HANDBOOK OF THERMOSET PLASTICS

HANDBOOK OF THERMOSET PLASTICS

Edited by

Sidney H. Goodman

Developmental Products Laboratory
Technology Support Division
Hughes Aircraft Company
El Segundo, California

and

Department of Chemical Engineering
University of Southern California
Los Angeles, California

NOYES PUBLICATIONS
Park Ridge, New Jersey, U.S.A.

Copyright © 1986 by Noyes Publications
No part of this book may be reproduced in any form
without permission in writing from the Publisher.
Library of Congress Catalog Card Number: 85-25932
ISBN: 0-8155-1054-3
Printed in the United States

Published in the United States of America by
Noyes Publications
Mill Road, Park Ridge, New Jersey 07656

10 9 8 7 6 5 4 3

Library of Congress Cataloging-in-Publication Data
Main entry under title:

Handbook of thermoset plastics.

Bibliography: p.
Includes index.
1. Thermosetting plastics. I. Goodman, Sidney H.
TP1180.T55H36   1986   668.4'22   85-25932
ISBN 0-8155-1054-3

To 'Pinky' and Debra for telling me things I don't always want to listen to but really need to hear.

To Adrienne, whose love and affection are reciprocated always, no matter how busy I get.

And most of all, to 'Maishe' whose few words of admonition many years ago, masked in much care and concern, gave me the guts to push on and reach this milestone, I sincerely dedicate this book.

Preface

Many years ago during the last part of my senior year in Chemical Engineering, my class participated in a series of plant tours of regional chemical companies. At one, our hosts made an enthusiastic presentation about this new plastic material, epoxy, with properties surpassing all but the elusive universal solvent. My unshared enthusiasm was tempered by an overwhelming desire to merely note the data needed to successfully (and I guess minimally) complete the written synopsis of the tours required of each student and a less than wholehearted interest in anything that smacked of organic chemistry, especially 'plastics.'

Some six months later, I found myself in the Technical Services Laboratories of a major paint company subsidiary whose major product line was epoxy resins and curing agents. Osmotically, my interest and career growth increased with time. For twenty years I have worked in nearly all phases of epoxy technology, expanded into urethanes, taught about phenolics, polyesters, silicones, and polyimides and other thermosets. Throughout this time I have always been conscious of thermoset technology's lesser yet significant role in the plastics industry.

The plastics industry distinguishes between linear polymers and those that are crosslinked. The former are generally cheaper, process easier and provide a broad, more-than-adequate property spectrum to meet the needs of designers and users. Thermoset resins, on the other hand, are perceived to be more expensive and harder to process, thus limiting their use to specialty applications where the inconveniences are tolerable and necessary. In the past, annually published sales/production data, which clearly demonstrates the dominance of thermoplastics over thermosets, would deign to include phenolics in their lists of 'engineering plastics' and relegate the remainder of thermoset resins to a specialty category or the ubiquitous 'miscellaneous.'

With time, however, the plastics industry has so burgeoned that its technology and technologists have elucidated the importance and contributions of thermosets. Sales and production figures routinely define unsaturated polyesters, urethanes, and epoxies as major commercial entities. The technical base has grown very large, accompanied by significant theoretical understanding of the chemical processes, manufacturing techniques and design of properties. In short, there is a lot more science and a lot less black magic involved today.

This volume has been produced to offer an up to date overview of this select segment of the huge plastics field. The contributors represent a combined experience of over 150 years in the field. Each contributor has made a career out of each of the subelements. They have presented not only traditional historical developments but the latest in technology.

Although grounded in polymer chemistry and science, the treatments presented here do not require expertise in these disciplines. It is sufficient for the reader to understand the general principles in the introduction, i.e., learn the jargon, and he/she can proceed to any specific chapter for the information sought. If there is to be a consistent thread throughout each chapter, it is that thermoset plastics are materials of construction, subject to strengths and weaknesses, as well as proper use and misuse. Marketing hyperbole is minimized, no 'universal solvent' can be found in these pages. The data and descriptions presented are for engineers, scientists, technicians, and students who routinely form judgements and take actions on the basis of informed analysis. It is our intent to help these readers make the right decisions and take the correct actions and thereby avoid the pitfalls our experience has uncovered.

The chapters on phenolics and amino resins emphasize the usage of these plastics as coatings and binders; a departure from the traditional stress on molding materials. The chapter on unsaturated polyesters includes the most recent developments in sheet and bulk molding as well as advancements in reinforced composites and decorative applications.

Thermosetting allyls, almost extinct five short years ago, have seen a resurgence that warrants their inclusion in this survey. New resins and curing agents commercialized in the last ten years, which have improved their usage in many traditional markets, are identified and described in the chapter on epoxy resins. For the first time, a clear description and explanation of urethane technology has been collated and presented in a practical "how" and "why" chapter. Polyimide resins have matured and are presented to reflect the intense interest in high temperature resistant plastics.

Although not hydrocarbon based, a review of recent developments in silicone technology is still important because of the resins' significant role in plastics applications. All too often, the crosslinking of thermoplastics is given short shrift in other treastises. Although relatively minor in sales volume, these materials satisfy a variety of specific needs, which warrants attention to their characteristics and properties.

As with any technical text, obsolescence is concomittant with publication. To minimize this effect and promote the readers interest in developments yet to come, a chapter on current research in thermosetting polymers has been included.

To each of the contributors, my personal thanks for your hard work, cooperation, and enthusiasm on this project. I speak for all of us in thanking that small 'army' of support who helped with the tough parts; typing, editing, proofreading, offering suggestions, and most importantly, their patient tolerance of our disorganization and mistakes. Finally, but no less important, our thanks to George Narita and his staff for making order out of chaos and keeping us in full view of our final goal.

Pacific Palisades, California
February, 1986

Sidney H. Goodman

Contributors

Sidney H. Goodman
Developmental Products
  Laboratory
Technology Support Division
Hughes Aircraft Company
El Segundo, California

Abraham L. Landis
Materials Science Dept.
Technology Support Division
Hughes Aircraft Company
El Segundo, California

Kreisler S.Y. Lau
Materials Science Dept.
Technology Support Division
Hughes Aircraft Company
El Segundo, California

Bernard Schneier
Reliability Engineering Dept.
Space and Strategic
  Engineering Division
Hughes Aircraft Company
El Segundo, California

Isao Shimoyama
DFC Company
Los Angeles, California

Arthur L. Wooten
Forest Products Utilization
  Laboratory
Mississippi State University
Mississippi State, Mississippi

Oscar C. Zaske
Silmar Division
Sohio Engineered Materials Co.
Hawthorne, California

Contents

PREFACE...................................vii
CONTRIBUTORS..............................xi

1. INTRODUCTION............................1
   Sidney H. Goodman
   History..................................3
   Definitions..............................4
   Crosslinking and Curing..................4
   Influence of Time, Temperature, and Mass.7
   Shelf Life and Pot Life.................10
   Curing..................................11
   Staging.................................14
   Stoichiometric Considerations...........14
   Prepolymerization and Adducting.........16
   Bibliography............................17

2. PHENOLIC RESINS........................18
   Arthur L. Wooten
   Introduction............................18
   Development.............................18
   Overview of U.S. Plastics Production and Phenolic
     Resins, 1983..........................19
   Raw Materials...........................20
     Production of Phenol..................22
     Formaldehyde Production...............23
   Reaction Mechanisms.....................23
     The Acid Catalyzed Reaction of Phenol-
       Formaldehyde........................24
     The Alkaline Catalyzed Reactions of Phenol-
       Formaldehyde........................25

xiv   Contents

        Chemistry of the Condensation Reaction 26
        Phenolics in Plywood 27
        Composition Boards 29
        Phenolic Molding Compounds. 31
        Thermal and Sound Insulation 34
        Decorative Laminates 36
        Industrial Laminates 37
        Foundry Resins. 38
        Friction Materials 39
        Miscellaneous Uses for Phenolic Resins 42
        Trade Names. 43
        References. 43

3. UREA, MELAMINE, AND FURAN RESINS 45
    Arthur L. Wooten
        Introduction 45
        Chemistry of the Urea Resins 47
        Chemistry of the Melamine Resins 48
        Fibrous and Granulated Wood Products 49
        Urea Bonded Plywood. 51
        Other Urea Resin Markets. 52
        The Furan Polymers 56
        Trade Names. 58
        References. 58

4. UNSATURATED POLYESTER AND VINYL ESTER
    RESINS. 59
    Oscar C. Zaske
        Unsaturated Polyesters 59
            History 59
            Chemistry 61
                General Concepts 61
                Functionality 62
                Polyesterification Reaction. 62
                Isomerization 63
                Polyesterification Reaction Speed. 64
            Processing 64
        Typical General Purpose (GP) Unsaturated Polyester
            Resin 65
        Common Resin Synthesis Raw Materials. 65
        Copolymerization of Unsaturated Polyester Alkyds
            with Monomers 66
        Processing Equipment and Manufacturing. 68
        Unsaturated Polyester Resin Alkyd Properties 69
        Styrenated Unsaturated Polyester Resin Liquid
            Properties. 70
        Monomers Used in Unsaturated Polyesters 71

    Unsaturated Polyester Properties and Chemical
       Composition....................................71
    General Purpose Resins...........................72
    Molar Ratio of PA to MA 72
    Flexibilization...................................72
    Isophthalic Resins...............................73
    Molecular Weight Comparisons 73
    Hydrolytic Stability and Chemical Resistance.........74
    Styrene Compatibility............................74
    Flame Retardance...............................75
Vinyl Ester Resins................................76
    Chemistry......................................76
    Basic Vinyl Ester Resin..........................77
    History..77
    Toughness and Chemical Resistance 77
    Vinyl Ester Resin Structure and Properties..........78
    Specialty Vinyl Ester Resins 78
        Vinyl Ester Resin Thickening for SMC 78
        Flame Retardant Vinyl Esters...................78
        One Step Vinyl Ester.........................79
        Rubber Modified Vinyl Ester 79
    Vinyl Ester Resins Overview 80
    Typical Styrenated Vinyl Ester Resin Liquid
       Properties....................................80
    Typical Styrenated Vinyl Ester Cast Resin
       Properties....................................80
Compounding of Unsaturated Polyester and Vinyl
  Ester Resins....................................81
    Overview.......................................81
    Curing Systems 82
        Room Temperature (RT) Curing Systems..........82
        Benzoyl Peroxide Catalyzed RT Cures 84
        Commercial Prepromoted Resins..................84
        Catalysts for RT Cobalt Promoted Resins..........84
        Heat Curing Systems 85
        Handling Catalysts and Promoters................85
    Ultraviolet Absorbers 85
        Types of UV Absorbers........................85
    Thixotropic/Flow Control Agents...................85
    Fillers 87
        Some Common Filler Applications in Resins........87
        Filler Dispersion and Mixing Equipment...........88
        Order of Mixing and Dispersion..................88
        Calcium Carbonate Fillers.....................89
        Clay Type Fillers............................89
        Talc Fillers 89
        Alumina Trihydrate Fillers (ATH)................89
        Pigments and Colorants........................90

Thickening Agents. 90
Fiber Reinforcements 91
Applicable Manufacturing Processes 91
Overview. 91
Hand Layup 91
Spray Layup. 93
Resin Transfer Molding (RTM) 93
Water Extended Polyester (WEP) 94
Casting 95
Acrylic Backup. 99
Matched Die Mat, Preform and Premix Molding 100
Pultrusion 101
Sheet and Bulk Molding Compounds (SMC and
BMC). 104
Bulk Molding Compound (BMC). 105
Recent Developments 107
Foamed Polyester 107
Urethane Hybrid Resins 107
Reduced Styrene Emission Resins. 107
**Manufacturers of Unsaturated Polyester and Vinyl
Ester Resins** 108
References. 109

5. ALLYLS 112
Sidney H. Goodman
Introduction 112
Chemistry 113
Polymerization and Processing 113
Formulation 115
Properties 116
Applications 129
Trade Names. 130
References and Bibliography. .130

6. EPOXY RESINS 132
Sidney H. Goodman
Introduction 132
Resin Types. 133
Diglycidyl Ether of Bisphenol A 133
Novolacs. 133
Peracid Resins. 137
Hydantoin Resins 138
Other Types 138
Curatives and Crosslinking Reactions 141
Stoichiometry. 141
Alkaline Curing Agents 144
Lewis Bases. 144
Primary and Secondary Aliphatic Amines. 144

Amine Adducts	145
Cyclic Amines	146
Aromatic Amines	147
Polyamides	149
Other Amines	150
Acid Curing Agents	**150**
Lewis Acids	150
Phenols	151
Organic Acids	151
Cyclic Anhydrides	151
Polysulfides and Mercaptans	155
Formulation Principles	**157**
Epoxy-Containing Reactive Diluents	158
Resinous Modifiers	160
Nonreactive Diluents	161
Fillers	162
Colorants and Dyes	163
Other Additives	164
Properties	**169**
Applications	**174**
Trade Names	**180**
References and Bibliography	**180**

7. THERMOSET POLYURETHANES 183
Isao Shimoyama

Introduction	**183**
Polyurethane Chemistry	**186**
What are Polyurethanes?	**186**
Polyurethane Raw Materials and Moisture	**188**
Handling of Polyurethane Components	**189**
Types of Polyurethane Systems	**190**
Castable Liquids	190
Thermoplastic Pellets or Millable Gums	192
Advantages of Adduction	**192**
Range and Types of Polyurethane Products	**193**
Polyurethane Uses	**194**
Polyurethane Coatings	**204**
Single Component Moisture Cured Polyurethane Coatings	204
Two Component Polyurethane Coatings	207
Components for Polyurethanes	**209**
Diisocyanates	210
TDI	210
MDI	211
Saturated Diisocyanates	211
Polymeric Isocyanates	211
Isocyanate Index (NCO/OH)	211
Catalysts	212

Special Situation Catalysts .212
Aromatic Amines 212
Conditioners 216
   Driers or Moisture Absorbers. 216
   Surfactants 216
   Adhesion Promoters 217
   Mold Release Agents. 217
   Fillers 217
   Extenders and Diluents. 217
   Amines 218
Polyols 218
   Simple Glycols 219
   Simple Triols. 219
   Other Polyols 219
   Polyols with Polyether (Polypropylene) Back-
     bones or Polyoxypropylene Derivatives of
     Trimethylolpropane, etc. 219
   Polyester Based Polyols. .220

Industrial Mathematics for Polyurethanes 220
Terminology 222
**Guidelines and Theories in Compounding Poly-
urethane Elastomers**. 223
Compounding of Thermoset Polyurethane Elastomers . . .228
   Prepolymers 228
   Prepolymers and Curing Compounds. 232
     Aromatic Amine (MOCA). 232
     Polyols 235
     Curing of MDI Base Prepolymers with Polyols 236
General Consideration. 237
   Other Polyurethane Ingredients. 238
     Castor Oil Derived Polyols. 238
     Polycaprolactone Polyols .238
     Graft Polyol 239
     Reaction Injection Molding (RIM) 239
   Recent Advances and New Products 239
     Internal Mold Release (IMR) Formulations. 239
     Recent Diisocyanates 239
     Recent Proprietary Catalysts. 242
Appendix 242
   General Polyurethane Math Development 242
   Toluene Diisocyanate (TDI) .248
   Methylene Diphenyl Diisocyanate (MDI) 250
   Methylene Dicyclohexane-4,4'-Diisocyanate
     (Saturated MDI). 251
   Isophorone Diisocyanate (IPDI) 252
   Determination of % Free —NCO 252
   Determination of Hydroxyl Numbers 254
   Determination of Isocyanate Equivalents 255

Method for Preparation of TDI Prepolymers..........256
Method for Preparation of MDI Prepolymers.........260
Raw Material Sources............................261
Pertinent ASTM Test Methods.....................264
References..264

8. COMMERCIAL POLYIMIDES..........................266
Abraham L. Landis
Polyimides from Condensation Reactions............268
Thermoplastic Polyimides..........................286
Addition Polyimides...............................304
References..316

9. SILICONES..318
Bernard Schneier
Silicone Fluids....................................319
Dimethyl Types..................................322
Methylphenyl Types..............................322
Silicone Rubbers..................................324
Room Temperature Vulcanizate....................324
One Component Systems.......................324
Two Component Systems.......................327
Condensation Cure...........................327
Addition Cure...............................328
Heat Cured Systems..............................330
Compounding.................................330
Curing......................................331
Silicone Laminates..............................331
Trade Names......................................333
References..334

10. CROSSLINKED THERMOPLASTICS....................335
Bernard Schneier
Crosslinking of Thermoplastics....................336
Effects of Crosslinking on Polymer................339
Polyethylene....................................339
Polypropylene...................................343
Polyvinyl Chloride..............................344
Chemical Crosslinking............................346
Polyethylene....................................346
Polypropylene...................................349
Rotational Molding...............................350
Post-Irradiation Effects..........................355
Acrylates...361
Trade Names......................................365
References..365

11. RESEARCH POLYMERS AND FUTURE DIRECTIONS.....367
Kreisler S.Y. Lau
- Nadimide-Terminated Thermosetting Polymers.........368
- Maleimide-Terminated Thermosetting Polymers........371
- Acetylene-Terminated Thermosetting Polymers........372
- Future Demands in Ultrahigh Temperature Resistant Polymers.......380
- Chemical Structures Suitable for Ultrahigh Temperature Use.......382
- Cure and Crosslinking Mechanisms at Ultrahigh Temperatures.......385
- References.......388

INDEX.....392

1
Introduction

Sidney H. Goodman

Developmental Products Laboratory
Technology Support Division
Hughes Aircraft Company
El Segundo, California

This book presents an overview of a major class of materials of construction: thermosetting plastics. Using the biological analogy, this class fits into the family of materials as shown in Figure 1-1.

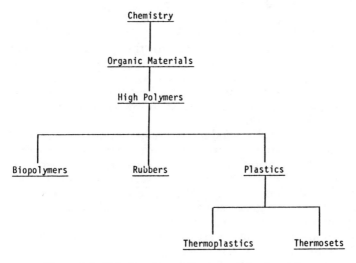

Figure 1-1: Relationship of thermosets in chemistry.

One popular definition of thermosets is:

> ... a polymeric material which can be formed by the application of heat and pressure, but as a result of a chemical reaction, permanently crosslinks and cannot be reformed upon further application of heat and pressure. (Goodman and Schwartz, p 9)

Another more rigorous definition is found in Whittington's *Dictionary of Plastics* (p 239):

> Resins or plastics compounds which in their final state as finished articles are substantially infusible and insoluble. Thermosetting resins are often liquid at some stage in their manufacture or processing, which are cured by heat, catalysis, or other chemical means. After being fully cured, thermosets cannot be resoftened by heat. Some plastics which are normally thermoplastic can be made thermosetting by means of crosslinking with other materials.

This leads to an interesting concept. All too often trade usage confers titles on classes of materials. These titles reflect a nomenclature or jargon that is fully comprehensible to those in the trade. Those new to the trade soon learn the meaning of the terms by association, osmosis, etc. At some point in the technology maturation, someone decides to establish a precise definition of the terms. The true definitions are quickly found to be elusive: no two practitioners define them exactly the same way: the definitions are not "scientific" enough; more exceptions to the rule exist than examples of the rule; and on and on. The term "thermoset" or "thermosetting plastics" is a classic illustration of this phenomenon.

This book is an attempt to collate and present the current practices and technology associated with a group of commercial polymeric materials called "Thermosets." Everyone who works with these materials has an intuitive understanding of the types of plastics that fall into this category. We know, for example, that chemical crosslinking must occur in order for the resultant product to be called a thermoset. We know that the monomeric precursors may or may not be polymeric in and of themselves, will undergo reaction when the chemical kinetics are right; that these precursors are commonly called thermoset resins because they will participate in a crosslinking reaction.

We also know that under the right conditions many of these resins can polymerize linearly and form a traditional thermoplastic polymer. Vulcanization is a form of crosslinking wherein a rubber is formed, yet rarely do technologists refer to rubber as a thermoset plastic. Biopolymers (amino acid/protein based) are known to crosslink (one theory suggests this as a root cause of aging) and we hardly think of animals as thermosetting plastics.

This book then will be structured based on the commonly perceived "definitions" of thermosetting resins. Both definitions stated earlier remain valid and useful.

This introductory chapter will include a series of basic terms and definitions that will be referred to throughout the individual chapters that follow. Many of the "definitions" will in fact be descriptions of the phenomena which best illustrate the sense of the terms, as opposed to a rigorous definition per se. That these explanations are "common usage" or "trade jargon," that they are not scientifically precise, does not compromise or lessen their meaning or value.

HISTORY

Goodyear's (and Hancock in England) discovery of the vulcanization of natural rubber in 1839 could be construed as the first successful commercial venture based on thermosetting polymers. The plastics industry dates the beginning of thermosetting plastics to the development by Leo Baekeland in 1909 of phenolics. In this instance, Baekeland not only produced the first synthetic crosslinked polymer, but as importantly, he discovered the molding process that enabled him to produce homogeneous useful articles of commerce. The Bakelite product line dominated plastics technology for years until the advent of alkyds in 1926 and the aminos in 1928. Table 1-1 lists a synopsis of the various historical milestones in thermosetting resin technology. Progress was made more often as a result of the economical commercialization of key precursor materials rather than as a conscientious result of a chemist's ability to tailor polymers for specific properties and characteristics. It must be remembered that the acceptance of Staudinger's heretical concept of macromolecules was not universally accepted until the late 1920s and early 1930s, long after products made from polymeric materials had reached commercial maturity.

Table 1-1: Historical Milestones of Thermosets*

```
1839 Goodyear discovered vulcanization of rubber.
1909 Baekeland granted his 'Heat and Pressure' patent for phenolic
     resins.
1926 Alkyd introduced.
     Aniline-formaldehyde introduced in U.S.
1928 Urea-formaldehyde introduced commercially.
1931 Hyde began research on organo-silicon polymers.
1933 Ellis patented unsaturated polyester resins.
1935 Henkel made melamine-formaldehyde resins.
1937 Automatic compression molding introduced commercially.
     Polyurethanes first produced.
1938 Melamine introduced commercially.
1939 First patent (in Germany) on epoxy.
1941 Urethane-polyester type-introduced in Germany.
1942 Dow Corning made silicone industrially.
1943 Castan's patent issued on epoxy.
1946 Polyurethane elastomers introduced.
1947 Epoxy introduced commercially.
1954 Polyurethane introduced in U.S.
1957 Urethane-polyether type-introduced in U.S.
1964 Polyimides introduced as a fabricated product.
```

*Extracted from SPE JOURNAL, 1967.

DEFINITIONS

The broad classifications of plastics—general purpose, engineering, and specialty—applies to thermosets as well as thermoplastics. *General purpose thermosets* are characterized by average (for thermosets) mechanical properties, lower resistance to temperature, higher coefficients of expansion, and low cost/commodity-like production and sales (tons/year). *Engineering thermosets* have higher mechanical properties and temperature resistance and they are perceived to be more durable. They are more expensive with a moderate production volume (pounds/year). *Specialty thermosets* are useful because of one or more highly specific and unusual property which offsets any lack of other "good" properties. They are usually very expensive and are produced in relatively small quantities (pounds/batch). Overlapping between the three categories often occurs—a general purpose phenolic is often competitive with an engineering polyimide. The individual families of plastics in this book can be loosely classed as shown in Table 1-2.

Table 1-2: Categories of Thermosets

General Purpose	Phenolics, aminos, polyesters
Engineering	Epoxy, polyurethane
Specialty	Silicones, allyls, high temperature thermosets, crosslinked thermoplastics

It is assumed that the reader has a reasonable understanding of the basic principles of polymer science and organic chemistry. These initial discussions therefore, are designed to highlight and review some of the basic concepts in order to establish the proper perspective for the material which follows.

CROSSLINKING AND CURING

A *linear polymer* is a long continuous chain of carbon-carbon bonds with the remaining two valence bonds attached primarily to hydrogen or another relatively small hydrocarbon moiety. Figure 1-2 shows a schematic representation of some linear polymer configurations.

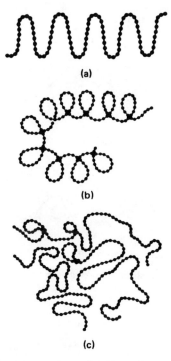

Figure 1-2: Simplified representation of various linear polymer configurations (Goodman & Schwartz, 1982).

A *network polymer* is formed as a result of the chemical interaction between linear polymer chains or the build-up from monomeric resinous reactants of a three-dimensional fish-net configuration [Figures 1-3(a) and 1.3(b)]. The process of interraction is called *crosslinking* and is the main distinguishing element of a *thermosetting* material. The "thermo" implies that the crosslinking proceeds through the influence of heat energy input, although, as will be seen in the individual chapters, much crosslinking occurs at room temperature (25°C, 77°F) and below. The "setting" term references the fact that an irreversible reaction has occured on a macro scale. The network polymer formed has an "infinite" molecular weight with chemical interconnects restricting long chain macromovement or slippage.

Molecular *functionality* (i.e., number of reactive moieties per mole of reactant) dictates the potential for a crosslinking reaction. A total average functionality between reactant elements greater than two suggests the potential for crosslinking independent of mechanism. In other words, the bifunctional C=C, would, via an addition reaction, normally produce a linear polymer. If, however, other unsaturation is generated or remains in the formed linear chain, crosslinking can yet occur (Figure 1-4).

6 *Handbook of Thermoset Plastics*

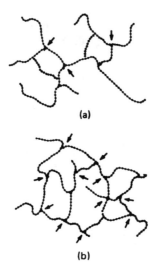

Figure 1-3: (a) Lightly crosslinked network polymer. (b) Highly crosslinked network polymer.

```
nC=C  →  --(C-C)n--        standard linear addition polymer
                           (polyethylene)

nC=C-C=C  →  --(C-C=C-C)n--
                     |
                     ↓
                     |
             --(C-C-C-C)--    crosslinked addition polymer
                     |
             --(C-C-C-C)--    (polybutadiene)
                     |
```

Figure 1-4: Linear chain formation and crosslinking via addition polymerization.

Similarly for a condensation reaction, a tri- or polyfunctional reactant will form a thermoset structure with a polyfunctional comonomer.

```
  O O                    O O
  ‖ ‖                    ‖ ‖
HOCRCOH + HOR'OH  →  --(OCRCOR')--        standard linear condensation
                                           polymer (linear polyester)

  O O                    O O    O O
  ‖ ‖                    ‖ ‖    ‖ ‖
HOCRCOH + HOR'OH  →  --(OCRCOR' OCRCO)--   crosslinked condensation
        |                    |
        OH                   OCRCO)--      polymer (polyester)
                             ‖ ‖
                             O O
```

Figure 1-5: Linear chain formation and crosslinking via condensation polymerization.

INFLUENCE OF TIME, TEMPERATURE, AND MASS

The temperature dependency of crosslinking reactions, for all intents and purposes, behaves in a traditional Arrhenius relationship. Thus ambient temperature strongly influences crosslinking rate. Since all commercial thermosetting reactions are exothermic, a mass effect also influences the rate of reaction. Monomer concentration effects are generally associated with stoichiometric balances between reactants as well as the normal free volume accessability of each of the reactants to each other.

Perceptually, it is easier to describe the events of crosslinking if we focus on the reaction between two low viscosity liquids. The principles, however, are valid whether the monomers are solids, liquids, gases or mixtures thereof.

Referring to Figure 1-6, we can track a polymerizing mixture of monomers by observing the viscosity change versus time at a given temperature. Beginning at t_0, the mixture has a viscosity η_0. The heat generated from the exothermic reaction produces a typical viscosity decrease (η_1). As the molecular weight of the mass increases, the resultant mixed viscosity increase outpaces and quickly surpasses any reduction caused by heat. The molecular growth continues over time until a perceptible macroscopic gel-like "lump" can be sensed. This is t_{gel}, the *gel point*, or more commonly, the *gel time*.

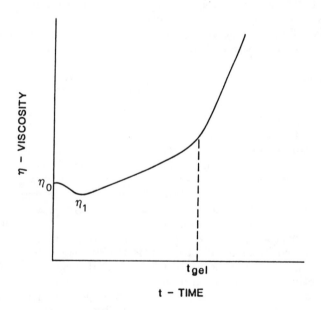

Figure 1-6: Viscosity vs time at constant temperature for a liquid thermosetting system.

From this point forward the viscosity goes to infinity, i.e., the polymeric mass becomes a macroscopic solid—a plastic. In some liquid systems the knee in the curve at the gel time is very hard to identify because the viscosity increase is very gradual over time. With solid molding powders, pressure and heat must be applied in order to generate a fluid condition so that the gel time can be determined.

Usually a wooden probe is sufficient to detect the gel point with a good deal of accuracy (± 0.5 minutes). Sophisticated equipment is available that automatically measures the gel point based on the length of flow of a molding powder, the increase in torque of an oscillating rubber-like mass, the change in dielectric constant of the crosslinking mass, as well as many others. However measured, accuracy less than minutes is rarely required.

The term most frequently used in the trade for this gelling phenomenon is *curing*. To cure a thermoset is to cause it to crosslink. Vulcanization is the rubber industry's term for curing.

Typically the coreactant monomers are referred to as the *resin* and *curing agent*. The resin is the resinous monomer from which the family name is derived; e.g., an epoxy plastic is an epoxy resin that has been crosslinked. The curing agent is the coreactant and goes by many names: curative, curing agent, hardener, catalyst, "Part B," etc.

Any crosslinking reaction is influenced by temperature. As the ambient temperature increases, the rate of reaction increases. Additionally these reactions are exothermic. Since all polymers are inherently thermal insulators, the exothermic heat cannot easily leave the curing mass and thus adds to the heat input for continued reaction. Figure 1-7 demonstrates this effect. Curve 1 represents a normal room temperature cure similar to Figure 1-6. With added heat ($T_2 > T_1$, Curve 2) the gel time decreases. Curve 3 ($T_3 < T_1$) shows the effect of decreasing the temperature, i.e., t_{gel} increases. Curve 4 ($T_4 \ll T_1$) describes a stable situation wherein the cure is arrested because the temperature is below the activation level necessary for inception of the reaction. Mixtures of resins and curing agents that are stored in such a stable or "latent" condition are called *one-can* or *single-package* systems. This latent storage is normally desired at room temperature although many applications can conveniently deal with refrigerated or frozen storage.

This latency can be useful in other ways. Resinous mixtures can be applied to supporting films or impregnated into cloth fabrics and then stored in their latent conditions. This is particularly useful in adhesive and laminating applications. The distinction between so-called "hot-melt" systems and latent one-can thermosets must be emphasized. The former are generally solid thermoplastics that can be liquified by melting. When coated on a film they return to a stable solid state. Latent thermosets are reactive. Whether applied to a surface from solution or as a melted liquid, the reaction will initiate

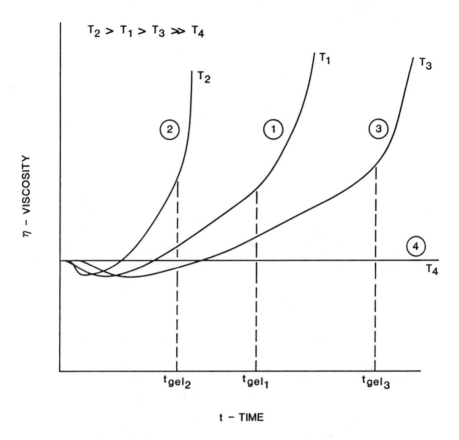

Figure 1-7: Influence of ambient cure temperature on the gel time of thermosets.

and proceed as a function of ambient temperature conditions. In other words, they have a finite shelf life (see definition below). The colder the storage conditions, the more extended the latency. Another process, B-staging, is different from these two conditions and will be discussed in more detail shortly.

The mass effect on gelation parallels the temperature effect. As a crosslinking mass increases in size, the ability to transfer the exothermically generated heat away from the reaction site decreases significantly because of the thermally insulative nature of polymers. Curve A in Figure 1-8 shows the typical gel profile for a given mass. Curve B represents the doubling of the mass, whereas curve C represents a halving of the mass (no Arrhenius relationship is suggested by doubling or halving). Curve D describes a condition where the mass is below some critical threshold size that arrests the crosslinking and generates an effective latency.

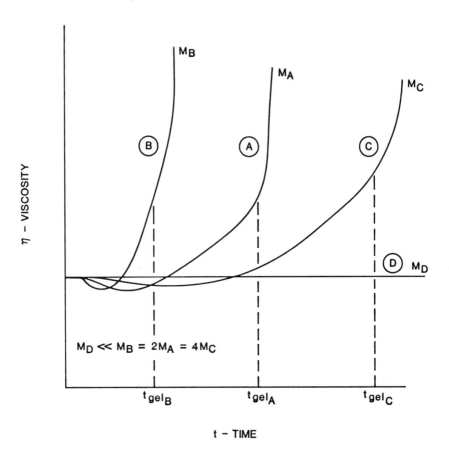

Figure 1-8: Influence of mass on the gel time of thermosets.

In practical terms, the temperature/mass dependency is very significant. For example, a 5-gallon mix of a urethane flooring varnish compound may gel in 20 to 30 minutes with an often violent exotherm. However, if the same mass is poured and spread over a cold floor within a few minutes of mixing, the gel time may extend itself to 4 to 8 hours. Similarly, an adhesive bonding two dissimilar metals will take longer to gel than if it is bonding two pieces of plastic: the metal acting as a heat sink, the plastic acting as an insulator.

SHELF LIFE AND POT LIFE

Shelf life is an arbitrary time for practical storage of a thermoset system. Shelf life derives from the storage concept; i.e., how long can a thermoset be left on the shelf before it becomes difficult or even impossible to use in the intended application? The term can refer to a one-can system (e.g., a phenolic molding compound must be molded within 1 year of compounding) or a two-can mix that must be set aside for a few hours before use. Shelf life is also used to describe the storage stability of unmixed components of a thermosetting resin system if there is some threat to their reactivity as a consequence of the storage. For example, some curing agents are very hygroscopic and will lose reactivity if airborne moisture were to penetrate the storage container.

Working life, or *pot life*, is the available time to process a reacting thermosetting formula. Once the ambient cure temperature is reached and the crosslinking reaction begins, pot life describes the time available before the mixture becomes intractable or otherwise difficult to process. For example, the pot life of a coating is the time during which the viscosity remains low enough to allow for easy brushing or spraying. In a molding compound, the working life represents the amount of residence time available in the molding machine before the material must be injected into the mold in order to have trouble-free molding and/or a defect-free part.

CURING

The establishment of a sequence of time, temperature, and pressure needed to produce a thermoset part is the *cure schedule*. A simple example of one such schedule is that found on tubes of household epoxy glue. Here residence at room temperature for 16 to 24 hours under slight contact pressure clearly defines the conditions needed to effect a sufficient bond. An example of the other extreme, an extended highly specialized schedule established for a polybenzimidazole laminate is shown in Table 1-3.

Table 1-3: Typical Processing Schedule for a Polybenzimidazole Laminate

Expose laminate prepreg in a press to $120°C$ with pressure increasing from contact to 200 psi. Increase temperature to $370°C$ and hold for 3 hours. Cool to $100°C$, remove from press and post cure under dry N_2 or vacuum as follows:

24 hours at $315°C$
24 hours at $345°C$
24 hours at $370°C$
24 hours at $400°C$
8 hours at $425°C$
3 hours at $370°C$ in air

In a multi-step curing sequence, the gel point most often occurs in the very early stages. From an efficient producibility point of view, the sooner a part can be handled after forming the better. This frees up the more expensive molding equipment and allows the cure to reach completion *(post-curing)* in a relatively low cost environment (e.g., an oven). The criteria used to establish this point are, generally sufficient "green" strength of the part (can be handled without deformation) and sufficient cure to minimize shrinkage (very little shrinkage stress or warpage will set in beyond this point).

A cure schedule is derived by plotting the change in the plastics' properties of interest against time at temperature and pressure. A sample curve is generated like the one shown in Figure 1-9. Practical considerations generally dictate that the cure time be chosen at some fractional level of the ultimate properties. This is because the time scale can often be logarithmic. Thus 90% of, say, ultimate tensile strength, may be achieved in a few hours at 25°C. The remaining 10% (often not needed for use) may require months to years for achievement.

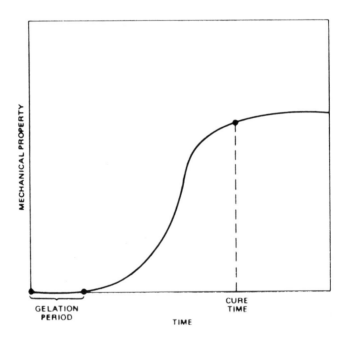

Figure 1-9: Mechanical property of a thermosetting polymer vs time.

Where more than one plastics property is important the cure schedule must reflect a reasonable time-temperature-pressure relationship that will yield an optimized combination of properties.

Figure 1-10 describes such a circumstance. A thermoset formulator must be aware of all these considerations in his design of a useful compound. In addition, heat-up and cool-down rates, volatiles release, part design, and many other factors influence the cure sequence ultimately designated to produce a part. An example of a ramped cure schedule for a polyimide composite is shown in Figure 1-11.

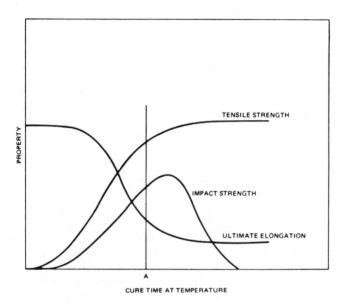

Figure 1-10: Optimization of cure schedule for thermosetting plastics. Optimum properties occur at point A.

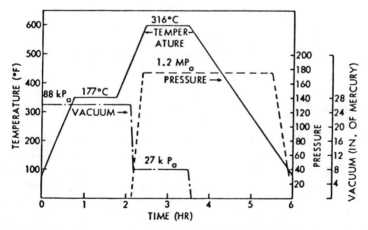

Figure 1-11: Autoclave cure cycle used for typical polyimide composite. (Delmonte)

STAGING

The influence of temperature on curing generates another practical production control. This is the concept of *staging*. The thermoset formula when first mixed (crosslinking has effectively not begun) is called the "A-stage." As time, and thus crosslinking, progresses the compound goes through the "B-stage." This is a time line as shown in Figure 1-12.

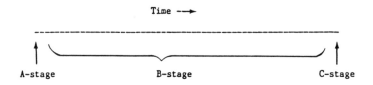

Figure 1-12: Staging time line for thermosets.

Many thermosets can have the reaction arrested at any point along this line. The simplest way to do this is to drop the temperature sufficiently below the reaction temperature to a point of latency. Depending on conditions, this latency period can be quite long (not uncommonly 6 to 24 months at room temperature and lower).

In addition, the polymer/monomer mixture at this point may well change its physical state to a more useful form. For example, a B-staged phenolic molding powder will harden and embrittle compared to the A-stage. It can be frangibly crushed into a non-sticking, free-flowing powder that is stable at room temperature for 12 or more months. Epoxy resins can be coated while liquid on a polymeric carrier film, B-staged to a fixed degree of tackiness, and stored under refrigeration for 6 to 12 months. This provides a useful tape-supported adhesive that only requires application to a substrate and subsequent heating for bonding.

The *C-stage* represents the fully crosslinked part in its final configuration.

STOICHIOMETRIC CONSIDERATIONS

One of the major responsibilities of the thermoset resin chemist is to balance the coreactants stoichiometrically. He does this by establishing a *mix ratio*, the weight to weight proportion of the resins and curatives. In theory, each functional group in each monomer must react on a 1/1 molar basis. The final crosslinked plastic should have no residual reactive sites if all reactants have been properly proportioned and subjected to optimum cure conditions.

In reality, many considerations drive the polymerization process away from the ideal. To begin with, as the molecular weight of the polymerizing mass increases, it becomes sterically less possible for reacting species to come together and react. Side reactions and chain stopping contaminants may reduce the calculated number of reactive sites. Although molecular movement never truly ceases in a crosslinked mass, the time span for "complete cure," i.e., to reach ultimate properties, may be logarithmic (see earlier discussion on cure schedules).

In practice, a resin formulator will calculate a theoretical stoichiometry for his intended formula. Then he will prepare samples under a given set of cure conditions and test for the change in selected properties of the resultant plastic versus change in stoichiometric ratio. This can be a long, cumbersome and expensive process. One short cut, common to the epoxy chemist, is to run a Soxhlet extraction in acetone. A curve like the one shown in Figure 1-13 is obtained.

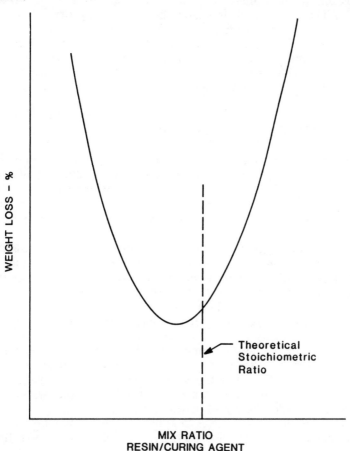

Figure 1-13: Optimizing mix-ratio using a Soxhlet apparatus.

Obviously that mix ratio which creates the least weight loss should represent the maximum integrity of the cured part, i.e., the best properties. Once established, the formulator can narrow his choice of mix ratios to those near this optimum and reduce the testing required to pinpoint the exact ratio that will provide the properties he is seeking. Examples of representative stoichiometric calculations can be found in the individual chapters on polyurethanes and epoxies.

Experience has demonstrated that it is not at all uncommon for mix ratios to depart from 1/1 molar by as much as 20 to 30%. The responsibility for insuring that the stoichiometric balance is maintained varies among thermoset types. The phenolic chemist is concerned during the initial manufacture of polymer. The user need only add heat and pressure to get a part. The polyester chemist establishes the balance when he makes the base resin. He then adds his crosslinking monomer and the user catalyzes the mix to effect the cure. The epoxy and urethane chemist, on the other hand, may not only do as the others, but he may also design a system that requires the end user to mix the reactants in the correct ratio. As will be described in later chapters, this may impose a significant constraint on the user to insure that he does not stray from the predetermined mix ratio tolerance.

PREPOLYMERIZATION AND ADDUCTING

Prepolymerization is a method of increasing the molecular weight of a forming polymer to some intermediate value. Prepolymerization is often confused with B-staging. During B-staging the polymer formation is arrested at some practical intermediate point. It is a random process which yields various molecular weight moieties and some crosslinking.

A prepolymer is normally formed under precisely controlled conditions to yield a stable polymer of specific molecular weight and configuration, most often without any crosslinking. For example, an isocyanate will be coreacted with a glycol at a stoichiometric ratio much greater than 1/1. A urethane prepolymer will be formed with sufficient residual isocyanate to further react in a curing environment.

There are many reasons for generating prepolymers. Among the more prevelant are: to increase the viscosity of monomer, to decrease toxicity and/or reactivity for control of gel time and exotherm, and to balance the mix ratio of a formulated system (compensate for the addition of additives and fillers).

When a monomeric resin is "capped" with a coreactant, the process is called *adducting*. The technique proceeds via the following schematic sequence.

Monomer A has 2 functional groups, F_A: $F_A R F_A$
Monomer B has 3 functional groups, F_B: $F_B \underset{\underset{F_B}{|}}{R'} F_B$

In a direct polymerization, the two monomers will react as in equation (1), i.e., a crosslinked polymer.

(1) $\quad 3nF_A R F_A + 2nF_B \underset{\underset{F_B}{|}}{R'} F_B \rightarrow -(RF_A F_B \underset{\underset{F_B F_A R-}{|}}{R'})-$

If F_A and F_B are olefinic bonds then the adducted moiety contains $F_A F_B$, i.e., C—C bonds, resulting from the typical addition reaction. If F_A is a carboxyl group, say, and F_B is an amine, then a condensation reaction will occur yielding an amide, $F_A F_B = -CONH-$, and H_2O.

The adduction process proceeds as in Equation (2).

(2) $\quad 3F_A R F_A + 1F_B \underset{\underset{F_B}{|}}{R'} F_B \rightarrow F_A R F_A F_B \underset{\underset{F_B F_A R F_A}{|}}{R'} F_B F_A R F_A$

The crosslinking is completed by stoichiometrically reacting the resulting pendant F_A groups with more F_B-containing reactant. The net result, equation (3), is polymer with essentially the same crosslinked structure as with a standard crosslinking, equation (1).

(3) $\quad mF_A R F_A F_B \underset{\underset{F_B F_A R F_A}{|}}{R'} F_B F_A R F_A + mF_B \underset{\underset{F_B}{|}}{R'} F_B \rightarrow -(RF_A F_B \underset{\underset{F_B F_A R-}{|}}{R'})_m-$

The reasons for adducting are essentially the same as for prepolymerization.

BIBLIOGRAPHY

Delmonte, J., *Technology of Carbon and Graphite Fiber Composites*, Van Nostrand Reinhold Co., New York (1981).
Morton, M., *Introduction to Rubber Technology*, Reinhold Publishing Corp., New York (1964).
Schwartz, S.S. and Goodman, S.H., *Plastics Materials and Processes*, Van Nostrand Reinhold Co., New York (1982).
Severs, E.T., *Rheology of Polymers*, Reinhold Publishing Co., New York (1967).
Whittington, L.R., *Whittington's Dictionary of Plastics*, Technomic Publishing Co., Stamford, Connecticut (1968).

2
Phenolic Resins

Arthur L. Wooten

Forest Products Utilization Laboratory
Mississippi State University
Mississippi State, Mississippi

INTRODUCTION

The phenol-formaldehyde polymers are formed by the reaction of phenol, or a mixture of phenols, with formaldehyde. In terms of volume, the commercial use of other phenols is very small; in terms of ability to achieve certain necessary technical properties, the other phenols may be very important. Formaldehyde is essentially the only aldehyde used. Furfural, in fact, is the only other aldehyde used commercially, at least to the author's knowledge.

DEVELOPMENT

Adolph Bayer[1] worked with phenol and formaldehyde in 1872. He could neither crystallize nor distill what we know now was a resulting polymer. The organic chemist, then as now, concentrated on the isolation, characterization, and identification of reaction products. The phenol-formaldehyde reaction received only sporadic attention. Efforts increased after Merklin-Lescam[2] began the commercial manufacture of formaldehyde in 1899. Blumer[2] introduced the first commercial phenol-formaldehyde resin called Laccain in 1902. This thermoplastic acid-condensed resin was intended as a shellac substitute. It was not a commercial success.

Leo Baekland has been called "the father of phenolic resins." As indicated above, he was not the inventor of phenolics. It is true, however, that "Baekland's heat and pressure" patents[3-5] of 1907 form the basis for almost all phenolic applications today.

Baekland, prior to his work on phenolics, had invented the fast photographic paper known as "Velox." The income from this invention allowed him to install a laboratory in the basement of his home in Yonkers, NY, and in 1905 he began devoting his full time to phenolic resin research. Baekland was gifted with not only a very superior technical mind but also a true sense of the commercially practical side of problems.

Prior to his 1907 patent, efforts to form molded objects failed in several ways. The low temperatures required to avoid steam evolution caused bubble formation and required production cycles that were too long. Also the cured resin was too brittle for most uses. Baekland overcame the latter by adding wood flour or other fillers. He overcame the former by using a closed mold and high pressure to suppress the evolution of gases and steam. These changes allowed a higher molding temperature and very short press cycle.

Baekland formed the General Bakelite Company in the USA in 1910 and similar companies in several other countries. In 1939 Bakelite Corp. became a subsidiary of Union Carbide Corporation. Bakelite took out more than 400 patents[2] and foresaw most of the current major applications with the obvious exceptions of space age usages such as ablation shields.

Phenolics are the oldest of the synthetic plastics. Cellulose acetate, which was in existence before Baekland's 1907 patent, is classed as "modified cellulose" or as a semi-synthetic. The continued importance of the phenolics is indicated by the symposium, "Phenolics Revisited, 75 Years Later," presented at the Washington meeting of the American Chemical Society in 1983.[6] U.S. production of phenolic resins during 1983 has been estimated at 1.1 million metric tons.[7]

OVERVIEW OF U.S. PLASTICS PRODUCTION AND PHENOLIC RESINS, 1983

In 1983 U.S. plastics production totaled 18.3 million metric (see Tables 2-1, 2-2 and 2-3).[7] As mentioned above, this total included 1.1 million metric tons of phenolic resins. This amount is about 6.1% of the total or approximately ten pounds of phenolic resin per man, woman, and child in the United States.

Most phenolic resins are sold in the form of solutions, either aqueous, strongly alkaline aqueous, or partially alcoholic. Table 2-4 shows the average resin content of various resin types as estimated by the *Chemical Economics Handbook* of the Stamford Research Institute.

Table 2-1: Phenolic Resins, Pattern of Consumption, 1983[7]

Market	1,000 Metric Tons	Percent of Total
Bonding and adhesive resins	295	26.5
Laminating	67	6.0
Plywood	580	52.0
Molding compounds	100	9.0
Protective coats	10	0.9
Export	8	0.7
Other	55	4.9
	1,115	100.0

Reprinted by permission of *Modern Plastics,* January 1984.

Table 2-2: Bonding and Adhesive Markets, 1983[7]

Market	1,000 Metric Tons	Percent of Total Phenolics Market
Coated and bonded abrasives	12	1.1
Fibrous and granulated wood	80	7.2
Friction materials	12	1.1
Foundry and shell moldings	26	2.3
Insulation materials	165	14.8
	295	26.5

Reprinted by permission of *Modern Plastics,* January 1984.

Table 2-3: Laminating Markets, 1983[7]

Market	1,000 Metric Tons	Percent of Total Phenolics Market
Building	17	1.5
Electrical/Electronics	12	1.1
Furniture	8	0.7
Other	30	2.7
	67	6.0

Reprinted by permission of *Modern Plastics,* January 1984.

RAW MATERIALS

Phenolic resins are the reaction product of one or more of the phenols with one or more of the aldehydes. The substituted phenols and the aldehydes other than formaldehyde are used in such small quantities that they will generally be omitted other than in this short summary.

Table 2-4: Average Resin Contents of Various Types of Phenolic Resins

	Conversion Factor	Remarks
Laminating	0.67	Normal product is 67% resin in water/alcohol solution.
Coated and bonded abrasives	0.75	Generally 50% of this product is shipped in 100% bonded abrasive resin. The remainder is 50% coated abrasive resin.
Friction materials	0.95	Dry product for 80% of total.
Insulation materials	0.46	10% dry resin for acoustical insulation; the balance is 40% solids.
Foundry and shell mold	0.60	Weight average of the resin systems in use.
Plywood	0.42	Generally shipped as 42% resin in water.
Fibrous and granulated wood	0.60	Generally shipped as 60% resins.
Rubber adhesives	0.60	Generally shipped as 60% resin for adhesive use; resins are 100% for processing and compounding use.
Molding compounds	1.00	Historically reported as product shipped. Generally 40% resin with fillers, colorant, etc.

Reprinted by permission of *Chemical Economics Handbook,* SRI International.

Phenol was originally obtained commercially from coal tar distillates. Cresols (monomethylphenols) and xylenols (dimethylphenols) were also obtained and for many years were cheaper than phenol. In the early days, this practice led to many uses of these chemicals in applications where phenol was equally or better suited. With the advent of cheaper synthetic phenol, many of these usages ceased. The cresols, especially, are still found in some electrical usages where their properties are better than those of straight phenolics.

The alkylated phenols are used in relatively specialized applications where the attached alkyl group offers special advantages. Resins based on p-tertiary butyl phenol are used as tackifiers in pressure sensitive tapes. The p-tertiary octylphenol resins are used as tackifiers in automobile and truck tires. Nonylphenol finds its widest use, not in resins, but as a hydrophobic tail on polyglycol condensates for nonionic surfactants. Prior to the advent of alkyd resins and the newer surface coating resins, para-substituted phenolics were widely used to modify the rosin-vegetable oil varnishes used in paints.

Resorcinol (m-dihydroxybenzene) is not used in large tonnage but its resins can be cured at room temperatures. This property makes it important in the manufacture of laminated beams for churches, boat keels, etc., which are too large and too specialized to fit in production hot press processes.

Among the aldehydes, furfural is the only one in commercial use other than the various forms of formaldehyde. Furfural-phenolic resins were of commercial importance in the early days of television. The early cabinets were molded, requiring a resin that had slow cure and extremely long flow in order to fill the large thin walled mold. They were soon replaced by other resins. The largest continuing usage for furfural-phenol resins is in the molded abrasives field. Sometimes a combination of liquid furfural is used to wet the grains of abrasives in order to help fix the powdered phenolic resin to the grain surface and to avoid resin segregation. Due to the presence of the furfural, the cured resin is softer and tends to abrade away. This exposes fresh abrasive grit and avoids plugging of the grinding wheel.

Production of Phenol

Phenol, as a commodity chemical, has an unusual history in that several different manufacturing processes have existed. Today, well over 90% of the world's production of phenol is by the cumene process.

Originally, phenol was derived commercially by the fractiontion of coal tar distillates. The oldest of the synthetic processes (about 1892) was the sulfonation processes. Benzene was sulfonated to benzenesulfonic acid and fused with sodium hydroxide to form sodium phenate plus sodium sulfite. The phenate was acidified with sulfur dioxide to yield phenol and form more sodium sulfite. The success of this process depended on stable markets in the kraft paper mills for the copious amounts of sodium sulfite that were produced. However, the ever increasing limitations of the regulatory agencies on trace amounts of phenol played a part in the loss of the sulfite market. Plant and process obsolescence were also factors contributing to the phasing out of this process around 1978.

The liquid phase hydrolysis of chlorobenzene with dilute alkali, known in the U.S.A. as the Dow process, was dominant for some years. Established in the 1920s, it produced diphenyl ether and p-phenylphenol as by-products. A eutectic mixture of diphenyl ether and diphenyl remains a standard high temperature heat transfer fluid (Dowtherm A) today. During World War II, varnishes based on p-phenylphenol provided the best of the marine varnishes. These have subsequently been replaced by better synthetic finishes.

The cumene process became commercial in 1952 and continues to add to its share of the market. This process, starting with benzene and propylene, proceeds as follows:

Cumene Process for Phenol

$$C_6H_6 + C_3H_6 \rightarrow C_6H_5CH(CH_3)_2 \xrightarrow{O_2} C_6H_5C(CH_3)_2OOH \rightarrow C_6H_5OH + (CH_3)_2CO$$

benzene   isopropylene   isopropyl benzene   cumene   phenol   acetone
                           or cumene   hydroperoxide

Formaldehyde Production

For many years, formaldehyde was produced only by the dehydrogenation of methanol over a silver catalyst. The dehydrogenation is endothermic so a portion of the evolved hydrogen is burned to maintain the necessary temperature. Careful control of the air-methanol ratio is required. The amount of heat from the partial burning of the hydrogen requires less air than the lower explosive limit of air-methanol, but an explosion can occur if the ratio increases due to control equipment failure or for other reasons. The silver is usually present as a bed of granules today. For many years the preferred form was silver gauze tightly wound into a plug. The plug was carefully inserted into a double flanged 6 by 1½ inch reactor tube. Excess air meant excessive temperature and melting or sintering of the gauze plug. A burner room with 400 to 500 individual gooseneck mounted reactors was not unusual. The individual burners were fed air-methanol from a manifold. This practice led to occasional operational problems inasmuch as the reactor tube temperature was the only individual reaction variable that could be monitored. The melted or plugged gauze was resold to the refinery for silver recovery. The conversion of methanol per pass was rather low, 70 to 80%. There are references[2] to 90% conversions being achieved by BASF.

In the late 1950s, Reichhold Chemicals commercialized the Formox Process. This process involves the direct oxidation (not dehydrogenation) of methanol. The advantages include complete methanol conversion at a single pass with formaldehyde yields above 95%. The disadvantages include a relatively high formic acid formation and a relatively larger air handling requirement. The formic acid is removed by an ion exchange unit at the end of the system. It is estimated that more than half of today's world production of formaldehyde is by the Formox Process.

REACTION MECHANISMS

The chemistry of the phenolic resin remains, relatively speaking, almost unknown. Until the advent of nuclear magnetic resonance (NMR) and gel phase chromatography there were literally almost no analytical tools with which to attack the scientific problems.

The problems involved in unravelling the concurrent, competitive, and consecutive reactions taking place in phenolic resins are orders of magnitudes greater than those for linear polymers, either vinyl or condensation types. The major phenolic resin manufacturers are doing some basic research, and they do fund a few academic studies. However, they also know that they have built a 2.5 billion pound industry with very little basic chemical knowledge but with many years of accumulated experience. In the author's opinion, it will be some time before chemistry of the phenolics is unravelled to any great extent. The discussion that follows is based on the recent works of Pizzi[8] and of Knopf and Schieb[2] and is strongly influenced by the 1972 work of Drumm and LeBlanc.[9]

The Acid Catalyzed Reaction of Phenol-Formaldehyde

We have chosen to introduce the phenol-formaldehyde condensations with the seemingly simple acid catalyzed type. Under acid conditions (pH less than 1) the reaction goes quickly to methylene linked phenols.

$$2C_6H_5OH + CH_2O \rightarrow HOC_6H_4-CH_2-C_6H_4OH$$

The phenol has an ideal reactivity of three and formaldehyde has a reactivity of two. The theoretical maximum combined formaldehyde in a fully cured resin is $3/2$ or 1.5 mol formaldehyde per mol of phenol. Novalacs are formulated with a small enough formaldehyde ratio that all chains are phenol terminated and thermoplastic. Additional formaldehyde must be added before curing. This is usually added as *hexa* (hexamethylenetetramine). At elevated temperatures hexa decomposes, reacts with the novalac, and causes the resin to become insoluble and infusible (thermoset).

At a mol ratio of 0.85 F/P the novalac is near gelation in the reactor. Commonly used commercial products have a ratio of 0.75 to 0.85. After reaction and distillation of water, the resin is vacuum treated to bring the melt point residual free phenol to the desired level. The resin is then flaked, crushed, pulverized, dissolved in organic solvents or put into aqueous suspension for customer use. Hexa may be added during this finishing step, or it may be added by the customer at a later stage.

The 0.85 critical mol ratio does not have a proven mechanistic explanation. At first glance, one would expect the branching and gelation equations of Flory[10] to apply. The reasons that they do not is that the novalacs do not meet the requirement that monomers and partially reacted products have equal reactivity. For a more detailed explanation see Drumm and LeBlanc.[9]

For many years it has been postulated that the initial reaction of formaldehyde with phenol in acidic solutions was to form methylolphenol. It was further postulated that this slow step was followed by a much more rapid reaction of the methylolphenol with phenol. This theory accounted for the absence of methylolphenol and was supported by work with model compounds. Knopf and Wagner[11] showed with NMR that a transient methylolphenol did, in fact, exist. This work is further supported by the presence of hydroxylbenzyl cations in acidic solutions of methylolphenols. The extreme reactivity of these cations with phenol or with the polymer accounts for the absence of methylolphenol in the novalac.

There is experimental evidence[9] showing the ratio of o-o', o-p', and p-p' isomers among the diphenolmethanes (DPM).

Experience has shown that the end groups in a novolac chain are much more reactive with methylolphenol than are the groups within the chain. For this reason, in acid condensations, branching is kept to a relatively low level. Later in the reaction when the concentration of polymer is high and that of the monomers is low, the situation changes and internal substitution occurs. This change may well be the cause of the gelation that occurs at a mol ratio of 0.85 to 0.87, but this theory has not been proven. The extractables in this initial gel are high indicating naturally a minimal gel. Considerable excesses of hexa crosslinking agent is necessary to obtain optimum properties.

The Alkaline Catalyzed Reactions of Phenol-Formaldehyde

During the first forty-five years of the phenolic resins, many technical papers were published on the chemistry involved. Good methods for free formaldehyde determination existed and were used. All other analytical methods were empirical in nature and measured neither structure, mechanisms, or kinetics of individual reactions. Methods such as gel time, time to turbidity, water dilutability, etc., offered comparative data but not structural information.

In 1954 Freeman and Lewis[12] published what has been referred to as the best single paper on phenolic chemistry. The initial step in the formation of an alkaline catalyzed phenolic resin is the addition of formaldehyde to the phenol molecule forming methylolphenols. There are five of these: two monosubstituted (o and p), two disubstituted (o-o and o-p) and one trisubstituted (o-o-p) products. Freeman and Lewis used paper chromatography to make quantitative studies on the rates of the seven reactions involved. These seven concurrent reactions are affected by many of the reaction conditions: temperature, pH, dielectric constant, and ionization constants of the intermediates are a few of these conditions. To interpret the experimental data would involve the solving of a dozen or more higher order simultaneous equations.

This work was done prior to the common availability of computers so that several mathematically simplifying adjustments were made. Sufficient alkali was used to insure conversion to the phenoxide ions and thereby eliminate the complications of the ionization constants of the substituted phenols involved. In each reaction the formaldehyde was adjusted to be equivalent to the reactive positions available. Room temperatures were involved both to slow the reaction sufficiently for the analyses to be made and to minimize dimolecular and higher condensations. Dilute solutions, 0.8 to 1.8 molar, were involved. Freeman and Lewis did show that the seven concurrent, competitive, and consecutive reactions could be untangled.

By 1968 computers, as well as gas chromatographic analysis of the methylolphenols, were available. Zavitsas, et al,[13,14] succeeded in including formaldehyde equilibria, temperature, pH, dielectric strength of the solvent, and other variables into the system of equations to be solved by the computer. Zavitsas' work vastly extends the work of Freeman and Lewis.

Chemistry of the Condensation Reaction

Under the usual commercial formaldehyde addition conditions, e.g., pH about 9, temperature about 60°C, relatively little intermolecular condensation takes place. The addition reaction is perhaps five times as rapid as the condensation reaction until at least 50% of the formaldehyde has reacted. Formaldehyde is more reactive with the methylolphenols than with phenol. One expects then to start the condensation process with a mixture of mono-, di-, and trimethylolphenol with free phenol also present. This concept is necessarily an oversimplification since any molecular kinetic energy distribution, Gaussian or otherwise, as well as the increased reactivity of the methylol phenols compared with phenol, means that some concurrent condensation is taking place during the addition step.

The kinetics and the mechanism of formation and dimerization of methylolphenols have been studied rather intensively since 1954. The clearest exposition is that of Drumm and LeBlanc[9] based primarily on the gas chromatographic study and computer work of Zavitsas, et al.[13,14] Prequantitative work prior to 1954 is documented by Martin[15] and Megson.[16]

There is a large gap in the technical literature between the dimers and the finished phenolic resins. Troughton and Rozon[17] indicated a gas chromatographic separation of isomers through the tetramers following trimethylsilylation. This technique has apparently not been pursued.

The phenolic resins, with their abundant hydroxyl groups, have hydrogen bonding, association, selective adsorption and other problems when GPC (gel permeation chromatography) is used. Many qualitative papers on phenolics using GPC have been presented. One of the more recent by Wellons and Gollab[18] using low angle laser light scattering as a detector is promising.

The associative and hydrogen bonding characteristics, both inter- and intramolecularly, together with the ability of alkalinity to supress these characteristics make phenolics a candidate for GFC (gel filtration chromatography) rather than GPC. (GFC is much slower than GPC, but allows the use of relatively highly alkaline aqueous samples.)

The direction that the effort to unravel the condensation steps of the phenolics will take will probably be GFC; or, hopefully, a modified GPC will lead the way; Carbon 13 NMR will contribute, as will magic angle FTIR. Infrared and ultraviolet spectroscopy can answer some questions regarding hydrogen bonding, especially with solvents. There is no immediate solution to the overall problem in sight.

PHENOLICS IN PLYWOOD

Exterior grade plywood is by far the largest consumer of phenolic resins in the U.S.A. In 1983, 580 thousand metric tons were used for this purpose. Plywood has been known since antiquity, although until this century plywood was primarily decorative rather than structural.

Commercial production of plywood in the U.S.A. probably began in 1904 when a plywood plant was started in Wisconsin to provide panels for doors. Until 1919 all U.S.A. plywood plants were producing door panels using a casein glue. By 1925 casein was being replaced by protein type glue with peanut meal as a principal ingredient. Peanut meal was soon replaced by the less expensive soybean meal. Much widespread use of plywood for interior uses was well underway.

New demands for exterior, weatherproof applications led to the importation about 1932 of the Goldschmidt glueline from Germany. This was a thin sheet of paper impregnated with phenolic resin best known here as "Tego film." Sheets were placed between each layer of veneer and hot-pressed at about $300°F$. There were obvious economies to be had by the use of water solutions of phenolic resins. The problems inherent in any new process were worked out and liquids were in commercial use by the early 1940s.

During World War II the demand for exterior plywood and the necessity for phenolic resins were greatly expanded. A large number of the U.S. mills added the necessary hot presses as did all of the British Columbia mills. In the ten years after World War II about one new Douglas fir plywood plant opened each month.

For years the industry, and particularly the Forest Products Laboratories, U.S.D.A., had studied the technology and the commercial feasibility of phenolic-bonded Southern pine. By 1963-1964, the technology and economics were right, and Temple Industries began commercial production in Diboll, Texas.

By 1973 in the U.S.A. and as early as 1969 in Canada, improved technology and lowered costs made phenolic-bonded plywood competitive with urea resin bonded plywood for interior applications. Today there is little urea used in plywood in the U.S.A.

The technical problems associated with the profitable production are so varied and interrelated that a full book would be required to discuss them adequately. The problem is made more difficult by the fact that wood bonding is an art, not a science. Wood, as a substrate, is a nonhomogeneous material even at the molecular level. Moisture content inevitably varies and must be taken into account. So-called smooth wood surfaces are very far from being smooth. Wood extractives can either cause excessive glue penetration or can prevent penetration. The adhesive can show insufficient tack, allowing the veneer to shift before being hot-pressed. The glue line can dry out before the hot press. The list of gluing factors could go on and on.

The phenolic resin formulated for plywood varies somewhat with the species of wood, the geographical area, and the season of the year. Further, the resin alone is not suitable for the commercial production of plywood. In order to control penetration and cure time, the phenolic resin is necessarily of a molecular weight high enough to be insoluble in water. This solubility is obtained by the presence of about 4.5 to 8.5% sodium hydroxide which forms the sodium phenate salt of the resin. The higher caustic levels are found in the northwest and especially in Canada. 5% caustic is more common with Southern pine plywood.

In the mixing of the phenolic resin to form a plywood adhesive, additional caustic as well as water is required. Both extenders and fillers are used. The extender, frequently wheat flour, is classically used to provide adhesive action and phenolic resin reduction. This concept was true with the urea-formaldehyde resin, but it is doubtful that the phenolic requirement is reduced. The extender is classically defined as an inert cost saver. It does cut costs but it also provides help in such functions as gap filling, etc. Both the extender and the filler are acted on by the caustic during the adhesive mixing operation, the extender more so than the filler. In the author's opinion, the extender is at least in part dispersed at the molecular level and is an integral part of the final bond. The filler may be slightly solubilized, but only slightly. A typical Southern pine adhesive mix will illustrate some of these points (see Table 2-5).

Table 2-5: Typical Southern Pine Plywood Adhesive Mix

Typical Resin Constants	
Percent nonvolatile	40.00%
Percent NaOH	5.11%
Percent free CH_2O	0.07%
Percent nitrogen	0.08%
pH	11.2
Viscosity, Brookfield at 25°C	789 cp
Specific gravity at 25°C	1.1770
121°C gel time, sunshine	6.8–7.1 minutes
Mix Information	
Water	1,200
Co-cob (100% oat hulls)	600
Wheat (Glu-X)	500
NaOH	270
Resin	6,300
	8,870
Percent resin solids	28.41%
Percent total solids	42.33%
Percent wheat on resin solids	19.84%
Percent NaOH on resin solids	5.36%
Typical Adhesive Constants	
pH	11.0
Percent nonvolatile	42.4%
Mix viscosity Brookfield at 25°C	2,900 cp
Specific gravity	1.1938

The added water is insufficient for the mix to be mixed properly after the caustic is added. For this reason, perhaps one-third of the total resin is added to the water, wheat flour, and Co-cob. The caustic is added and while mixing is continued, the exothermic heat of dilution of the caustic causes the mix to become warm. During the empirically determined mixing time the wheat flour becomes solubilized and the Co-cob somewhat so. The remaining resin is added to chill the mix. In careful laboratory work, the adhesive mixes are frequently allowed to stand overnight before testing.

This mix should be a good starting point for roller coating or for curtain coating. It would be too viscous for spraying. It is not unusual for the technician to make 4 to 6 mixes with a new resin before getting the adhesive constants necessary for application.

The adhesive is applied to the Southern pine veneer at about 40 to 42 pounds per 1,000 square feet of single glue line. For $3/8$-inch plywood the veneer is three ply at $1/8$-inch per ply. The center ply is cross-grain to the outer plies. Characteristically, plywood will have an odd number of plies to maintain a balanced, nonwarping structure.

The molecular weight of the phenolic resin is one of the most important variables. Today the manufacturer controls this and all variables in an effort to reproduce exactly the minute details of each step in the phenolic resin cooking operation. If the process is reproduced, the viscosity of the resin is a measure of its molecular weight.

The lack of a characterization method or test has been a continuing headache to both the producer and to the user of phenolics for 75 years.

COMPOSITION BOARDS

Composition board is used in this chapter to designate those wood-resin boards other than plywood. These include particleboard, fiberboard, waferboard, strandboard, and others.

The production of low density insulating board was begun about 1914. Equipment similar to papermaking equipment was used, but a thick layer was formed rather than thin paper. This mat was oven dried and could be used for interior applications such as ceiling tiles.

Wet process hardboard was invented by H. Mason in 1924. He was using exploded wood fiber. Chips were loaded into a high pressure retort. The pressure was raised with steam and water and suddenly released. The explosive release of pressure produced bundles of fibers with the hemicelluloses rendered water soluble. It was found that under high pressure and about 400°F the product, since known as Masonite board, was formed. The effective bonding agent was the humidified lignin present. Resin binder was not used in the early boards. A small amount is used today in order to produce better machinability without edge chipping.

The origins of other composition board products usually preceded their commercial production by long lead times. As early as 1887 a proposal was made to utilize waste sawdust mixed with glue and compressed with heat. This is almost a literal description of our modern particleboard but it was not commercialized until 1941. Particleboard, almost alone among the composition board products, utilizes essentially nonplanar particles. For this reason they retain little of the strength properties of the cellulose fiber. It is anticipated that future major growth will be in the direction of other composition board products rather than in plywood or particleboard.

About 1960 the massive expansion of particleboard began in the U.S.A. Plant capacities increased to 2,000 tons of board per day. Breakthroughs into the furniture core and floor underlayment markets were made. Particleboard was, and is, primarily a urea resin bonded product. Some 415 thousand metric tons of urea resins were used in 1983.[7]

By 1970, the waferboard with its broad thin chips, sometimes from planer shavings, was combined with a phenolic binder for structural use. This market has continued to expand.

The oriented strandboard utilizes a three-inch by half-inch flake of thin wood. By orienting the strands, either mechanically or electrostatically, a board is obtained that offers improved structural characteristics. Again, phenolic resin is the binder.

In wet process hardboard, mentioned above, the fibers are deposited from a slurry onto a moving screen and partially dewatered by vacuum. This process is a modification of low density fiberboard and its associated papermaking background. The American Plywood Association through its research arm, the Plywood Research Foundation, spent years working on a water-free hardboard process. The process involves cooking, adding resin and wax, and mechanically grinding the wood chips to fiber bundles, which are air conveyed through heated drying ducts and stored in cyclone collectors. The fibers are metered onto cauls and hot pressed. Because the fibers have much less chance to become entangled than in the wet process, phenolic resin is required as the binder. Operating conditions can be adjusted to control moisture content. With moisture content below 10%, the hardboard can be pressed between cauls to produce smooth-two-sides. Above 10% moisture content the mat must be pressed with a screen for venting steam on one side yielding, as in the wet process, boards smooth-one-side.

Developed in the 1960s, medium-density fiberboard is usually thicker than dry process fiberboard and finds its market as core stock in the furniture industry. It is almost always thicker than $3/8$ inch.

Wood flour molding found a good American market primarily for toilet seats. Wood flour is blended with about 10% of a powdered novolac-hexa resin. The separate seats and tops are formed in heated metal molds and are molded at about 300°F and about 1,200 psi. This process yields a quality product. Plastic molded seats continue to make inroads on this industry.

PHENOLIC MOLDING COMPOUNDS

Leo H. Baekland applied for his "heat and pressure" patent in 1907. His work is briefly summarized early in this chapter. Equally important, but not so widely remarked, was his introduction of wood flour and other fillers to improve strength and other properties. By 1910 Baekland had formed companies in Germany, U.S.A. and several other countries to produce Bakelite. The use of phenolic molding compounds in the U.S.A. in 1983 was about 100 thousand metric tons.

Some molding compounds are made from one-step resins or resoles, but the great majority are made from two-step novolac resins. Powdered hexa is added as a curing agent. Lubricants and fillers and colors are blended in. The mixture is homogenized and fused on heated rolls or by passing through heated extruders.

The overall market for phenolic molding is being eroded by the newer engineering plastics. The all-around performance of phenolics including cost/value insures that a substantial market will remain.

Four major classes of phenolic molding powders are generally recognized:

(1) The *general purpose* compounds are low cost and have good electrical properties.

(2) *Heat resistant* low voltage electrical compounds. The indiscriminate use of asbestos as a filler in this class is no longer permissible but good high temperature phenolics are still readily available.

(3) *High impact* types usually have fabric, glass, or rubber as the filler.

(4) The *special electrical* grades usually contain mica or glass as filler. These increase the humidity resistance so that their good electrical characteristics are maintained over a wide range of environments. They are used in such applications as resistor cases and special electrical switches.

All phenolic molding compounds contain fillers, otherwise they would be too brittle for general use. The physical properties of molding compounds are usually greatly improved by the addition of the filler. Special properties can be enhanced by careful selection of fillers.

A typical general purpose molding compound recipe is shown in Table 2-6. Table 2-7 lists typical properties of the four types of phenolic molding compounds. Figure 2-1 illustrates a use of an electrical grade phenolic.

Table 2-6: Typical General Purpose Molding Compound

Novalac resin	100
Hexa	12.5-15.0
Wood flour	110-125
Lime or MgO	2.0-2.5
Lubricants	2.0-2.5
Colorants or pigments	2.0-5.0

Table 2-7: Properties of Phenolic Molding Compounds

	General-purpose	High-impact-strength	Heat-resistant	Special Electrical Insulation
Water absorption 1/8-in. thickness in 24 hr, %	0.8-1.5	1.5-1.75	0.2-0.5	0.07-0.1
Flexural strength, psi	7,000-9,000	7,000-9,000	6,000-9,000	8,000
Tensile strength, psi	3,500-6,500	4,000-6,000	4,000-4,500	4,000-4,500
Izod impact strength, ft-lb/in. of notch	0.24-0.46	0.8-4.0	0.23-0.64	0.3
Dielectric constant	4-10	4.4-10	0.5-50	5.0-6.5
Dissipation factor	0.03-0.3	0.03-0.6	0.1-0.8	0.01-0.07
Flexural modulus of elasticity, psi	1,000,000		2,000,000	
Relative cost/cu in., based on minimum list prices	1.0	1.2	1.3	2.1

[a]Specifications according to ASTM D 700-63T

Each commercial molding resin is usually available in three flows: short, medium, and long. The flow test by which it is determined is designed to measure the length of flow of the molten compound before gelation stops it. Obviously a thin-walled, long mold will require a resin with longer flow than a spherical mold.

The almost endless variety of end uses makes a logical discussion of properties almost impossible. The result always resembles a listing instead of a discussion.

Phenolics excel in rigidity but are moderate in strength. Their stiffness to weight ratio is high but so also is their density. They can, by incorporating fibrous fillers, be made quite impact-resistant but at the expense of moldability. They retain good electrical properties at elevated temperatures. The dimensional stability and creep resistance are excellent.

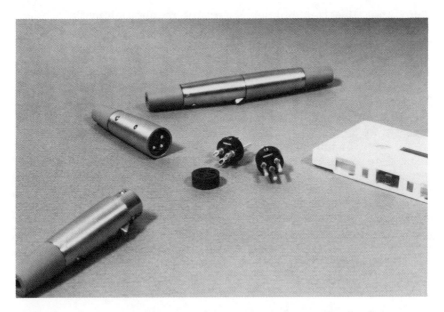

Figure 2-1: A one-step phenolic molding compound from Rogers Corp. was selected for lead insulation in these audio cable connectors for Switchcraft™. The compound resists creep and has excellent electrical properties. (Courtesy of the Rogers Corporation.)

Phenolics are affected by alkalis and by oxidizing acids. They are resistant to weak acids, solvents, detergents, and hydrocarbons. Extended exposure to weathering and U.V. can cause failure.

There are three principle molding methods in use with phenolic compounds; compression, transfer, and injection molding processes.

Compression molding is the original method and continues today. Pelletizing and radio frequency preheating is used with compression molding. Compression molding yields the strongest product in that there is less damage to any fibrous filler compared with the two other processes.

The transfer molding process adds a heated transfer chamber from which the preweighed, preheated compound is pressed by a plunger through a gate into the mold. The transfer process is especially good in molding parts with both thick-walled and thin-walled sections.

The operation of an injection molding phenolic operation resembles an injection thermoplastic. The loose material is fed from a hopper to a screw feed mechanism. Heat is generated primarily by friction, and water cooling may be used. When the preheated compound has built up in the nozzles, the screw becomes a ram, and the compound is injected into the mold by hydraulic pressure.

The three different molding processes require special types of compounds. Injection molding, especially, requires a long flow resin.

Both organic and inorganic fillers are used with molding compounds. The more widely used organic fillers generally have poorer water resistance than the inorganics. Their other properties cause them to be chosen.

Wood flour is the least expensive and the large usage filler. Characteristics include good flow and appearance, good tensile strength, and low heat conductivity. Wood fillers lack good impact strength.

In terms of impact strength, this property improves as the filler changes to cotton floc, chopped fabric, and to chopped twisted cord. The flow becomes progressively poorer until chopped twisted cord is reached; it has good flow but poor water resistance.

The inorganic fillers are reserved for specialty uses. Asbestos offers good heat, chemical, and fire resistance. Diatomaceous silica gives heat resistance and good electrical properties, but it is very abrasive. Mica offers good electrical properties, dimensional stability, and heat resistance. It does not wet well, and it causes sticking in the mold. Graphite improves frictional properties. Fiber glass gives good electrical properties, heat resistance, and impact strength.

THERMAL AND SOUND INSULATION

Insulation is widely used in the home and industry to prevent loss (or gain) of heat. In the home this has become more and more the domain of fiber glass at the expense of the cheaper rock wool. In industry, molded pipe and boiler insulations with the more heat-resistant rock wool is still in use. For low temperature insulation, moisture permeation is frequently a problem, and fiber glass with its superior moisture resistance dominates this industry.

The various ingredients for glass (blended to obtain the desired ratio of silicon, alumina, lime, magnesium oxide, borates, sodium and potassium) are charged to a cupola and melted. In general, the higher the silica, the higher the softening point of the resulting glass. The lower the sodium and potassium, the better the moisture resistance. The molten glass flows through small openings in a platinum die and is hit with blasts of steam or compressed air. Adjustments in steam pressure produces fibers of the desired length and thickness in the forming chamber. At the same time diluted aqueous phenolic resin is sprayed into the chamber. Exhaust fans draw the cooled resin coated fibers onto a moving perforated belt at the bottom of the chamber. The bulky batt emerges and is compressed by moving metal belts to the desired thickness as they pass through the horizontal curing ovens. The cured slabs emerge and go to further processing. Frequently foil or plastic vapor barriers are attached to the matt surfaces.

Rock wool and slag fibers were the forerunners of glass fiber.

Tradition has it that slag fibers were first noted in the gaps between bricks in blast furnaces early in the nineteenth century. The first rock wool plant was built in the U.S.A. in 1870. Slag wool disappeared some time ago and rock wool is down to about 10 to 15% of the glass fiber industry today.

Rock wool, properly formulated, has a higher maximum operating temperature than glass fiber: 725°F versus 500°F. Rock wool fibers are much thicker, hence heavier than glass wool. The yield per cupola charge is much lower. One of the reasons is the much larger percentage of shot (small pellets of nonfibrous material) in the product. The overall fiber yield may be as low as 40%.

The manufacturing process differs in principle in the fiber forming operations. The molten product is fed onto rapidly spinning discs at about 4,000 rpm. Centrifugal force slings the molten material out to form fibers. The remainder of the operation is similar in principle to that for fiber glass.

Powdered one-step phenolic resins were at one time used with fiber glass for molded under-the-hood automobile sound insulation. A veil of rather loose fiber glass matting descended past a device that blew powdered one-step resin against it. Exhaust ducts immediately behind the descending veil drew the powder through and conducted the excess to cyclone separators for recycling. The fiber glass-resin was molded to form the hood lining for one of the major automobile producers.

The water soluble phenolic resins used in the conventional fiberglass insulation blankets are usually low solids (50%), high combined formaldehyde, high dilutability resins. The alkaline catalyst is neutralized with an acid to produce soluble ash resins. Lime is usually used as catalyst and precipitated with sulfuric acid to give a low ash version for higher humidity resistance. The insulation producer further dilutes the resin to low solids so that he prefers a resin with very high initial water dilutability to simplify his storage problem. Even here, refrigerated storage of the resin is considered a necessity. The term resin efficiency is used to quantify the amount of resin solids that stay on the glass. Due to the large volume of air and high initial temperatures in the forming chamber much of the lower molecular weight molecules are lost up the stack. The resin manufacturers' solution is to increase the combined formaldehyde to decrease the volatility and to increase the cross link density of the curing resin.

Vinsol, lignin, and other extenders are sometimes added to the brown matts for cost reduction. Ammonia is usually used to give a yellow and/or pink colored insulation. Urea may be added both as a cost cutter and as a formaldehyde scavenger, but the humidity resistance of the insulation can suffer.

The fiber glass insulation usually has about 7% binder solids on it and mineral wool has about 3%.

An estimated 165 thousand metric tons of phenolic resin were used on insulation products in 1983.[7] Probably 80% was applied to glass fiber with perhaps 10% to rock wool and 10% to nonglass sound insulation, carpet backings, etc.

DECORATIVE LAMINATES

The 1983 consumption of phenolic resin in laminates has been estimated to be about 67 thousand metric tons.[7] Today at least 75% of laminate production is of the form called "decorative laminate." Perhaps the best known example of these is "Formica" produced by the Formica Corporation, a subsidiary of American Cyanamid Company. There are several other major producers and many smaller ones. The other class of laminates is "industrial laminates." These are engineering composites and have enhanced special properties such as strength, electrical properties, heat resistance, chemical resistance, etc.

A laminate is produced by impregnating a web of substrate to the desired resin content. The web is then passed through an air-circulating oven to remove the solvent and to advance the resin to the desired degree of B-stage. The cooled web is wound into rolls and stored until needed in the pressing operation. The webs are cut to size and stacked to give the desired final thickness. Caul plates are used, bottom and top, to give the desired surface finish. Newer presses are described with 20 to 22 openings and 8 to 12 laminates per opening.[2] Pressing is done at a controlled rate of heating and subsequent cooling. A typical top temperature would be about 375°F and 2,000 psi pressure. Finishing operations are performed as required by the desired end use. Thicknesses vary from fractions of a millimeter to inches, with corresponding differences in curing conditions.

The production control of the B-staging process is very important to the quality of the finished laminate. The resin solids pickup, expressed in percent pickup based on dry paper, varies from 30 to 70%. In decorative laminates this is usually 32 to 34%. The resin pickup is affected not only by the resin properties, type and thickness of paper and time of immersion in the resin, but also by the temperature and moisture content of the incoming new rolls of paper. Pickup can be controlled to a certain extent by increasing or decreasing the depth to which the paper is submerged in the dip tank. It can also be varied by the resin temperature. Line speed is seldom used to vary pickup. If determined manually, resin pickup is tested by die cutting perhaps ten samples of untreated paper. Then the same number of samples of B-staged paper are cut. After thorough curing, the weights of the two sets allow the calculation of resin pickup. On a high speed computerized line the measurement is made electronically, recorded and displayed immediately on a CRT tube.

The nonvolatiles are expressed as a weight percentage of the uncured B-staged resin. A suitable sized sample of B-staged paper is weighed and suspended in a circulating air oven until thoroughly cured. The weight loss is the percent volatiles. The nonvolatiles are kept below 3%. Some years ago when the primary solvent for decorative laminates was still alcohol, this specification was necessary as a safety precaution. Today a large part of the solvent has been replaced by water. The precaution is still observed not only for safety, but also because it has a definite effect on the resulting flow test.

The flow test is equally important for quality production. As the heated hydraulic press closes on the B-staged paper, the partially cured resin melts and flows to fill any voids. If the flow is too high, resin is pressed out of the laminate onto the caul plates. Manually this test is performed by pressing a stack of B-staged die cut samples. The cured specimen is removed and the "flash" or extruded resin is scraped off the edges of the laminate. A flow of perhaps 3 to 4% is normal.

The surface of a decorative laminate consists of a print sheet with the printed pattern. This is an alpha-cellulose sheet with perhaps 50 to 60% B-staged melamine. Over this is the overlay sheet, also alpha-cellulose, filled melamine. The print sheet, under properly controlled processing, acts as a barrier sheet to prevent the dark phenolic resin from bleeding through and marring the surface appearance. The overlay sheet adds to the gloss, alcohol resistance, scratch resistance, hardness and abrasion resistance of the laminate.

European practices differ somewhat in that they frequently add a balance sheet of melamine to the back of the laminate to help prevent warping from unequal expansion with humidity change. This back sheet is seldom found on U.S.A. laminates. Careful control of resin formulations and processing conditions can adequately reduce warping.

INDUSTRIAL LAMINATES

Industrial laminates are produced by basically the same process as the decorative laminates. Being specialized, premium quality, engineering composites there is much more variety in the webs, resins, B-staging conditions, and pressing conditions.

The material forming the web in industrial laminates varies with the specialized requirements dictated by the end use. Kraft paper is the more usual base with its low cost and good strength. Alpha-cellulose paper gives better electrical characteristics, good punchability, and for printed circuits, it can be impregnated in one pass, where Kraft requires two. Creped Kraft was used in the past where post-forming of the finished flat stock (kitchen bench tops) was required. Creped paper has reduced production for some years now. Cotton canvas is used for impact resistance (gears). Asbestos is used to a lesser extent today than previously to impart heat and flame resistance and will continue to be replaced with new asbestos-free formulations. Glass cloth is used when temperature resistance, electrical resistance, or moisture resistance is required. Synthetic plastics such as nylon fabric offer excellent moisture and humidity resistance.

Today phenol is used for most of the industrial laminates. In the past, cresols were primarily used and were felt to have superior electrical characteristics. Rosin also was commonly used to modify the phenolic resin for both industrial and decorative types. When rosin and cresols became more expensive then phenol, better phenol based resins were quickly developed. Canvas based laminates are sometimes made in thicknesses of inches rather than millimeters. At these thicknesses, the exothermic nature of the phenol resin curing process can lead to actual carbonization in the center of the laminate. Ortho- and para-cresols are slower curing and may be substituted for part of the phenol in the resin. Electrical grade punching stock resins (NEMA XXP and especially XXXP) are frequently plasticized with tung oil (China wood oil) to avoid shattering on the impact of punching. The XXP cold punching requirement is especially severe. Rosin is sometimes required to achieve compatability with the phenol and the oil.

FOUNDRY RESINS

Casting of metals can be done in either permanent molds or lost molds. The permanent molds are used in such applications as die castings of low melting alloys in mass production. Lost molds are primarily either plaster of paris (lost wax process) for the precision molds of items such as jewelry and of items such as jet turbine blades of titanium. Almost all ferrous casting is made in sand molds.

Many castings require cores to produce a desired metal free volume in the final casting. Examples are the holes required for cylinders in an engine block or the water jacket section of a cylinder head. The core binder has one more strict requirement than the mold binder. The core binder must decompose after the metal solidifies. Sometimes the opening through which this sand must flow is quite small and lumps of bonded sand can not be tolerated.

Rough ferrous castings are usually made in green sand molds. These are made of sand, clay, and water. They are very inexpensive molds but their accuracy and finish are poor. Almost all green sand castings are machined before use.

The shell molding (Croning Process) was introduced to the U.S.A. just after World War II. Before 1950 one of the major automobile companies was producing crank shafts, cam shafts, and engine valves by this process. There was said to be a 90% reduction in the post-machining processing.

This initial shell molding was done by blending about 8% novolac containing about 14% hexa with sand. The sand-resin was dumped onto a hot metal pattern for about 30 seconds. The box and hot pattern were inverted to remove the excess sand and resin. The shell was cured for several minutes at about 450°F. Two shells, of course, were required for the complete mold. These were joined and backed with support material for the heavier castings. The metal solidified in the 0.3 to 0.5-inch-thick mold without burning through. The remnants of the shell were removed on the shakeout line and the casting was ready for further processing.

Within a short time the cost advantages of hot coating the sand were realized and put into production. Flake novolac resin without hexa was added to a muller containing hot sand. The resin melted and coated the individual grains of sand. After a short coating period, a carefully weighed amount of water was added to cool the mixture. The necessary hexa was frequently added with the water. Release agents such as calcium stearate were added. When the muller had completed the breaking up of lumps, the muller was discharged through screens to the production line.

During this same period two further events modified the existing shell molding processes: warm sand coating was introduced in which water-alcohol solutions of novolacs were used and minute amounts of added silanes were found to greatly increase the effectiveness of the bonding process. These led to further reductions in the amount of expensive phenolic resins required.

The discussion to this point has been limited to molds. During this same period, processes starting with the introduction of coated sand technique for bonding cores have also been developed. Prior to shell molding, the cores were almost universally bonded with core oils. These were cheap vegetable oils or triglycerides of tall oil. The core mixture was water and cereal and the core oil. The water-cereal provided a cold gluing action (green strength) so that the blown core would hold its shape in the oven until the oil could polymerize.

The coated sand hot box core process has declined with the introduction of three newer processes. The furane or furfuryl-alcohol phenolic system and the furfuryl alcohol-urea system are referred to as no-bake systems. The necessary acidic catalyst for initiation of the furfural alcohol resin systems is achieved by injection of sulfur dioxide fumes. Special ventilation is required. The system is also used with conventional acids which are premixed into the system in the muller, limiting the useful life of the mixture.

The cold box process uses polyisocyanates and polyols. The polyol is usually a phenolic resin. After the core is blown, a small amount of gaseous amine is blown through the core causing the conversion of the binder into a polyurethane, which bonds the sand.

FRICTION MATERIALS

Brake linings, clutch facings, grinding wheels, and emery paper are generally classed as friction materials. The production volume of phenolic resins for these usages has been estimated at 12 thousand metric tons for 1983.[7]

Brake linings are used by the automotive, aircraft, railroad, and trucking industries. They also are used in such areas as oil well drilling rigs and in construction cranes. Industry standards are not used; each user of brake linings develops his own standard specification, and the lining manufacturer develops a product to fit. The SAE (Society of Automotive Engineers) has standard test methods but no product standards. Some individual states are moving in the direction of independent automotive standards.

There are other variations in requirements; for example, the higher speed limits in Europe require higher temperature formulations. The variation in requirements by end use are normal and to be expected. A family car requires a relatively soft lining for easy brake control, quiet action, and good friction. A taxicab or bus requires long lasting linings. A heavy duty truck requires excellent heat resistance and long life.

The formulations of the supplier are kept very confidential and usually contain about twenty ingredients. A typical formula might include 15% resin, 55% asbestos and 30% fillers.[2] The phenolic may be resole or novolac and powder or solution depending on the formulation, fabrication process, and end use.

Asbestos imparts strength and heat resistance. The phenolic resin imparts strength and bonding action. Its action is frequently modified with rubber, epoxy resin, etc. The different inorganic and organic fillers are used to improve friction.

There are at least three different manufacturing processes for brake linings: the impregnation method for asbestos fabric for linings and clutch facings, the wet dough process for drum brake lining, and the dry process for disk brakes as well as drum linings.

The impregnation process involves elements of the laminates B-staging as well as homogenizing calendaring such as is used in making molding compounds. Clutch facings are made by this process.

The wet dough process involves mixing the ingredients and any predissolved rubber in a sigma blade mixer. The dough-like mass is extruded to shape, cut to length, and dried at a low temperature. These preforms are slowly cured in clamped molds or rapidly by compression molding.

The dry process first masticates the rubber and asbestos in order to open the fibers. The other ingredients are then added in a regular blender and thoroughly mixed. Disk brakes can be molded directly onto the metal plate or an asbestos-phenolic precoat can be molded on to increase shear strength. Drum linings can be made in a similar fashion. Frequently the mix is B-staged into sheets and cut into strips. These can be shaped into molds and slow cured or compression molded rapidly.[2]

In 1909 Baekland gave a speech during which he displayed the first Bakelite bonded grinding wheel. Today about half of all grinding wheels are phenolic bonded. They have, to this extent, replaced the ceramic wheel, largely because the phenolic bond can withstand more thermal and mechanical shock. Also the phenolic wheel with more tensile strength can run at a faster speed and remove metal more efficiently.

The layman may think of his home workshop or some local industry when he thinks of abrasives. Actually, the great majority of consumption is in the steel producers plants and in the fabricators shops. A cut-off wheel, for instance, 18 inches in diameter and 0.1 inch thick can slice a 1 inch bar of steel in a few seconds leaving a mirrored finish.

The manufacturing process is fairly well standardized. The grit, fillers, and wetting agents are tumbled together to wet the abrasive grains and to cause the powdered novolac-hexa resin to adhere. The mixture is weighed into molds and pressure produces the preforms. These can be very slowly dried and cured as with the brake linings. They are more often hot pressure molded into the higher density varieties.

Wheels are designated by hardness and by grade. The grade relates to the abrasive grain, bond volume, and porosity. The bond volume directly affects the grade of hardness. This is not the hardness of the individual grain but rather the ease of removal from the matrix. A Norton scale grades hardness of these wheels from E to Z.[2]

There are no standardized performance tests for industrial grinding wheels. Each major user works with the supplier to develop the best wheel for his particular operation. Fillers contribute to toughness, stability, flexural strength and heat resistance of the wheel.

Sandpaper, the home workshop light colored sheet of grit bonded to heavy paper with animal hide glue has been largely replaced in industry by the phenolic bonded coated abrasives.

The manufacturing process is relatively simple although requiring exacting care. Usually a specially made paper is used as a backing. The cellulose fibers are frequently oriented in the line of motion to provide additional strength. The paper is usually modified with polymers or rubber during its manufacture to increase moisture resistance. Aluminum oxide and silicon carbide are the preferred grits for high performance usage.

First a make coat of perhaps 15,000 cp at 75% solids is applied to one side of the backing. The backing is passed over the grit with an electrostatic force being applied. The nature of the electrostatic force assures that the grain will embed in the make coat with its longer axis vertical to the backing. This vertical axis is the desired orientation. The magnitude of the electrostatic field controls the closeness of the embedded grit. The make coat is partially cured in a low temperature festoon dryer and a light size coat of resin is applied very thinly onto the relatively rigid grit. The coated abrasive goes through the festooned final cure and is cut to size.

MISCELLANEOUS USES FOR PHENOLIC RESINS

The gluing of laminated beams with cold setting resins is a small but important use for phenolics. Resorcinol is meta-dihydroxybenzene. Its reaction with formaldehyde is so fast that it can be used as a two-part system at room or slightly elevated temperatures. The phenolic laminated beam industry got its start with keels for PT and other wooden boats during World War II. When cost became very important after the war, the resin research labs learned how to graft resorcinol onto the ends of resole molecules. This product gave a slightly slower cure but at much lower cost. Church beams are a typical usage today.

Phenolic foam insulation has been much discussed for 25 years. Phenolic foam has good fire resistance, low smoke generation, good thermal insulating properties, and good acoustical insulating properties. Aqueous resoles are used, and a surfactant is added as are blowing agents. Strong acid catalysis generates a controllable exotherm with gelation before foam collapse.

Sound insulation mats for automobiles, offices, auditoriums, etc., are made from novolac-hexa resin and shredded rags. A minimum amount of resin is used so the mats are, in effect, spot welded.

Oil, air, and gasoline filters for automobiles are made from B-staged paper laminates. The resin-paper is still soft enough to be pleated or corrugated into shape before curing. Some air filters are white rather than brown; these white filters are bonded with melamines.

Laminated tubes based on phenolic-fabric and on phenolic-paper are found in widespread small applications. Generally the base material is B-staged, slit into narrow widths and then wound onto heated mandrels with heated backup pressure rolls. Final cure is usually in an oven.

Battery separators between electrodes in automotive batteries are required both as spacers and as electrical insulation. A top grade of thick paper with controlled porosity is impregnated, B-staged, and rolled up until used. The B-staged paper is cut to size and grooved with corrugated rolls; the ribs are reinforced with a kiss-coat of additional resin and cured. There are two basic requirements for the resole: it must be low in salt content to prevent electrical problems in the battery, and it must be primarily monomeric. This requirement is made so the molecules will be small enough to pass through the cellulose cell walls. If the resin molecule is too large, it will lie on the exterior of the fiber, and the battery acid will eventually eat away the cellulose core, leaving a drinking straw effect. The battery is ruined, of course.

At one time varnishes employing phenolic resins and particularly para substituted phenol had a wide acceptance in the paint industry. These have, for the most part, been replaced by newer polymers.

Alkylated phenolic novolacs have a place in the automotive tire industry as tackifiers. These make the rubber just sticky enough to hold the individual plies in place until the mold is closed and the tire vulcanized.

Phenolic-polyvinyl butyral combinations have found use as adhesives. They, as well as phenolic-nitrile adhesives, are used as structural adhesives.

Para substituted hard phenolic resoles are used in both neoprene cements (shoes) and in pressure sensitive tapes such as masking tapes.

Some chemical equipment for corrosive, high temperature applications are phenolic-asbestos based.

Can linings, drum linings, tank and tank core linings are still in use. A pioneer in this field was Heresite. A good part of these markets has been captured by newer plastics, the epoxies, in particular.

TRADE NAMES

Arochems	Ashland Chemical Co.
Bakelite	Union Carbide Corp.
Beckacite	Reichhold Chemicals, Inc.
Catacol	Ashland Chemical Co.
Durez	Durez Plastics Division
Genal	General Electric Co.
Micarta	Westinghouse Electric Co.
Plenco	Plastics Engineering Co.
Plyophen	Reichhold Chemicals, Inc.
Resinox	Monsanto Co.

REFERENCES

1. Bayer, A., *Ber.* 5:25 (1872).
2. Knopf, A. and Sheib, W., *Chemistry and Applications of Phenolic Resins*, p 2, Springer-Verlag, New York (1979).
3. Baekland, L.H., U.S. Patent 942,699; July 13, 1907.
4. Baekland, L.H., *Journal of Ind. and Eng. Chem.* 1:149-161 (1909).
5. Baekland, L.H., *Journal of Ind. and Eng. Chem.* 6:90 (1914).
6. Abstracts of Papers, 186th ACS National Meeting, Washington, DC, Aug. 28- to Sept. 2, 1983. American Chemical Society, 1155 16th St., N.W., Washington, DC 20036.
7. The Statistical Story of Industrial Recovery, *Modern Plastics* 61(1):57-66 (1984).
8. Pizzi, A., *Wood Adhesives, Chemistry & Technology*, (A. Pizzi, ed.), Chapter Three, Phenolic Wood Adhesives, Marcel Dekker, New York (1983).
9. Drumm, M.F. and LeBlanc, J.R., *Step-Growth Polymerizations* (D.H. Solomon, ed.), Vol. 4, Chapter 5, The Reactions of Formaldehyde with Phenols, Melamine, Aniline and Urea, Marcel Dekker, New York (1972).

10. Flory, P.J., *Principles of Polymer Chemistry*, Cornell Univ. Press, Ithaca, New York (1953).
11. Knopf, P.W. and Wagner, E.R., *Polymer Sci., Polymer Chem. Ed.* 11:939 (1973).
12. Freeman, J.H. and Lewis, C.W., *J. Am. Chem. Soc.* 76:2080-2087 (1954).
13. Zavitsas, A.A., *J. Polymer Sci.*, Part A-1, 6:2533-2540 (1968).
14. Zavitsas, A.A., Beaulieu, R.D., and LeBlanc, *J. Polymer Sci.*, 2541-2559 (1968).
15. Martin, R.W., *The Chemistry of Phenolic Resins*, John Wiley & Sons, Inc., New York (1956).
16. Megson, N.J.L., *Phenolic Resin Chemistry*, Butterworth, London, England (1958).
17. Troughton, G.E. and Rozon, L., *Wood Science* 4(4):219-224 (1972).
18. Wellons, J.D. and Gollob, L., *Symposium on Wood Adhesives*, Madison, Wisconsin, September 23-25 (1980).
19. Staff, *Modern Plastics* 6(1):57-67 (1984).

3

Urea, Melamine, and Furan Resins

Arthur L. Wooten
*Forest Products Utilization Laboratory
Mississippi State University
Mississippi State, Mississippi*

INTRODUCTION

The first synthesis of urea was made over 150 years ago by Wohler.[4] This incident remains noteworthy because before this experiment, the scientific world considered that an insurmountable barrier existed between the chemistry of living things and that of inanimate matter. By 1884 Tollens[3] and others had noted the sticky resinous substance obtained under some reaction conditions with urea and formaldehyde.

The early industrial work was dominated by CIBA and I.G. Farbenindustrie with the control being based on Pollak's patent pool of the 1920s. Urea molding compounds were the initial center of interest and adhesives did not become important until the 1930s. New British and American processes for the manufacture of urea broke the monopoly of I.G. Farbenindustrie in 1936 and resin interest increased rapidly.[2]

In 1918 Johns obtained a patent on urea condensation products capable of commercial utilization. With the acceptance of the commercial importance of the resins the veil of secrecy dropped. Much of the data were kept as trade secrets and never patented or published. This tendency towards secrecy remains characteristic of the resin industry today except in those areas where absolute novelty allows very strong patent protection (American Cyanamid, melamines;

Shell Chemical, epoxy resins). Probably the clearest summary of this era of urea resins was given by Pollak and Ripper in 1924.[1]

The best recent summary of the history of the urea resins is that in 1979 by Meyer.[2]

To summarize, much of the initial commercial interest in the urea resins was as a potential glass substitute. When the problem of moisture sensitivity and subsequent fogging together with lack of scratch resistance proved insurmountable, interest shifted to molding compounds about 1928. The use of urea resins as adhesives began a few years later. When melamines were introduced in 1939, they quickly dominated a large share of the urea molding compound market. The ureas then encountered (1960-1980) a very fast growth as interior grade wood adhesives. This usage continues today particularly in particleboard and medium-density fiberboard.

Tables 3-1, 3-2 and 3-3 show production and usage for urea and melamine resins.

Table 3-1: Urea and Melamine Pattern of Consumption

Market	...1000 Metric Tons...	
	1982	1983
Bonding and adhesive resins for:		
Fibrous and granulated wood*	345	415
Laminating	12	17
Plywood*	27	20
Molding compounds	28	34
Paper treatment and coating resins	30	33
Protective coatings	29	35
Textile treatment and coating resins	27	31
Export	8	10
Other	6	7
Total	512	602

*Includes melamine modified urea resins.

Reprinted by permission of *Modern Plastics,* January 1984.

Table 3-2: Urea Molding Powder Markets

Market	.....1000 Metric Tons.....	
	1982	1983
Closures	4	5
Electrical	11	14
Other	n*	1
Total	15	20

*n = negligible.

Reprinted by permission of *Modern Plastics,* January 1984.

Table 3-3: Melamine Powder Markets

Market	1000 Metric Tons 1982	1983
Buttons	1	1
Dinnerware	10	11
Sanitaryware	1	1
Other	1	1
Total	13	14

Reprinted by permission of *Modern Plastics,* January 1984.

CHEMISTRY OF THE UREA RESINS

Urea is the diamide of carbonic acid. It is manufactured under high pressures from carbon dioxide and ammonia.

$$2NH_3 + CO_2 \xrightarrow[\text{Heat}]{\text{Pressure}} \underset{\text{urea}}{NH_2CONH_2} + H_2O$$

Urea has become a large tonnage commodity chemical. Perhaps 10% of production is used to make urea resins. The primary use of the remainder is as fertilizer.

Urea and formaldehyde react in a pH controlled reaction to form condensation polymers. The initial formaldehyde addition is usually conducted under alkaline conditions at a pH of perhaps 8.0.

$$NH_2COCNH_2 + CH_2O \xrightarrow{pH\ 8} \underset{\text{monomethylol urea}}{NH_2CONHCH_2OH}$$

$$NH_2CONHCH_2OH + CH_2O \xrightarrow{pH\ 8} \underset{\text{dimethylol urea}}{HOCH_2NHCONHCH_2OH}$$

$$HOCH_2NHCONHCH_2OH + CH_2O \xrightarrow{pH\ 8} \underset{\text{trimethylol urea}}{(HOCH_2)_2NCONHCH_2OH}$$

The relative rates of the three successive reactions are approximately: 1:1/3:1/9. Tetramethylol urea has been detected in solutions at high formaldehyde ratios by using ^{13}C NMR. It has not been isolated. Compared to other formaldehyde condensation resins, the methylol ureas are quite stable on the alkaline side. There is relatively little condensation. In fact, crystallization of dimethylol urea was somewhat of a nuisance in the early days.

After the initial formaldehyde step, the pH is usually dropped to 5.0-5.5 and the condensation step begins. This is of two types, methylol ether formation and methylene bridge formation. Obviously the methylol ether must split out formaldehyde if the final methylene bridge structure is to be reached. There is no evidence, however, to indicate that the methylol ether is a necessary intermediate step to the methylene bridge formation taking place during the acid condensation step.

$$2\ U-CH_2OH \xrightarrow{pH\ 5} U-CH_2-O-CH_2-U$$

$$2\ U-CH_2OH \xrightarrow{pH\ 5} U-CH_2-U-CH_2OH$$

$$U-CH_2-O-CH_2-U \xrightarrow[\text{cure}]{\substack{pH \leqslant 1 \\ \text{final}}} U-CH_2-U + CH_2O$$

where U = mixture of mono-, di- and trimethylol urea.

The average urea resin is a mixture of the first type of reactions. For simplicity, the reacting methylol urea has been shown as the monocondensate. It is actually a mixture of mono-, di-, and trimethylol urea.

During the initial stages of condensation, the resin is very low in viscosity. The viscosity increases as chain length grows and network formation begins. Viscosity is the usual commercial control variable. At the predetermined endpoint the resin is cooled and the pH adjusted to 7.0-8.0 for maximum stability. Buffers and/or latent catalysts may be added at this point if needed for customer convenience.

Urea resins may be catalyzed to cure at room temperatures or to cure in a matter of minutes by hot pressing. The rate of cure is dependent on low pHs, and the pH may vary from 1-4 dependent on the processing conditions.

CHEMISTRY OF THE MELAMINE RESINS

Melamine is the triamide of cyanuric acid,

(2,4,6-triamino-1,3,5-triazine). Its production was patented by Ciba A.G. in 1936. The process converts calcium cyanamide to cyanamide and causes the product to react with ammonia under pressure to produce melamine.

The melamine analogues of urea resins appeared concurrently with the commercial availability of melamine. The addition of formaldehyde to melamine occurs more readily and more completely than with urea. Commercial products are available containing anywhere from two to six methylols per melamine molecule. The cured melamine resins are more resistant to water than are the ureas; they are harder and more chemically resistant. They are more expensive and cannot be cured at room temperature. Plates and dinnerware constitute a major market. They comprise a market of about one-fifth or more of that of the ureas. Melamine fortified urea resins for particleboard, medium-density fiberboard, and plywood are typically reported together with the ureas without being separately itemized.

FIBROUS AND GRANULATED WOOD PRODUCTS

These products, often referred to as composition boards, account for about 70% of urea resin consumption. Particleboard and medium-density fiberboard (MDF) are the largest markets in this area. The particleboard industry consumes approximately four times as much resin as does the relatively new MDF board (Table 3-1).

Particleboard consists of discrete particles of wood bonded under heat and pressure with about 7% urea resin solids (based on bone dry wood weight). Particle board was commercialized around 1946. The initial wood source was waste wood: planer shavings, plywood scrap, and even sawdust. It soon became apparent that the geometry of the particle had a significant effect on the physical properties of the finished board. This realization, together with changing manufacturing methods by the plywood and lumber producers that produced less scrap, caused the particleboard manufacturers to increase their use of low quality, locally available logs as raw materials. This, in turn, allowed the design of chippers to produce a flake with more nearly optimum geometric shape. This continuing evolution in composition boards has led to the profusion of types available today. The significance of this sequence is apparent in the market areas penetrated by particleboard. The major markets are all for interior use since the moisture induced breakdown of urea resins prevents its use in exterior, damp applications. In the decade 1963-1973 (which coincided with the emergence of the southern pine composition board industry) the annual growth rate of the industry was probably about 20%.

At the leading edge of today's technology are waferboard and oriented strandboard (OSB). Both are phenolic bonded rather than urea bonded and are aimed at the exterior and structural phenolic plywood market. Urethane bonded boards are being made although it is too early to anticipate their long range commercial success.

MDF boards are made from fiber rather than from discreet particles of wood. This process, introduced by Celotex in 1965, uses dry fibers and about 9% urea resin solids. The increase in resin content is required by the increased surface area of the individual fibers. Again, heat and pressure are used to consolidate and cure the resin-fiber mixture. MDF is somewhat more expensive than particleboard, but it quickly found a place because of its increased strength, extreme smoothness, and tight edges. Particleboard must be edge-banded to hide the rough interior. Again, as with any urea bonded composite, it finds its market in interior use. Furniture is an obvious and major outlet for MDF.

Figure 3-1: Fabricated items made from particleboard. Particleboard has superior structural strength and is somewhat less expensive than medium-density fiberboard. (Courtesy of Weyerhaeuser.)

Figure 3-2: MDF (medium-density fiberboard) is much smoother and has tighter edges than particleboard. It is not used in structural applications. (Courtesy of Weyerhaeuser.)

UREA BONDED PLYWOOD

Urea bonded plywood is a relatively small part of the urea adhesive market. Urea is restricted to interior applications and, in the U.S., is found primarily in the hardwood decorative panels. The much larger structural market has to a large degree shifted to phenolic resins even for interior uses. This change is largely induced by product economics and the size of the commodity type structural plywood market.

Urea plywood resins, like the particleboard resins, are catalyzed by acids to provide the needed gel time. The cure time can be controlled by the supplier with latent temperature-sensitive catalysts or by the customer adding ammonium chloride or other catalysts.

The urea plywood system, unlike the particleboard system, is usually highly extended, up to 100% with wheat flour.

Urea bonded plywood is, in the U.S., a stable or declining market. About 20 thousand metric tons of urea and melamine resins were consumed in 1983 (Table 3-1). Imports of urea bonded hardwood plywood, primarily from Asia, have continued to increase. Today they have well over one-half the market. Decorative paneling is the principal end use.

OTHER UREA RESIN MARKETS

Laminated beams can be bonded by ureas catalyzed to cure at room or at slightly elevated temperatures. These are too large for presses and ovens.

Etherified urea and melamine resins have traditionally been used as crosslinking agents in industrial baking finishes. Ureas have continued to decline, and melamines probably account for 70% of today's consumption.

The resins are etherified for solubility in organic solvents. Butanol was for years the favorite; inroads were made by isobutanol on a cost basis. In recent years, methyl alcohol modified resins have increased their share of this market because of their water solubility. The drive, spearheaded by the regulatory agencies, to eliminate organic solvents in favor of water solvents or 100% nonvolatile coatings continues unabated.

The alkylated resins are quite compatible with alkyds, cellulosics, and epoxies. They improve the cure rate, hardness, and chemical resistance of the binder system.

Until about 1978, alkyd-ureas dominated the factory-finished plywood panel market. The base coat is generally about 25% urea and the topcoat about 35%. Vinyl coatings and waterborne acrylics have penetrated the base coat market. In 1981, alkyd-amino and alkyd-acrylic solvent based topcoats still held about 80% of the topcoat market. The waterborne systems are slower curing. Kitchen cabinets and low cost furniture use etherified ureas as crosslinkers with alkyds. They have lost their dominance to newer systems, but are expected to continue to be a factor in the market.[5]

A major market for alkyd-urea combinations continues to be the baked enamel metal coating industry. Here again, melamines are used where the maximum resistance is needed. This market will decline with the increase in high solids coatings, waterborne coatings, and electrodeposition.[5]

Probably 70 to 80% of beverage cans contained amino resin crosslinkers in 1978. Waterborne epoxies probably predominate in this field today.[5]

The use of ureas in automotive topcoats has been replaced by the harder melamines. These, in turn, are being replaced by the newer systems.

About 30 thousand metric tons of urea resins at about 35 to 40% nonvolatile were consumed by the paper industry in 1983. A small amount of this went into surface coatings but most of it was used as wet strength additives. A low molecular weight urea resin modified with relatively small amounts of a polyamine (e.g., diethylenetriamine) is a cationic polymer. These charged molecules are selectively adsorbed onto the cellulose from the dilute pulp. Small amounts contribute significantly to the wet strength of the resulting paper (e.g., butcher wrap paper, paper toweling, etc.). The pH of the pulp is dropped by the addition of alum, not only to set the resin size but also to catalyze the cure of the urea. This market is shared with colloidal dispersions of melamine resins and also with polyamide-epichlorohydrin polymers which offer the sometimes advantage that they can be cured at neutral or slightly alkaline pH.

Production of urea molding compounds increased considerably in 1983. Whether the trend is significant remains to be seen. The market has been relatively static since about 1968 (see Table 3-1). Urea molding compounds have good color, heat resistance, and electrical properties. Seventy percent of the use is in electrical applications, much of it in wall switches and electrical receptacles. For molding, ureas are almost always compounded with alpha-cellulose. Colors, release agents, etc., are added during the compounding. Table 3-4 shows some selected properties of alpha-cellulose filled urea molding compounds.

A more recent market for modified ureas has been found in the "tarpaper shingle" market. Traditionally, these shingles were made from a woodpulp base, saturated with a sprinkling of mineral particles. Around 1977 it was found that modified urea resins could bond fiberglass mats which, after curing, could be substituted for the tarpaper. These shingles are rendered essentially fireproof by the elimination of the paper base and decreased tar content, and carry an improved fire risk insurance rating.

The original crease-resistant, wash-and-wear clothing treatments were based on urea resins. The low molecular weight resins were able to penetrate the individual cellulose cell walls so that the fabric was not stiffened by the treatment. It soon became apparent that the urea molecule was not sufficiently resistant to chlorine based bleaches. Ethylene urea was considered superior for some time, then melamine was shown to be superior to ethylene urea. Today this is a decreasing market both because of new processes and new synthetic fabrics.

Foamed-in-place housing insulation has been used for decades in Europe, particularly in the Scandinavian countries. This market began in the U.S. in the early 1970s. Improper formulations, untrained applicators, and poor performance caused resistance to the use of foamed urea-formaldehyde. Additionally, the regulatory agencies have prohibited this usage because of alleged toxicological problems.

The use of melamine resins for print sheet and for overlay sheet

Table 3-4: Properties of Amino (Urea, Melamine, Furan) Molding Compounds

			Urea	Melamine formaldehyde	
	Properties	ASTM test method	Alpha cellulose-filled	Cellulose-filled	Glass fiber-reinforced
Processing	1. Melting temperature, °C. T_m (crystalline)		Thermoset	Thermoset	Thermoset
	T_g (amorphous)				
	2. Processing temperature range, °F. (C = compression; T = transfer; I = injection; E = extrusion)		C: 275-350 I: 290-320 T: 270-300	C: 280-370 I: 200-340 T: 300	C: 280-350
	3. Molding pressure range, 10^3 p.s.i.		2-20	8-20	2-8
	4. Compression ratio		2.2-3.0	2.1-3.1	5-10
	5. Mold (linear) shrinkage, in./in.	D955	0.006-0.014	0.005-0.015	0.001-0.006
Mechanical	6. Tensile strength at break, p.s.i.	D638	5500-13,000	5000-13,000	5000-10,500
	7. Elongation at break, %	D638	<1	0.6-1	0.6
	8. Tensile yield strength, p.s.i.	D638			
	9. Compressive strength (rupture or yield), p.s.i.	D695	25,000-45,000	33,000-45,000	20,000-35,000
	10. Flexural strength (rupture or yield), p.s.i.	D790	10,000-18,000	9000-16,000	14,000-23,000
	11. Tensile modulus, 10^3 p.s.i.	D638	1000-1500	1100-1400	1600-2400
	12. Compressive modulus, 10^3 p.s.i.	D695			
	13. Flexural modulus, 10^3 p.s.i. 73° F.	D790	1300-1600	1100	
	200° F.	D790			
	250° F.	D790			
	300° F.	D790			
	14. Izod impact, ft.-lb./in. of notch (⅛-in. thick specimen)	D256A	0.25-0.40	0.2-0.4	0.6-18
	15. Hardness Rockwell	D785	M110-120	M115-125	M115
	Shore/Barcol	D2240/ D2583			
Thermal	16. Coef. of linear thermal expansion, 10^{-6} in./in./°C.	D696	22-36	40-45	15-28
	17. Deflection temperature under flexural load, °F. 264 p.s.i.	D648	260-290	350-390	375-400
	66 p.s.i.	D648			
	18. Thermal conductivity, 10^{-4} cal.-cm./sec.-cm.^{2}-°C.	C177	7-10	6.5-10	10-11.5
Physical	19. Specific gravity	D792	1.47-1.52	1.47-1.52	1.5-2.0
	20. Water absorption (⅛-in. thick specimen), % 24 hr.	D570	0.4-0.8	0.1-0.8	0.09-1.3
	Saturation	D570			
	21. Dielectric strength (⅛-in. thick specimen), short time, v./mil	D149	300-400	270-400 175-215 @ 100° C	130-370

Materials	Properties		ASTM test method	Melamine phenolic Woodflour- and cellulose- filled	Furan Asbestos-filled
Processing	1.	Melting temperature, °C. T_m (crystalline) T_g (amorphous)		Thermoset	Thermoset
	2.	Processing temperature range, °F. (C = compression; T = transfer; I = injection; E = extrusion)		C: 300-350 I: 350-400	C: 275-300
	3.	Molding pressure range, 10^3 p.s.i.		5-20	0.1-0.5
	4.	Compression ratio		2.1-4.4	
	5.	Mold (linear) shrinkage, in./in.	D955	0.009-0.010	
Mechanical	6.	Tensile strength at break, p.s.i.	D638	6000-8000	3000-4500
	7.	Elongation at break, %	D638	0.4-0.8	
	8.	Tensile yield strength, p.s.i.	D638		
	9.	Compressive strength (rupture or yield), p.s.i.	D695	26,000-30,000	10,000-13,000
	10.	Flexural strength (rupture or yield), p.s.i.	D790	8000-10,000	600-9000
	11.	Tensile modulus, 10^3 p.s.i.	D638	800-1700	1580
	12.	Compressive modulus, 10^3 p.s.i.	D695		
	13.	Flexural modulus, 10^3 p.s.i. 73° F.	D790	1000-1200	
		200° F.	D790		
		250° F.	D790		
		300° F.	D790		
	14.	Izod impact, ft.-lb./in. of notch (⅛-in. thick specimen)	D256A	0.2-0.4	
	15.	Hardness Rockwell	D785	E95-100	R110
		Shore/Barcol	D2240/ D2583		
Thermal	16.	Coef. of linear thermal expansion, 10^{-6} in./in./°C.	D696	10-40	
	17.	Deflection temperature under flexural load, °F. 264 p.s.i.	D648	285-310	
		66 p.s.i.	D648		
	18.	Thermal conductivity, 10^{-4} cal.-cm./ sec.-cm.^{2}-°C.	C177	4-7	
Physical	19.	Specific gravity	D792	1.5-1.7	1.75
	20.	Water absorption (⅛-in. thick specimen), % 24 hr.	D570	0.3-0.65	0.01-0.2
		Saturation	D570		
	21.	Dielectric strength (⅛-in. thick specimen), short time, v./mil	D149	220-325	

Reprinted by permission of *Modern Plastics Encyclopedia*, McGraw-Hill, Inc.

continues to offer substantial markets. The decorative laminate, most widely known as Formica (trade name of Formica Co., Division of American Cyanamid) is widely used as a covering for kitchen counters, cabinets, etc. There are a variety of alternates being offered at lesser cost, and the long-term market for melamine is expected to decline because of cost.

Melamine molding powders are more expensive than urea-formaldehyde compounds but have superior properties. Systems filled with alpha-cellulose are more thermally and chemically resistant, are harder and have considerably less water absorption (50 to 75% less depending on formulation). Because melamines can withstand long term exposure to 100°C (212°F) they can be repeatedly washed and sterilized.

This latter property, along with improved gloss and stain resistance, makes melamines ideal for low cost institutional and household tableware. However, their scratch resistance and stain resistance to many foodstuffs is not the best overall, even with many recently developed overcoats. Other molded parts made from melamine include lavatory units, knobs, electrical appliance housings, automotive ignition parts, terminal blocks and connectors.

Table 3-4 shows selected properties of some general types of melamine molding compounds.

Melamines plasticized with oil-modified alkyds have been used in coatings that produce better chemical, heat and water resistance than urea-based coatings. They must be heat cured and require as much as two times more plasticization to get equivalent hardness, flexibility and adhesion to substrates.

THE FURAN POLYMERS

Furan polymers are not large volume resins in comparison to the other plastics. For this reason they are frequently grouped with other resins, in this case—the ureas. This classification is made in the belief that the urea-formaldehyde-furfuryl alcohol resins for foundry usage constitute the largest market for furans.

The term furan polymer or resin is a loosely defined term. It can be used to denote polymers based on furfural (I), furfuryl alcohol (II), or furan (III)

I   II   III

Furfural is the starting material for all of these compounds. Commercialized by Quaker Oats Company just after World War I, it is obtained by acid hydrolysis and steam distillation of various agricultural residues. Corn cobs and rice hulls have frequently been used.

Furfural readily forms novolac resin with phenol. Contrary to the normal formaldehyde based phenolics for novalac molding compounds, the furfural resins are prepared with an alkaline catalyst. The molding compounds made from these resins have a characteristic long flow for a given gel time. They were in considerable demand for molding TV cabinets in the early days of TV. A very long flowing resin was required to properly fill the large thin walled mold. The phenol-furfural resins have excellent chemical resistance but their sales volume has decreased substantially in recent years, primarily because of competitive resin advantages. Table 3-4 shows selected data for an asbestos-filled furan.

A considerable amount of furfural is used in the resin bonded abrasive wheel industry. The abrasive grit is wet with a small amount of furfural before the powdered hexa/phenolic resin is added to the mixer. The furfural causes the particles of resin to coat the grains and prevents the powder from segregating. During the heat curing cycle about 90% of the furfural chemically combines with the phenolic resin.

Furfuryl alcohol is manufactured by hydrogenating furfural. Furfuryl alcohol resinifies readily in the presence of acids. In addition to the normal methylol reaction, it can also polymerize by reaction at the unsaturated ring bonds and by ring cleavage. It can also copolymerize with urea and phenolic resins.

Furfuryl alcohol urea resins find their largest market in the foundry industry as no-bake core binders. The urea resin containing perhaps 20 to 50% furfuryl alcohol is usually mixed with sand already coated with a predetermined amount of acid catalyst. After mixing, the sand resin mixtures are formed into the desired core shape and allowed to stand until the resin has cured. The time is determined by the amount of acid catalyst present and may for large cores, be overnight. The working life of the sand resin mixture naturally decreases as the catalyst is increased. No-bake binders can also be made with a combination of furan resin-furfuryl alcohol rather than urea formaldehyde-furfuryl alcohol.

The chemical resistance of furan resins has been used for many years to advantage in chemical cements.

After carbonizing, furfuryl alcohol impregnated graphite gives a very low permeability graphite that finds application in nuclear reactors.

Tetrahydrofuran, obtained by the hydrogenation of furan, can polymerize to long chain butane diol based polyethers. The initiation is cationic, and chain polymerization occurs.

TRADE NAMES

Trade Name	Type of Resin	Company
Beetle	U/F	American Cyanamid
Cata-	U/F, M/F and Furan	Ashland Chemical
Cymel	M/F	American Cyanamid
Fabrez	U/F	Reichhold Chemicals
Melamine	M/F	Fiberite Corp.
Melatine	M/F	Ciba-Geigy Corp.
Melmac	M/F	American Cyanamid
Melurac	M/F	American Cyanamid
Permalite	M/F	Ciba-Geigy Corp.
Plaskon	U/F and M/F	Allied Corp.
Plaspreg	Furan	Furane Div., M&T Chemical
QuaCorr	Furan	Quaker Oats Co.

REFERENCES

1. Pollak, F., and Ripper, K., *Chem. Zeit.* 48,569 (1924).
2. Meyer, B., *Urea Formaldehyde Resins*, Addison-Westey Publishing Company, Inc., Reading, MA (1979).
3. Tollens, B., *Ber.* 17,653 (1885).
4. Wohler, F., *Pogg. Annalen der Physik* 12,253 (1828).
5. *Chemical Economics Handbook*, Amino Resin, SRI International, Menlo Park, CA (1982).

4

Unsaturated Polyester and Vinyl Ester Resins

Oscar C. Zaske
Silmar Division
Sohio Engineered Materials Company
Hawthorne, California

UNSATURATED POLYESTERS

History

  The laboratory preparation of polyesters probably first occurred in 1847 with Berzelius cooking a saturated polyester from tartaric acid and glycerine.[1] The earliest record of chemical work with unsaturated polyesters is the study of glycol maleates by Vorlander in 1894.[2]

  Polymer chemistry as a science did not really develop until the first half of the twentieth century. The 1920s witnessed the brilliant and pioneering work of Wallace Carothers which included studies on polyesters among which were unsaturated types prepared from ethylene glycol and unsaturated acids and anhydrides such as fumaric acid and maleic anhydride.[3]

  It was soon discovered that unsaturated polyesters, although possessing reactive double bond unsaturation, were sluggish in reacting with themselves or homopolymerizing. Although responsive to catalysis with peroxide catalysts, relatively high temperatures and rather long times were required to obtain a complete curing reaction. These resins were also totally unlike our present day low viscosity liquid unsaturated polyester (UP) resins in that they were either solids or very high viscosity, relatively immobile liquids.

60 *Handbook of Thermoset Plastics*

The key which made possible the modern unsaturated polyester resins of today was the discovery by Carlton Ellis that the addition of liquid unsaturated monomers such as monomeric styrene gave mixtures which would copolymerize at rates twenty to thirty times faster than the homopolymerization rate for unsaturated polyesters by themselves.

Further, if a low viscosity liquid monomer such as styrene were used, the resultant mixture could be an easily handled liquid that could be readily cast or molded without the need for high molding pressures.

Styrene monomer, although an optimum monomeric diluent for unsaturated polyesters, was in this time period still a relatively costly chemical. The economics of unsaturated polyesters based on such high cost styrene would have, very likely, retarded any broad based applications of UP.

The second World War provided the last two developments which propelled UP into the applications and economic position they occupy today. Early during the war it was found that styrenated polyester could yield high strength, low weight structures when reinforced with glass fibers. The fabrication of such structures could be accomplished using very low molding pressures making it possible to mold very large structures, if necessary, in relatively light weight, low cost tooling.

It was also found that such fiber glass reinforced UP composites had excellent electrical properties including low loss factor properties which gave them relatively high transparency to the radar beams which were to play an ever increasingly important role as the war progressed.

Concurrently styrene monomer became increasingly available and cheaper as styrene monomer moved from the status of a lab chemical to that of a chemical commodity. This was a result of large styrene plants being built with U.S. Government assistance to provide the styrene necessary for the production of styrene-butadiene synthetic rubber. Prior to the entry of the United States into World War II, the U.S. Government foresaw the development of natural rubber shortages due to the spread of the war into the Far East and established The Rubber Reserve Corporation to stockpile natural rubber and pursue synthetic rubber research and development. This research work resulted in the development of GR-S (Government Rubber Styrene) rubber, a styrene-butadiene copolymer. By the end of the war five styrene production plants had been built with government financing to supply the styrene for the production of GR-S rubber. These styrene plants, by 1955, were all sold to private industry.[4]

The war years saw the production techniques for making polyester fiber glass radomes developed to a fine art. With the end of the war, commercial development proceeded rapidly with materials and molding methods research moving in all directions. Resin types proliferated, applications multiplied and new raw materials for resin manufacture became available. Hosts of compounding materials, such as fillers, pigments, reinforcements, light stabilizers, curing catalysts and promoters were introduced for use with unsaturated polyesters.

UP today are a billion pound per year business which grew steadily from a production level of about 19 million pounds in 1952.[5] The number of UP producers also markedly increased from those early days until by May of 1984 there were fifteen major domestic producers with a total production capacity estimated to be 1,775,000,000 pounds per year. A recent U.S. projection estimated a 1985 UP production of about 1.4 billion pounds compared to 1.3 billion plus in 1984 and about 1.1 billion in 1983.[6]

Consumption patterns for unsaturated polyesters were estimated in a recent Jan. 1985 article in *Modern Plastics* magazine, "Materials '85" (p. 63) to be as follows:

Table 4-1: Unsaturated Polyester Applications

Market	1,000 Metric Tons 1983	1984
Reinforced polyester*		
Molded, filament-wound, pultruded, etc.	300	350
Sheet, flat and corrugated	74	80
Surface coating	6	7
All other**	111	127

*Resin only.
**Autobody putty, furniture, cultured marble, etc.

Similar data for vinyl ester resins is difficult to obtain or estimate but some market surveys have estimated volume to be in the neighborhood of twenty to thirty million pounds per year.

The history of vinyl ester resins development is mainly chemical and will be covered in the section on the chemistry of vinyl ester resins later in the chapter.

Chemistry

General Concepts: Unsaturated polyesters are condensation polymers formed by the reaction of polyols and polycarboxylic acids with olefinic unsaturation being contributed by one of the reactants, usually the acid. The polyols and polycarboxylic acids used are usually difunctional alcohols (glycols), and difunctional acids such as phthalic and maleic. Water is produced as the by-product of the esterification reaction and is removed from the reaction mass as soon as it is formed to drive the polyesterification reaction to completion. All of the materials used must be at least difunctional to make the polyesterification reaction possible.

Unsaturated polyesters differ from saturated polyesters such as the polyethylene terephthalate which constitutes the polyester films and fibers of commerce in that acids or glycols having double bond unsaturation are included in the formula to provide reactive olefinic unsaturation in the unsaturated polyester alkyd. This reactive unsaturation can then be used to form thermosetting, crosslinked polymers with monomers such as styrene and methyl methacrylate which contain olefinic double bonds. The term "alkyd" originating from the early days of coating resins technology is often also used for polyester resins capable of crosslinking with themselves or copolymerizing with monomers.

Functionality: As noted above the acids and polyols used to make unsaturated polyesters must be at least difunctional. The fundamental work underlying this understanding was done on the types of alkyds used for protective coatings. A prominent worker in coating alkyds, R.H. Kienle, proposed the functionality theory[7,8,9] in papers during the 1930s.

This theory simply states that monomer molecules must have at least two reactive groups to be able to form a polymer. Monofunctional reactants such as ethyl alcohol and acetic acid can react (esterify) to form an ester but are incapable of forming a polymer (polyester). Two difunctional reactants, such as propylene glycol (dihydroxy functionality) and maleic acid (dicarboxylic functionality) can esterify with the esterification reaction continuing to form a long chain polyester which can contain many repeating units of the basic mer, propylene glycol maleate.

Polyesterification Reaction: Polyesterification is the most important reaction in the preparation of unsaturated polyesters. Side reactions also take place. These have been enumerated by E.E. Parker[10] as:

1. Isomerization of maleate to fumarate.
2. Addition of glycol to maleate and fumarate double bonds.
3. Oxidative destruction of double bonds.
4. Loss of glycol.

The general chemistry of unsaturated polyesters can be illustrated by the following representation of the synthesis of a general purpose propylene glycol, maleic anhydride, phthalic anhydride polyester.

$$HO-CH_2-\underset{\underset{OH}{|}}{CH}-CH_3 \;+\; \underset{CH-C}{\overset{CH-C}{\underset{\|}{}}}\!\!\overset{O}{\underset{O}{\diagup\!\!\diagdown}}\;O \;+\; \underset{}{\bigodot}\!\!\overset{C=O}{\underset{C=O}{\diagup\!\!\diagdown}}O \;\xrightarrow[\substack{400\text{-}450°F \\ (204\text{-}232°C)}]{\text{Inert Gas Sparge}}$$

Propylene Glycol   Maleic Anhydride   Phthalic Anhydride

$$HO\!\!\left[\underset{\underset{CH_3}{|}}{CH}-CH_2-O-\underset{\underset{O}{\|}}{C}-CH=CH-\underset{\underset{O}{\|}}{C}-O-\underset{\underset{CH_3}{|}}{CH}-CH_2-O-\underset{\underset{O}{\|}}{C}-\bigodot-COOH\right]_n + HOH\uparrow$$

1:1 MA:PA-PG   GP Unsaturated Polyester Alkyd

Isomerization: The esterified maleic is shown as the trans isomer, fumarate, since under the conditions of esterification more than 90% of the maleate ester isomerizes from the cis maleate isomer to the trans fumarate isomer.

$$\begin{array}{cc} \text{CH}-\text{C}\overset{O}{\underset{O}{\diagdown}} & \text{CH}-\text{COOH} \\ \parallel \quad \quad \diagup & \parallel \\ \text{CH}-\text{C}\overset{}{\diagdown_O} & \text{COOH}-\text{CH} \end{array}$$

cis   trans

The trans isomer is a lower energy level, less strained structure than the cis isomer configuration and the isomerization of maleate to fumarate occurs readily in the processing of most unsaturated polyesters. This is of great commercial importance because maleic anhydride is the least costly source for unsaturation in unsaturated polyesters. Fortunately, fumarate unsaturation is much more reactive in crosslinking reactions than maleate unsaturation and is the isomer preferred for practical reactivity levels. Fumarate ester in the form of ethyl fumarate has been cited as being up to forty times more reactive with vinyl monomers than ethyl maleate.[11]

The polyols used in the manufacture of unsaturated polyesters are generally dihydric alcohols or glycols. Propylene, ethylene, dipropylene and diethylene glycols are frequently used members of the glycol family.

Use of secondary glycols such as propylene glycol in the unsaturated polyester cook will favor a high isomerization of maleate to fumarate. Isomerization will be less with the use of primary glycols such as ethylene glycol.

V. Szmercsanyi, Marcos and Zahran investigated the effect of different glycols on the maleate-fumarate isomerization in unsaturated polyesters. Table 4-2 shows the isomerization variations incidental to the use of different glycols they found.[12]

Table 4-2: Isomerization vs Glycol Type

Reaction Temperature Constant = 356°F (180°C)

Glycol	Type	Isomerization, %
1,2-Propylene	Secondary	96
Ethylene	Primary	64
Diethylene	Primary	53
1,6-Hexamethylene	Primary	36

The extent of maleate to fumarate isomerization also depends on reaction or cooking temperature and time, higher temperatures and longer times generally favoring greater isomerization.

V. Szmercsanyi et al,[12] in their investigation of the kinetics of maleate-fumarate isomerization, measured the percent isomerization in the cooking of poly(propylene glycol maleate) at temperatures ranging from 221° to 356°F (105° to 180°C). These investigators found a direct dependence of isomerization on reaction temperature as shown in Table 4-3.

Table 4-3: Isomerization vs Reaction Temperature

Reaction Temperature, °F (°C)	Isomerization to Fumarate, %
221 (105)	ca 34
257 (125)	ca 56
284 (140)	ca 75
356 (180)	ca 96

From an overall standpoint it was concluded that the rate of isomerization was mainly a function of the glycol structure and the temperature of the condensation since after a time, with all glycols and temperatures, the isomerization approached a constant value.

Polyesterification Reaction Speed: The speed of the polyesterification reaction can be increased in several ways all of which effect more efficient removal of the water produced as a by-product of the reaction. Unsaturated polyesters are cooked under a blanket of inert gas to minimize oxidative degradation at cook temperatures. Inert gas is also usually introduced below the surface of the cook (sparging) to increase the liquid/gas interfacial area for mass transfer of water from the cook and to assist with the agitation. Increasing the inert gas sparge rate or agitator speed, by increasing mass transfer area, increases the rate of water removal and hence reaction rate. The use of vacuum can also increase reaction rate by increasing the partial pressure of the by-product water.

The reaction can also be accelerated by the introduction of esterification catalysts. Among these are mineral acids such as sulfuric, aryl sulfonic acids such as p-toluene sulfonic acid, tin compounds such as dibutyl tin oxide and titanates such as tetrabutyl titanate.

Most esterification catalysts introduce disadvantages such as darker colored resins, haze in the resin or the necessity for their removal from the resin on completion of reaction.

In unsaturated polyesters some of the tin compounds such as dibutyl tin oxide and butyl stannoic acid offer the best compromise between reaction acceleration and undesirable side effects.

Processing

It is usually necessary to use a stoichiometric excess of glycol in cooking unsaturated polyesters because of some glycol loss along with the reaction water and decomposition of some of the glycol.

This glycol excess can often by reduced with the use of some of the more efficient esterification catalysts.

On completion of the polyesterification reaction the unsaturated polyester alkyd can be discharged from the reaction kettle as a molten mass into drums or onto flaking equipment such as a water cooled stainless steel belt or drum or it can be mixed with a liquid monomer such as styrene to give the liquid resin which is the most commonly used form of unsaturated polyesters.

As noted previously, the homopolymerization rates of unsaturated polyesters are very sluggish and they are usually used in combination with unsaturated monomers, the most common of which, by far, is monomeric styrene.

In the manufacture of a typical styrenated unsaturated polyester the molten unsaturated polyester alkyd, on completion of the cook, is transferred from the cooking kettle to a thinning or styrenation tank containing styrene monomer where the alkyd and styrene are blended with vigorous agitation. Since the mixture of unsaturated polyester alkyd and monomer are very coreactive, polymerization inhibitors must be present in the styrene or alkyd before thinning to prevent copolymerization from occurring during the thinning operation. Hydroquinone or substituted hydroquinones such as toluhydroquinone or monotertiary butyl hydroquinone are commonly used.

Typical General Purpose (GP) Unsaturated Polyester Resin

Table 4-4 shows the reactor charge for a typical GP polyester resin.

Table 4-4: Reactor Charge for a Typical GP Polyester Resin

Material	Mols	Pounds	Pounds Per Pound of Resin .....@ 60% NVM*......
Phthalic Anhydride	1.5	222.0	0.2774
Propylene Glycol	2.7	205.2	0.2564
Maleic Anhydride	1.0	98.0	0.1225
Hydroquinone	–	–	0.0001
Styrene Monomer	–	–	0.4000
Total materials to make 1 lb. of thinned resin at 60% NVM (40% styrene content)			1.0564

Reaction water = approximately 45 pounds.
Alkyd cook yield = approximately 480.2 lbs. = approximately 91.43%.
Overall thinned resin yield = approximately 94.66%.

*Nonvolatile Matter.

Common Resin Synthesis Raw Materials

Common raw materials used for cooking unsaturated GP and specialty polyesters are shown in Table 4-5.

Table 4-5: Common Raw Materials for Polyesters

	Contributes
Glycols:	
Propylene Glycol (PG)	Low cost, styrene compatibility
Ethylene Glycol (EG)	Low cost, rigidity
Dipropylene Glycol (DPG)	Flexibility, toughness
Diethylene Glycol (DEG)	Flexibility, toughness
Neopentyl Glycol (NPG)	UV, water and chemical resistance
Trimethylpentanediol (TMPD)	Water and chemical resistance
Cyclohexane Dimethanol (CHDM)	Electrical properties
Propoxylated Bisphenol A (PBPA)	Water and chemical resistance
Hydrogenated Bisphenol A (HBPA)	Water and chemical resistance
Dibromoneopentyl Glycol (DBNPG)	Flame retardance
Acids:	
Phthalic Anhydride (PA)	Low cost, styrene compatibility
Maleic Anhydride (MA)	Lowest cost unsaturation
Adipic Acid (AA)	Flexibility, toughness
Isophthalic Acid (IPA)	Toughness, water and chemical resistance
Terephthalic Acid (TPA)	Higher heat deflection point
Fumaric Acid (FA)	Maximum reactive unsaturation
Glutaric Acid	Flexibility, toughness
Dimer Acids	Flexibility, toughness
Azelaic Acid	Flexibility, toughness
Chlorendic Acid	Flame retardance, chemical resistance
Tetrabromophthalic Anhydride	Flame retardance, chemical resistance
Tetrachlorophthalic Anhydride	Flame retardance, chemical resistance
Endomethylenetetrahydrophthalic Anhydride	Air drying properties

Copolymerization of Unsaturated Polyester Alkyds with Monomers

Unsaturated polyesters copolymerize with monomers having olefinic unsaturation much more rapidly than they homopolymerize so most unsaturated polyesters are used as mixtures with reactive, usually liquid, monomers. Of such monomers styrene is by far the most used monomer. Commercially the term unsaturated polyester usually refers to the combination of an unsaturated polyester alkyd with styrene monomer.

With number average molecular weights usually in the range of 800 to 3,000, unsaturated polyester alkyds are not high polymers but rather may be considered to be reactive low molecular weight prepolymers. Copolymerization with an unsaturated monomer such as styrene in a subsequent curing reaction forms the three-dimensionally crosslinked polymer that is a cured unsaturated polyester resin. Unless noted otherwise, subsequent reference to polyester resin will refer to the liquid resin made by the solution of unsaturated polyester alkyd in styrene monomer. This type of resin, being a liquid system, offers all of the advantages of liquids such as ease of handling with little or no pressure being required for mixing with other ingredients or casting and molding operations.

Styrene is an ideal monomer in most respects for use in polyester resins because it is low in cost, can give low viscosity resins at reasonable monomer levels and copolymerizes readily with unsaturated polyester alkyd in either room temperature or elevated temperature curing systems.

The copolymerization chemistry of unsaturated polyester alkyds and unsaturated monomers is usually initiated by free radicals generated by the decomposition of peroxides, azo compounds or the generation of free radicals by the use of medium to high energy radiation such as ultraviolet light or electron beams.

With styrene as the monomer, the copolymerization reaction involves the addition of styrene monomer across the fumarate/maleate double bonds in the unsaturated polyester alkyd chains as pictorially depicted below.

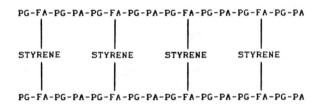

PG-FA = Propylene Glycol Fumarate in alkyd chain

PG-PA = Propylene Glycol Phthalate in alkyd chain.

The above is a simplified schematic; in commercial resins more styrene is employed resulting in a ratio of styrene double bonds to alkyd double bond greater than one. Workers such as Haman, Funke and Gilch early found that the number of fumarate double bonds participating in the crosslinking reaction with styrene increases with increasing styrene content in the unsaturated polyester resin.[13] In an unsaturated polyester prepared from 3.6 mols of FA, 2.4 mols of AA and 6.6 mols of 1.6 hexane diol it was found that maximum cure occurred at a ratio of about 1.5 to 2.0 styrene double bonds per fumarate bond. Lower styrene contents resulted in no reaction of an appreciable number of the unsaturated polyester alkyd double bonds.[14]

Typical general purpose resins have styrene to alkyd double bond ratios ranging from two to three or even higher. This means, of course, that some of the styrene crosslinks in the cured resins can have more than one styrene molecule forming the crosslink and that some styrene molecules may simply add a branch on an alkyd chain formed from two or three styrene molecules which have no attachment to another alkyd chain. It has been claimed that a small amount of low molecular weight polystyrene can be formed under some conditions.[15] It has also been claimed, on the basis of degradative and spectroscopic studies, sol-gel analysis and dynamic measurements that in addition to styrene-fumarate copolymerization, a fumarate-fumarate crosslinking reaction occurs at low styrene concentrations.[16]

In the polymerization or cure of a typical styrene containing unsaturated polyester the liquid resin reaches gel-like consistency (gel point) when less than 5% of the original carbon to carbon double bonds have been converted and a physically rigid state is achieved when only 50% of the double bonds of the unsaturated polyester alkyd and styrene have been used up.[17]

Cure of the resin system, of course, normally continues on beyond the 50% double bond conversion point to a fairly high degree of conversion, usually above 90 to 95%.

The curing behavior of an unsaturated polyester resin can be illustrated by the time versus exotherm temperature curve generated by the SPI (Society of The Plastics Industry) Gel Time Test, a very useful tool for assessing the curing rates and exothermic characteristics of a resin.

Processing Equipment and Manufacturing

Typically unsaturated polyesters are cooked in 304 stainless steel kettles from 1,000 to 5,000 gallon capacities. The kettles are equipped for heating via internal coils and heat transfer fluids. The same coils and fluids, cooled, are also used for cooling the kettle contents when necessary. The kettles are equipped with heavy duty agitators. The reaction water is taken off the kettle through either packed distillation columns or partial condensors to more efficiently separate the glycol which comes off with the reaction water so the glycol can be returned to the kettle. Coming off with the reaction water are reaction by-products from glycol decomposition such as aldehydes which are more volatile. These reaction by-products, together with the reaction water and small nonrecoverable amounts of the glycols and acids used in the charge, must then be dealt with in conformance to regulations that regulate the handling and disposal of industrial wastes.

The cooking operation begins with charging raw materials into the kettle by the use, usually, of a jacketed weigh hopper for the materials handled as liquids such as molten PA, molten MA, PG, EG, DEG, and DPG. Nonliquid materials such as AA and IPA can manually be "bagged" in via the kettle charging port or by handling equipment for powdered and granular materials.

The liquid glycols are usually charged first followed by the granular and then the molten materials such as PA and MA. Agitation can be started once a liquid "heel" has been established in the kettle. The kettle is then slowly brought up to temperature at a programmed rate which is designed to allow for the significant exothermic heat released when the MA and PA anhydride rings open on reacting with the glycols in the charge.

Once the kettle contents have been brought up to cooking temperature the contents are sampled periodically during the cook for the measurement of acid number and process viscosity. Cooking is continued until the desired acid number and process viscosity are achieved in the resin batch. A deviation from the desired acid number-viscosity curve can usually be detected early enough in the cook so that the resin batch can be brought back to the correct acid number-viscosity relationship by the addition of a small, calculated amount of extra glycol or acid.

Typical resin cooking temperatures are in the range of 400° to 450°F (204° to 232°C). Cooking times range from 8 hours to 28 hours depending on the raw materials used, the degree of condensation or molecular weight desired and the cooking temperature employed.

Mixing the completed unsaturated polyester alkyd with a liquid monomer, usually styrene, is accomplished by transferring the finished molten alkyd to a thinning tank previously charged with styrene containing polymerization inhibitors to prevent copolymerization of the unsaturated polyester alkyd and styrene. The molten alkyd is added gradually with vigorous agitation in the thinning tank. The thinning tank, also usually made of 304 stainless steel, is equipped with cooling coils through which cooling water is circulated to keep the thinning tank contents below a safe temperature, usually 180° to 200°F (82° to 93°C). The alkyd must be transferred to the thinning tank at a rate at which the sensible heat being added by the alkyd does not exceed the ability of the cooling system to keep the tank contents at a safe temperature level.

Thinning tanks, naturally, must be larger than their corresponding cooking kettles.

On completion of the thinning operation the thinned unsaturated polyester is cooled down to or near room temperature and further compounded to make the various types of unsaturated polyester resins of commerce.

Unsaturated Polyester Resin Alkyd Properties

The 100% alkyd from a typical general purpose resin cook is a pale yellow solid at room temperature having the general properties shown in Table 4-6.

The preceding alkyd properties would change as functions of the raw materials used and the molecular weight to which the resin is cooked.

Substitution of either glycols such as diethylene glycol in place of propylene glycol and adipic acid in place of PA can give an alkyd that is a viscous liquid at room temperature rather than a solid resin.

Cooking to a much higher molecular weight, as is often done with resins using isophthalic acid in place of phthalic anhydride, can raise the melting point of the alkyd, for example, to well over 200° to 240°F (104° to 116°C). Employing halogenated intermediates in

Table 4-6: GP Unsaturated Polyester Alkyd Properties

Specific gravity	1.13–1.15
Solubility	Insoluble H_2O, soluble ketones and aromatic solvents
Melting point (Durran Mercury Method)	140°-170°F (60°-77°C)
Acid number	30-40
Viscosity of 60% solution in methyl Cellosolve	G-H (Gardner Holdt)
Molecular weights by gel permeation chromatography:	
Number average	900
Weight average	2,400
Z average	6,800
Dispersity	2.7

place of phthalic to build in fire retardance can increase the specific gravity to 1.3 to 1.5 depending on the amount of halogen, such as chlorine or bromine, introduced into the alkyd.

Although some of the unsaturated polyester alkyd produced is sold as alkyd for specialized uses such as prepreg or molding compounds, most of the unsaturated polyester alkyd is sold and used as a solution in styrene monomer.

Styrenated Unsaturated Polyester Resin Liquid Properties

Liquid unsaturated polyester resins range in viscosity from thin 50 centipoise liquids to quite viscous fluids with viscosities of 4,000 to 6,000 centipoises and higher. Liquid resin colors can range from a very pale yellow to dark amber. These basic resin colors can further be affected by the presence of color contributing additives such as curing promoters.

Typical liquid resin properties obtained by the styrenation of a typical general purpose unsaturated polyester alkyd are shown in Table 4-7.

Table 4-7: Styrenated GP Polyester Liquid Resin Properties

Styrene content, by weight	32%
Viscosity, centipoise	1,100
Specific gravity	1.14
Color	Pale yellow
SPI gel time	5-7 minutes
Peak exotherm time	6-8 minutes
Peak exotherm temperature	340°-360°F (171°-182°C)
Flash point (tag closed cup)	87°-95°F (31°-35°C)

As mentioned previously, styrene is by the far the most widely used monomer employed for making liquid unsaturated polyester. With the use of styrene a reactive liquid resin can be created which is relatively low in cost compared to other reactive liquid resin systems such as epoxies and polyurethanes. Further the styrenated unsaturated polyester can be made to a very wide range of viscosities and can be made compatible with curing methods ranging from ambient temperature cures all the way to high temperature curing conditions.

The styrenated unsaturated polyester resins do not present the allergy and high toxicity problems associated with some liquid reactive epoxy and urethane systems. On the other hand styrene monomer is to an extent volatile and the use of unsaturated polyester resins must be carried out in a manner that will keep the styrene in the air and workplace below the levels recommended/mandated by advisory or regulatory bodies (50 to 100 ppm depending on the agency).[18,19] The styrene monomer in an unsaturated polyester results in the liquid resin having a flash point below 100°F (37.8°C) and such unsaturated polyesters are classified as flammable liquids which require a Department of Transportation (DOT) Red Label. Other less volatile monomers having olefinic unsaturation, such as vinyl toluene or p-methylstyrene, can be used in place of styrene to increase the flash point of the resin or to reduce monomer emission levels.

General purpose styrene containing polyester resins of this type find uses in compounding resins for glass fiber reinforced composites such as boats and other marine applications, translucent fiber glass reinforced building panels, cast synthetic stones and general purpose fiber glass reinforced molded articles.

Monomers Used in Unsaturated Polyesters

Monomers other than styrene used in unsaturated polyesters are shown with their common abbreviations in Table 4-8.

Table 4-8: Nonstyrene Monomers Used in Polyesters

Monomer	Application
Methyl methacrylate (MMA)	Enhanced weather resistance
Butyl acrylate (BA)	Enhanced weather resistance
Butyl methacrylate (BMA)	Enhanced weather resistance
Alpha methyl styrene (AMS)	"Cooler" cure, reduced exotherm
Vinyl toluene (VT)	Less volatility, higher flash point
Para-methyl styrene (PMS)	Less volatility, higher flash point
Diallyl phthalate (DAP)	Very low volatility, prepregs
Diallyl isophthalate (DAIP)	Very low volatility, prepregs
Octyl acrylamide (OAA)	Solid monomer, molding compounds
Trimethylol propane triacrylate (TMPT)	UV and electron beam cures
Triallyl cyanurate (TAC)	High heat deflection
Triallyl isocyanurate (TAIC)	High heat deflection
Diallyl maleate (DAM)	High heat deflection
Diallyl tetrabromophthalate	Fire retardance

Unsaturated Polyester Properties and Chemical Composition

The physical properties of cast cured unsaturated polyester will largely depend on the raw materials used for manufacture and to a lesser extent on the degree of condensation or molecular weight to which the alkyd is cooked.

General Purpose Resins

Rigid general purpose (GP) resins are cooked using chiefly propylene glycol (1,2-propanediol), maleic anhydride and phthalic anhydride. The molar ratio of phthalic anhydride to maleic anhydride will usually range from 1:1 to 2:1. Rigid GP resins of this type can be used for most types of casting, molding and laminating. These are the so-called GP ortho resins of commerce, so named for the ortho isomer of phthalic used in their manufacture. (Phthalic acid can exist in three isomeric forms: the ortho isomer is always referred to as phthalic acid or anhydride, the meta isomer as isophthalic and the para isomer as terephthalic).

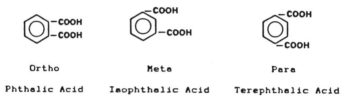

| Ortho | Meta | Para |
| Phthalic Acid | Isophthalic Acid | Terephthalic Acid |

Molar Ratio of PA to MA

The molar ratio of PA to MA will affect such cast resin properties as cured hardness, tensile elongation, heat deflection point or softening temperature, cast resin refractive index and reactivity. Increasing unsaturation, by using more MA, will increase cured hardness, heat deflection point and reactivity. Increased PA on the other hand will give resins with higher refractive index and decreased reactivity.

Flexibilization

Greater tensile elongation, flexibility and toughness can be built into unsaturated polyester resins by substitution of ether glycols such as dipropylene or diethylene glycol for part or all of the propylene glycol and by using flexibilizing long chain aliphatic acids such as adipic acid for all or part of the phthalic used in the GP resin formula.

$$CH_3-\overset{OH}{CH}-CH_2-O-CH_2-\overset{OH}{CH}-CH_3 \qquad HO-CH_2-CH_2-O-CH_2-CH_2-OH$$

Dipropylene Glycol   Diethylene Glycol

$$COOH-CH_2-CH_2-CH_2-CH_2-COOH$$

Adipic Acid

Such flexibilized resins will be softer when cured and will have their heat deflection points reduced in proportion to the amount of flexibilization employed. Since the flexibilization is introduced into the alkyd "backbone" of the resin the flexibilization in the cured, crosslinked resin is permanent unlike that which might be created by the addition of a fugitive external plasticizer such as dibutyl phthalate. A drawback of flexibilization is reduced hydrolytic stability.

Isophthalic Resins

Resins made with phthalic anhydride can be practically cooked up to only a relatively modest molecular weight before decarboxylation by loss of PA places a practical limit on the extent of condensation. Ortho-phthalic based unsaturated polyester alkyds will generally be condensed to (by GPC) 800 to 1,000 number average molecular weights. Substitution of the meta-phthalic isomer, isophthalic acid, for the ortho PA is the most common commercial route to cooking higher molecular weight unsaturated polyester resins. Resins with (by GPC) number average molecular weights of 1,500 to 2,000 can be easily made.

The higher molecular weight isophthalic resins can give higher physical properties with respect to toughness, heat deflection point, hydrolytic stability and chemical resistance compared to their lower molecular weight ortho-phthalic cousins.

Molecular Weight Comparisons

A typical GP ortho-phthalic resin compares in molecular weight characteristics to a high molecular weight isophthalic resin as shown in Table 4-9 below when tested by size exclusion chromatography (Gel Permeation Chromatography).

Table 4-9: Molecular Weights of Ortho- and Isophthalic UP*

	GP Ortho	High MW Iso
Number average molecular weight	910	1,520
Weight average molecular weight	2,430	10,160
Z average molecular weight	6,800	26,550
Dispersity (MWw/MWn)	2.7	6.7

*(Run on a Waters Associates Liquid Chromatograph[20] equipped with Varian Associates[21] Micropak TSK 1000H and 3000H Columns at 2 ml/min, tetrahydrofuran carrier solvent with data reduction by a Waters Data Module microprocessor.)

GPC is an excellent characterization technique for determining the molecular weight characteristics of unsaturated polyester resins and is especially useful if calibration data is available from an absolute correlative detector such as a Low Angle Laser Light Scattering (LALLS) detector.

GPC can give considerably more data than the process viscosity and thinned resin viscosity measurements which also are a measure of molecular weight. An extensive literature is available on the application of GPC to polymer characterization.[22,23,24]

Hydrolytic Stability and Chemical Resistance

Significant improvement in hydrolytic stability, chemical resistance and resistance to yellowing on exposure to ultraviolet radiation can be obtained by substituting neopentyl glycol (NPG) for propylene glycol as follows:

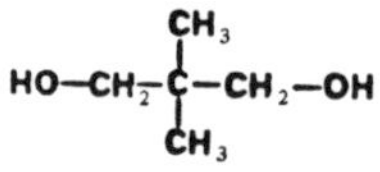

$$HO-CH_2-\underset{\underset{CH_3}{|}}{\overset{\overset{CH_3}{|}}{C}}-CH_2-OH$$

Neopentyl Glycol (NPG)

Neopentyl glycol confers these advantages in both ortho- and isophthalic unsaturated polyester resins. NPG, because of its linear primary glycol structure, will tend to give a more linear alkyd structure. This can result in alkyd-styrene incompatibility problems in high maleic content resins or if the alkyd structure is made more linear by the use of significant amounts of a straight chain acid such as adipic in the formulation.

Styrene Compatibility

The styrene compatibility limitations of such NPG formulations can be greatly reduced or eliminated by reducing the stereoregularity of the alkyd chain by replacing part of the NPG with a secondary glycol such as propylene glycol.

For the same styrene compatibility reasons ethylene glycol is very seldom used as the sole glycol in an unsaturated polyester.

$$OH-CH_2-CH_2-OH$$

Ethylene Glycol (EG)

Cyclohexanedimethanol, a cycloaliphatic primary glycol, will also tend to give stereoregular alkyds if cooked with fumaric acid alone. With proper alkyd formulation, resins having good electrical properties can be made with this diol.

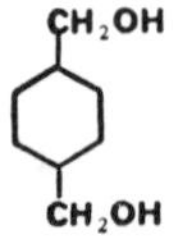

Cyclohexanedimethanol (CHDM)

Flame Retardance

Flame retardance can be built into unsaturated polyesters by the use of halogenated intermediates in the cook charge. The halogens used are either chlorine or bromine.

Part or all of the saturated acid in the unsaturated polyester cook can be replaced with tetrabromophthalic anhydride (TBPA), tetrachlorophthalic anhydride (TCPA) or chlorendic acid (CA).

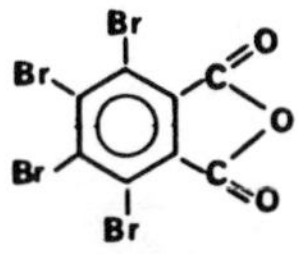

TBPA 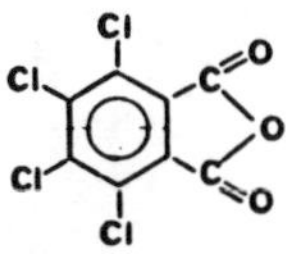   TCPA 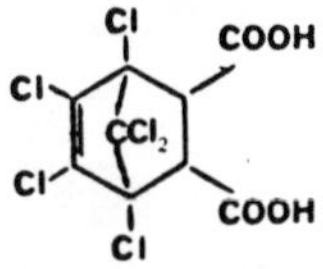   CA

Bromine can also be built into the unsaturated polyester backbone by the use of dibromopentyl glycol as the diol in the cook.

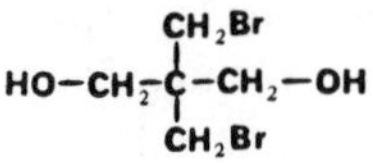

$$HO-CH_2-\underset{CH_2Br}{\underset{|}{\overset{CH_2Br}{\overset{|}{C}}}}-CH_2-OH$$

Dibromoneopentyl glycol

Improved chemical resistance can also be obtained by the use of aromatic and cycloaliphatic diols such as propoxylated bisphenol A and hydrogenated bisphenol A in place of the conventional diols such as propylene glycol.

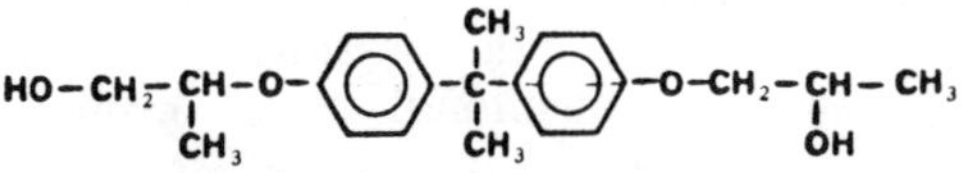

Propoxylated bisphenol A

An equilibrium is reached in the propoxylation of bisphenol A in the formation of primary and secondary hydroxyl groups. According to Kerle, Connolly and Rosenfeld the ratio of primary to secondary hydroxyls is 15:85.[25] The same investigators also claimed that the percentage of para, para isomer in the commercial bisphenol A used could vary from 97.15 to 100.00% depending on the source.

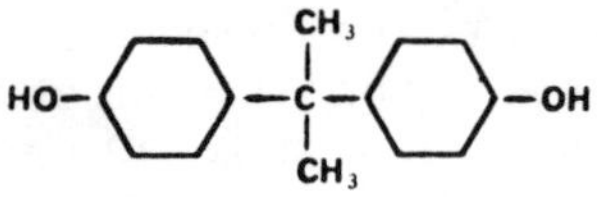

Hydrogenated bisphenol A

Unsaturated polyesters can give fairly good chemical resistance to aqueous acids and salts but are vulnerable to attack by high pH alkaline media on the unsaturated polyester alkyd ester linkages. Use of the bisphenol A derived diols introduces considerable steric shielding of the ester linkages and can extend the use of unsaturated polyesters into higher pH alkaline environments.

2,2,4-Trimethyl-1,3-pentanediol (TMPD) can also give improved hydrolytic and chemical resistance.

$$HOCH_2-\underset{\underset{CH_3}{|}}{\overset{\overset{CH_3}{|}}{C}}-CH-\underset{\underset{OH}{|}}{\overset{\overset{CH_3}{|}}{CH}}-CH_3$$

TMPD

The methyl groups of this glycol sterically shield ester linkages formed by its hydroxyl groups from hydrolytic and chemical attack. The same steric shielding makes it a bit more difficult to synthesize resins with TMPD and also reduces curing reactivity somewhat. These problems have been addressed by the manufacturer of the glycol with the development of techniques for formulating and synthesizing TMPD resins and room temperature curing promoter systems for resins made with this glycol.[26] TMPD resins' cured specific gravities tend to be lower than those of conventional resins.

VINYL ESTER RESINS

Chemistry

Vinyl esters are the reaction products of epoxy resins with ethylenically unsaturated carboxylic acids. Simple diepoxide resins, such as the diglycidyl ether of bisphenol A or bisphenol A extended higher molecular weight homologues thereof, can be used as well as brominated analogues for flame retardance or epoxy novolac resins for special properties such as higher heat resistance. Vinyl ester resins typically have terminal unsaturation except for some special types such as those designed for thickening with Group II metal oxides and hydroxides in the manufacture of sheet molding compound (SMC). The terminal unsaturation will react to give crosslinking either by homopolymerization of the vinyl ester resin with itself or by copolymerization with unsaturated monomers such as styrene.

The most common vinyl esters are made by esterifying a diepoxide resin with a monocarboxylic unsaturated acid such as methacrylic acid or acrylic acid. Such epoxy methacrylates or epoxy acrylates can be used in free radical curing reactions alone or can be dissolved in unsaturated monomers such as styrene to give liquid resins which can be used very much like styrenated unsaturated polyester resins.

Basic Vinyl Ester Resin

A basic vinyl ester preparation employing a simple diglycidyl ether of bisphenol A epoxy resin and methacrylic acid can be represented as:

$$CH_2\overset{O}{\triangle}CH-CH_2-O-\bigcirc-\underset{CH_3}{\overset{CH_3}{C}}-\bigcirc-O-CH_2\overset{O}{\triangle}CH-CH_2 + HO-\overset{O}{\overset{\|}{C}}-\underset{CH_3}{C}=CH_2 \quad \xrightarrow[\text{Catalyst*}]{250\text{-}300°F\ (121\text{-}149°C)}$$

$$CH_2=\underset{CH_3}{C}-\overset{O}{\overset{\|}{C}}-O-CH_2-CH-CH_2-O-\bigcirc-\underset{CH_3}{\overset{CH_3}{C}}-\bigcirc-O-CH_2-CH-CH_2-O-\overset{O}{\overset{\|}{C}}-\underset{CH_3}{C}=CH_2$$
                                           OH   OH

*Benzyl trimethyl ammonium chloride catalyst

Generally, onium salts, tertiary amines and phosphines are effective catalysts for the acid-epoxide reaction which proceeds readily at only moderately elevated temperatures.

Methacrylic acid is most commonly used for vinyl ester resins intended for composites applications while acrylic acid is favored for resins intended for application in coatings. The use of other unsaturated acids such as crotonic and cinnamic acids have been reported.[27]

History

Some of the earliest vinyl ester resins resulted from efforts by the U.S. Commerce Department to find a better dental adhesive with which to bond acrylic dental prostheses to teeth. R.L. Bowen esterified glycidyl methacrylate and acrylate with bisphenol A in these efforts.[28,29] Unfortunately these resins were too reactive to afford a practical working life and only further vinyl ester development succeeded in providing commercially useful dental bonding resins. Other workers such as Fekete and his associates worked on resins intended for use in electrical insulation and chemically resistant composites.[30,31]

Toughness and Chemical Resistance

Vinyl ester resins offer toughness and chemical resistance properties which are generally superior to unsaturated polyesters.

The epoxy resin backbone used in making vinyl ester resins confers toughness and greater tensile elongation properties to these resins. The molecular weight of the vinyl ester resin can be varied by the choice of the epoxy backbone employed. For the most common vinyl ester resins used for composites two mols of the diglycidyl ether of bisphenol A are chain extended with one mol of bisphenol A to form the epoxy backbone.

In this manner, molecular weight and backbone structure dependent properties such as tensile strength and elongation, heat deflection point and reactivity can be varied for different applications.

Vinyl Ester Resin Structure and Properties

The superior chemical resistance (compared to unsaturated polyesters) of vinyl ester resins is in part due to the absence of ester linkages in the epoxy backbone in those sites where the polymer units are connected with phenyl ether linkages. These latter moities are much more resistant than ester linkages to degradation in many chemical environments and especially in high pH alkaline situations. The ester linkages in a vinyl ester resin are present only at the end of the molecule which minimizes the number of ester linkages that can be chemically attacked. Further, if the vinyl ester resin molecule is terminated with methacrylate groups the spatially large methyl group pendant on the methacrylate group sterically shields the ester linkage from chemical attack.

With the vinyl unsaturation present on the ends of the molecule, vinyl ester resins can be made to be very reactive. They have the ability to cure rapidly with fast green strength development either as homopolymers or as copolymers with monomers such as styrene.[27]

Specialty Vinyl Ester Resins

Specialty vinyl ester resins can be made based on the use of epoxy novolacs (epoxy resins based on phenol formaldehyde novolacs) for the epoxy resin backbone. Heat deflection points of 270° to 300°F (132° to 149°C) can be achieved while still maintaining excellent chemical resistance. Flame retardant vinyl ester resins can be made by using diepoxide resins based on the use of tetrabromobisphenol A.

Vinyl Ester Resin Thickening for SMC: Conventional vinyl ester resins do not respond well to the divalent Group II metal oxides and hydroxides used for thickening sheet molding compounds (SMC). Good thickening response can be obtained by introducing carboxyl group functionality on the vinyl ester molecule. R.J. Jackson described acid modification of a vinyl ester resin by replacing part of the monocarboxylic methacrylic acid with a dicarboxylic acid, maleic acid.[32] An example from Jackson's patent illustrates such a formula change and the improvement in thickening response. This is shown in Table 4-10 below.

Thickening responses of the two resins were quite different on having 2.5 PHR (parts per hundred of resin) of magnesium oxide thickener dispersed in them.

Table 4-11 shows the thickening of acid modified vinyl ester resin.

Flame Retardant Vinyl Esters: Flame retardant vinyl ester resins are usually made using an epoxy backbone incorporating tetrabromobisphenol A in place of bisphenol A.

Table 4-10: Maleic Acid Modified Vinyl Ester SMC Resin

	Standard Vinyl Ester	Acid Modified Vinyl Ester
Stage I:*		
Epoxy Resin (DGEBA)**	795	795
Bisphenol A	237	237
Trimethyl ammonium chloride	0.825	0.825
Stage II:*		
Glacial methacrylic acid	195.2	171.3
Hydroquinone	0.45	–
Trimethyl ammonium chloride	6.6	6.6
Maleic acid	–	32.25
Styrene	970	970

*Stage I was processed for 1 hour at 340°F (171°C). Stage II was processed for 3 hours at 240°F (116°C) with a nitrogen-air sparge.
**DGEBA: Av. MW 350, equiv. wt. about 170-190.

Table 4-11: Thickening of Acid Modified Vinyl Ester Resin

	Brookfield Viscosity, cps.	
	Standard Vinyl Ester	Vinyl Ester Acid Modified
Viscosity		
Initial	500	1,230
After 24 hours	500	85,000
After 48 hours	500	2.15×10^6
After 7 days	500	4.00×10^6

One Step Vinyl Ester: An early novel "one step" approach to the preparation of vinyl ester type resins was described by C.A. May, who using a kettle charge of bisphenol A, epichlorohydrin and methacrylic acid, first reacted the epichlorohydrin with the methacrylic acid and then, adding sodium hydroxide, reacted the bisphenol A with the methacrylated epichlorohydrin residue.[33] Other variants were explored by May. Among these was the modification of a vinyl ester resin by reacting the secondary hydroxyls of the resin with isocyanate to give a urethane modified vinyl ester resin.[34]

Rubber Modified Vinyl Ester: More impact resistant vinyl ester resins were made by D.J. Najvar by replacing up to 20% of the unsaturated monocarboxylic acid, such as methacrylic acid, with a functionally equivalent amount of a liquid carboxy terminated polydiene rubber.[35]

As an example the following were reacted to an exotherm of 374°F (190°C) as shown in Table 4-12.

Table 4-12: Rubber Modified Vinyl Ester Resin Synthesis

	Parts by Weight
DGEBA epoxy equiv. wt. (EEW) 189 (Dow 331)	455
Bisphenol A	154
t-Butyl phosphonium acetate	0.5

This gave a polyepoxide resin with EEW ca 600.
This was then reacted at 248° to 266°F (120° to 130°C) with:

Methacrylic acid	76
DMP-30 catalyst	1.2
B.F. Goodrich Hycar CTBN (COOH terminated butadiene acrylonitrile rubber with 2.5% COOH)	228
Hydroquinone	0.17

The above was reacted to 1.15% COOH content.

Vinyl Ester Resins Overview

The epoxy backbone and methacrylic/acrylic acid used to make vinyl esters make them significantly more costly than conventional unsaturated polyester resins. Vinyl esters, however, do offer advantages over unsaturated polyester resins which have been very well summarized by Anderson and Messick:[36]

1. Excellent reactivity due to terminal vinyl unsaturation in either homopolymerization or copolymerization reactions.
2. With methacrylate termination, increased hydrolysis resistance due to ester linkage shielding by the methacrylate methyl group.
3. 35 to 50% fewer hydrolysis prone ester linkages than conventional unsaturated polyester resins.
4. Better wetting and bonding to glass reinforcements due to secondary hydroxyls on the vinyl ester resin molecule.
5. Improved elongation and toughness conferred by the epoxy resin backbone whose ether linkages give superior acid resistance.

Typical Styrenated Vinyl Ester Resin Liquid Properties

Table 4-13 shows the typical styrenated vinyl ester resin liquid properties.

Typical Styrenated Vinyl Ester Cast Resin Properties

Table 4-14 shows the typical styrenated vinyl ester cast resin properties.

Table 4-13: Styrenated Vinyl Ester Resin Liquid Properties

Styrene content, %	50-45
NVM, %	50-55
Viscosity, cps, 77°F (25°C)	80-600
Color, Gardner	2-3
Wt. per gallon, lbs.	8.6-8.7
Flash point, tag open cup, °F (°C)	95 (35)
Reactivity	
Gel time at 77°F (25°C) min.*	20-25
SPI gel time, min.	10-19

*Promoted with 0.5 PHR 12% cobalt octoate and catalyzed with 1.0 PHR MEK peroxide catalyst.

Table 4-14: Physical Properties of Cast Vinyl Ester Resin

Tensile strength, psi, 77°F (25°C)	10,000-11,000
Tensile modulus, psi, 77°F (25°C)	$0.36\text{-}0.44 \times 10^6$
Tensile elongation, %	5.3-5.8
Flexural strength, psi, 77°F (25°C)	18,000-19,000
Flexural modulus, psi, 77°F (25°C)	$0.45\text{-}0.46 \times 10^6$
Heat deflection temperature, °F (°C)	190-210 (88-99)
Notched Izod impact - ft.lbs./in. of notch	0.5-0.55
Barcol hardness	40-50

COMPOUNDING OF UNSATURATED POLYESTER AND VINYL ESTER RESINS

Overview

Most commercial unsaturated polyester resins contain styrene as the crosslinking monomer. Some exceptions are resins for:

Structural panels which may also contain some acrylic monomer for improved outdoor weathering.

Mine Bolt Resins which usually contain enough vinyl toluene in addition to styrene to raise the flashpoint to over 100°F (37.7°C).

Resins for the manufacture of prepreg and molding compounds which may be sold as monomer-free 100% unsaturated polyester alkyd or as monomer-free alkyd dissolved in acetone.

The liquid styrenated resins which form the volume of commerce can be compounded for use with:

Curing Systems
    Promoters for room temperature cures.
    Promoters for elevated temperature cures.

Thixotropic Agents
: To give flow control and prevent sagging in vertical lamination and coating.

Fillers
: To reduce cost, reduce curing shrinkage and impart special properties such as flame retardance.

Pigments
: For coloration.

Thickening Agents
: To give the compound thickening necessary for sheet and bulk molding compounds (SMC and BMC).

Fiber Reinforcements
: To give high strength composites as in laminates, SMC and BMC.

Wetting Agents
: Facilitating wetout of fillers and reinforcements.

Bubble Release Agents
: To enhance air bubble release in laminating or casting.

Internal Mold Release Agents

Catalyst Indicators
: Which indicate catalyst addition by a color change.

Most of the compounding techniques and materials used with polyesters can also be used in the compounding of vinyl ester resins. Vinyl esters, being more costly than polyesters, are generally not used in as broad a range of applications as polyesters. Vinyl esters owe most of their applications to their superior chemical resistance properties and higher physical properties of composites made with them.

Curing Systems

Styrenated unsaturated polyester resins can be cured by either room temperature (RT) or heat curing methods. Other monomers such as vinyl toluene, methyl methacrylate and para-methylstyrene can also be used, generally with styrene, to give room temperature or heat curing systems. Diallylphthalate or isophthalate monomers do not respond well in RT cures and are generally only used in such heat curing applications as prepregs and molding compounds. Vinyl esters can also be cured by either room temperature or elevated temperature curing routines, with styrene being the most common monomer in use.

Room Temperature (RT) Curing Systems: RT curing systems are usually based on the use of transition metal soaps as the primary promoters and ketone peroxides as the catalysts. Cobalt soaps such as cobalt naphthenate, octoate or neodecanoate are the most popular metallic promoters providing RT curing resins which give good curing behavior and package life.

Other primary metallic promoters that have been used are compounds of manganese, vanadium, tin, calcium and barium.

Vanadium can give very fast curing RT systems with ketone peroxide catalysts but imparts a pronounced yellow color to the cured resin with the additional drawback of shelf life problems with the promoted resin unless special stabilization methods are used.

Manganese is primarily useful with cumene hydroperoxide which is attractive as a lower cost catalyst.

Calcium and barium have been cited as being able to speed up the cure of promoted systems employing very low cobalt concentrations to minimize cured resin color.

An example of a cobalt single promoted RT curing resin is shown in the following Table 4-15.

Table 4-15: Single Promoted Polyester Resin Formulation

	Parts by Weight
Polyester resin containing 40% styrene	100
Cobalt octoate 12% (containing 12% cobalt as metal)	0.25

On catalyzation with 1 PHR of MEK peroxide catalyst (Lupersol DDM-9, a trademark of Lucidol Pennwalt Corp.) reactivity at 77°F (25°C) would be:

50 gram resin sample
Gel time*	14-20 min.
Peak temperature interval**	25-30 min.
Peak exotherm temperature***	240°-270°F (116°-132°C)

*Time from catalyst addition to gelation.
**Time from gelation to maximum temperature in the curing resin mass.
***Maximum temperature reached in the resin mass.

Many commercial resins are double promoted in that they contain not only metallic primary promoters but also secondary promoters which can:

1. Speed up the cure after gelation.
2. Help stabilize the gel time drift on aging that often occurs with single cobalt promoted resins.

Although cobalt compounds are the major primary promoters, there are many different types of secondary promoters in use in commercial RT curing resins.

Some common secondary promoters are shown below in Table 4-16.

Table 4-16: Secondary Promoters

Tertiary Amines
  Diethylaniline
  Dimethylaniline
  Dimethylethanolamine
  Phenyldiethanolamine
  Dimethyl-p-toluidine
Acetamides
  N,N-dimethylacetoacetamide
  mono-N-methylacetamide
Acetoacetates
  Methyl acetoacetate
  Ethyl acetoacetate
Quaternary Salts
  Benzyltrimethylammonium chloride
  C_{12}-C_{24} quaternary ammonium chlorides
  Triphenylsulfonium chloride

As an example, conversion of the single promoted resin shown earlier to a double promoted system could be accomplished by the addition of 0.25 PHR of diethylaniline. The reactivity characteristics using the same catalyzation are shown in Table 4-17 below.

Table 4-17: Double Promoted Polyester Reactivity

Gel time	8-12 min.
Peak temperature interval	8-12 min.
Peak exotherm temperature	290°-320°F (143°-160°C)

Benzoyl Peroxide Catalyzed RT Cures: Fast RT curing systems can also be obtained by the use of aromatic tertiary amines such as diethylaniline and dimethylaniline as promoters and an acyl peroxide such as benzoyl peroxide for the catalyst. These systems are most often used by the user of unpromoted resins who will very often employ them in a "2 pot method" where half of the resin is promoted with the amine promoter and the other half of the resin is catalyzed with the benzoyl peroxide. Equal parts of promoted and catalyzed resin portions are then mixed at point of use. This approach suffers from limited pot life of the catalyzed resin portion. It has the advantage that it is much less sensitive to inhibition by moisture than the cobalt promoted ketone peroxide systems.

Commercial Prepromoted Resins: Most commercially available resins prepromoted for RT cure employed cobalt-based single and double promoter systems designed for use with ketone peroxides. Resins of this type can be easily catalyzed by the addition of liquid ketone peroxide either by batchwise addition and mixing or by the convenient catalyst injection equipment available for spray-layup, casting and laminating equipment.

Catalysts for RT Cobalt Promoted Resins: Methyl ethyl ketone peroxide is probably the most widely used catalyst for the room temperature curing of prepromoted polyester and vinyl ester resins. This peroxide catalyst is not a pure compound and may contain varying ratios of the peroxide and its dimer as well as hydrogen peroxide and free water. Pentanedione peroxide catalyst is sometimes used in conjunction with MEK peroxide catalyst or alone when faster room temperature curing systems are necessary. Many varieties of MEK peroxide catalysts are commercially available as well as related catalyst compounds. Vinyl ester resins are more commonly sensitive to the peroxide dimer content than polyesters.

The cobalt promoted resins are sufficiently stable for adequate package life and with proper compounding can have minimal variation in curing behavior as they age after manufacture.

Heat Curing Systems: Heat curing systems are used in processes such as matched die molding, building panel lamination, press molding and pultrusion. Organic peroxides having higher decomposition temperatures, such as t-butyl peroctoate, benzoyl peroxide and t-butyl perbenzoate, are used.

Handling Catalysts and Promoters: Catalysts and promoters must never be allowed to directly contact each other because a violent reaction can occur which may result in fire, explosion or injury. Promoters, if used, must be thoroughly mixed into the resin before the addition of any catalyst. Catalysts and promoters should be stored and handled in a manner which will eliminate the possibility of accidental contact. Good housekeeping is especially important with catalysts and promoters. Catalyst suppliers can supply excellent instructions on this subject.

Ultraviolet Absorbers

Cured unsaturated polyester resins are quite susceptible to degradative attack by the ultraviolet radiant energy in daylight and sunlight. Such degradation results in discoloration of the resin to a yellow and in severe cases a brownish coloration. Some halogenated unsaturated polyester resins such as those based on tetrachlorophthalic and tetrabromophthalic anhydrides are more UV sensitive than non-halogenated types.

Ultraviolet light absorbers can be added to unsaturated polyester resins to greatly improve the resistance to UV degradation.

Types of UV Absorbers: UV absorbers most commonly used in unsaturated polyester resins fall into two classes as shown in Table 4-18.

UV absorbers are typically used in polyester resins at levels ranging from 0.1 to 1.0 PHR. The higher levels would be typical in the more UV susceptible resin systems such as the halogenated types.

Thixotropic/Flow Control Agents

In some processes in which polyester resins are used, such as open mold hand layup, spray layup or gel coating, it is essential that

Table 4-18: Ultraviolet Absorbers

Chemical Type	Commercial Examples
Substituted benzophenones	Cyasorb UV-9 (American Cyanamid)
	Uvinol 400 (Ciba-Geigy)
Substituted benzotriazoles	Tinuvin P (Ciba-Geigy)
	Tinuvin 327
	Tinuvin 328
	Cyasorb UV-5411

the resin does not drain down from inclined or vertical surfaces before gelation takes place. The desired behavior is much like that in a high grade drip-resistant paint which will exhibit minimal downward flow or sagging on application to a vertical surface. Thixotropic agents can be dispersed in polyester resins to develop the necessary thixotropic or psuedoplastic rheology.

With ideal thixotropic properties, the polyester resin would have a storage viscosity high enough to keep the thixotropic agent in perfect suspension and yet become a relatively low viscosity liquid on the application of liquid shearing forces so it can be readily handled, sprayed and impregnated into reinforcements. The thixotropic "false" viscosity should then quickly redevelop to prevent drainage from a coating or from reinforcement. Not surprisingly, real life thixotropic resin systems approach but do not completely achieve such ideal behavior.

Thixotropic agents commonly impart these characteristics by the formation of shear labile hydrogen bonds between their own particles to form microchains between their particles and the resin. Being shear labile, the thixotropic viscosity generating three dimensional structure that is formed is readily broken on exposure to liquid shearing forces such as occur on mixing, spraying or impregnation of reinforcements. The hydrogen bonds quickly assert themselves on removal of shear with regeneration of the higher thixotropic viscosity. Examples of these thixotropic agents are shown below in Table 4-19.

Table 4-19: Thixotropic Agents

Type	Typical Commercial Examples
Pyrogenic silicas	Aerosil 200 (Degussa Corp.)
	Cabosil PTG (Cabot Corp.)
Precipitated silicas	Sylox TX (Davison Chemical Division)
Modified bentonite clays	Claytone PS (Southern Clay Products Co.)
Hydrogenated castor oil	Thixin E (Baker Castor Oil Co.)

Excellent dispersion of the thixotropic agent is usually necessary to obtain optimum thixotropy and maintenance of this property as the resin ages. Thixotropic agent levels of 0.4 to 1 PHR are commonly used in laminating resins while gel coats will generally require higher levels such as 1.5 to 2 PHR. Small (0.1 to 0.3 PHR) additions of polar hydroxylic additives such as simple glycols, glycerin or surfactants such as Tween 20 (Atlas Chemical Industries) are often used in conjunction with silica thixotropic agents to obtain enhanced thixotropic behavior.

Thixotropic resin viscosities are commonly measured using rotational multispeed viscometers such as the Brookfield Synchroelectric Viscometer.[37] Viscosities are usually measured at a higher and lower rotational speed such as 50 and 5 RPM and 20 and 2 RPM. A thixotropic resin will give a higher viscosity reading at the lower RPM than at the higher RPM. The low RPM viscosity reading divided by the high RPM viscosity reading is called the Thixotropic Index.

Fillers

Liquid polyester resins can accept relatively large filler loadings. Most of the fillers used are inorganic and most are of mineral origin.

The addition of mineral fillers to polyester resins usually produces the following effects.

1. Viscosity of the liquid resin system is increased.
2. Curing shrinkage is decreased.
3. Peak exotherm on curing is decreased since the filler acts as a heat sink.
4. Specific gravity of the resin-filler mixture is higher than pure resin if a filler denser than resin is used.
5. Cured hardness of the resin-filler mixture is increased (except with cellular fillers).
6. Moduli are increased.
7. Tensile elongation is decreased.
8. Impact strengths are decreased.
9. Curing behavior may be affected. Gel time may be accelerated or retarded.
10. Raw material cost of the filler-resin composite is usually lower on a weight or volume basis than the pure cured resin.

Specialized low density fillers can also be used to reduce the specific gravity of the resin-filler composite. Hollow, very small diameter, glass or ceramic microballoons may be used for this purpose. Plastic microballoons made from polyvinylidene chloride, epoxy or phenolic resin have also been used for achieving lower density composites.

Some Common Filler Applications in Resins: Common filler applications are shown in Table 4-20.

Table 4-20: Filler Applications

	Typical Applications
Calcium carbonates	Synthetic marble, SMC, BMC, matched die molding, building panels, autobody putty, mine bolt grouts
Clays	Molding compounds, SMC, BMC
Talcs	Autobody putty, gel coats
Alumina trihydrate	Synthetic onyx, flame retardant SMC, BMC and molding compounds, flame retardant construction composites
Dolomites	SMC, BMC, matched die molding
Glass and ceramic microballoons	Synthetic marble, bowling ball cores, low density SMC, deep submergence vehicles
Phenolic microballoons	Deep submergence vehicles
Glass microspheres	SMC, gel coats
Nepheline syenite	Building panels
Silica sand	Polymer concrete

Filler Dispersion and Mixing Equipment: In compounding polyester resins with fillers care must be taken to completely wet out the fillers with resin. Adequate mixing equipment must be used. Medium shear mixers of the Cowles or Myers type are effective for medium to high filler loading mixing. Dough mixers of the Hobart type have also been used, especially in synthetic marble and onyx manufacture.

High filler content mixes which must also incorporate reinforcing fibers such as BMC compounds are most commonly made using double arm sigma blade mixers.

Hollow fillers such as inorganic and organic microballoons require special care in mixing because the relatively thin walls of the microballoons can be fractured by mixing action that is too vigorous. Microballoon types will vary in their sensitivity in this respect. Suppliers and their literature should be carefully consulted for mixing recommendations and cautions.

Order of Mixing and Dispersion: The order of addition of ingredients in filled compositions will vary somewhat with the application. With heat curing systems where a long catalyzed pot life can be expected, the polyester resin is often first mixed with the catalyst. Typical of these are filled systems for:

1. SMC
2. BMC
3. Filled systems for matched die molding with glass mat and glass fiber preforms.

On thorough dispersion of the catalyst in the resin the rest of the additives such as internal mold release, wetting/bubble release agents, pigments, etc., are added. A variant on this sequence can occur with shrink controlled SMC and BMC where the shrink control/low profile agent may precede the catalyst. On mixing of the preceding ingredients, the filler is added slowly and thoroughly dispersed. Propeller type mixers can be used for low filler content systems such as those for mat and preform molding where 40 or so PHR of filler is used. SMC mixes, which are termed pastes and on the average contain 150 PHR filler, require more powerful mixers of the Cowles or Myers type.

Calcium Carbonate Fillers: Calcium carbonate fillers of either the ground limestone type or the precipitated chalk type are the most widely used fillers for the preparation of filled compositions for mat and preform molding and SMC and BMC. The calcium carbonate fillers are low in cost and generally have low oil absorption numbers which allow considerable freedom in filler loading levels while keeping the viscosity of the filled system at a usable level. Calcium carbonate fillers have minimal effect on the curing properties of the filled systems.

One of the largest applications for calcium carbonate fillers is in the manufacture of synthetic or, as the industry prefers to call it, cultured marble. The carbonate filler levels used here are quite high, ranging from 300 to 400 PHR. Relatively coarse particle size fillers ranging from 30 to 100 mesh are used to achieve such high filler loading levels.

Two compounding sequences are common in the use of calcium carbonate fillers in marble. In the first sequence, resin is catalyzed and then mixed with the filler. The common production methods depend on room temperature curing systems so pot life on catalyst addition is very short. A second sequence, called the "master batch" method, mixes the filler with the resin first in larger batch amounts than with the first sequence. Portions of this "master mix" are then catalyzed just before use as required.

Clay Type Fillers: Clay type fillers are also used even though they generally have higher oil absorption numbers and tend to increase mix viscosity more than the calcium carbonates at equivalent filler loading levels. The incorporation of clay fillers can give good electrical properties, enhance flow in molding and being more chemically inert than carbonates, enhance chemical resistance. Clays also find use in pigmented gel coats.

Talc Fillers: Talc fillers generally have high oil absorption numbers and consequently give high viscosities on addition to resin. Auto body putties used for auto body repair commonly contain some talc filler. Talcs are also used in pigmented gel coats.

Alumina Trihydrate Fillers (ATH): Among the earliest applications for ATH in polyesters was the improvement of electrical tracking resistance and flame retardancy.[38]

Currently, ATH is used largely for imparting flame retardance to polyester composites and, because its refractive index approaches that of cured polyester resin, for the manufacture of synthetic/cultured onyx.

The similarity between the refractive indices of ATH and the cured polyester resins permits the manufacture of ATH filled cultured onyx composites which have the depth of translucence of natural onyx. Special resins which on cure give almost colorless castings have been developed for the manufacture of cultured onyx.

The flame retardance conferred by ATH is due to the energy absorption that occurs as three water molecules are liberated from each molecule of ATH at combusion temperatures. By the use of 100 PHR of ATH GP laminating resins can give glass fiber reinforced laminates which can yield moderate flame retardant properties. ATH can also enhance the flame retardance of flame retardant halogenated resin systems to give more cost effective systems.[39,40,41,42]

Development of flame retardance by the use of ATH can also give lower smoke densities than flame retardant systems which depend on the use of halogenated resins alone.[43]

SMC and BMC employ filler loading levels of from 150 to over 200 parts of filler per 100 parts of resin. Partial or complete replacement of these fillers with ATH can give such molding compounds good flame retardance without the use of halogenated flame retardants. Zaske, Wang and Wuh investigated the effects of different ATH fillers on the thickening of SMC paste and came to the conclusion that the major factors affecting thickening were the particle size distribution of the ATH and the water content in the paste.[44]

Pigments and Colorants: Very often pigments and colors are added to polyester resins in the form of dispersions rather than as pure dry colors. The use of dispersions is convenient and permits the use of low shear mixing equipment such as propellor mixers with the assurance that the color will be excellently dispersed in the resin mix. Pigment dispersions are prepared by suppliers from pure colors and special pigment grinding vehicles using high shear dispersion equipment such as three-roll mills and sandmills. Early pigment dispersions employed low molecular weight saturated polyesters as vehicles. The pigment grinding vehicles used in modern dispersions are often special liquid unsaturated polyesters which contain no styrene or other diluent monomer and which being unsaturated, can crosslink with styrenated polyester during the curing process.

Inorganic pigments such as titanium oxides and organic pigments such as the phthalocyanine pigments are used as well as dyes. Care must be taken in using colorants in polyester resins to ensure that the pigments or dyes will not interfere with proper cure of the resin. With peroxide catalyzed systems, the colorants must also be able to resist the oxidizing action of the peroxide.

Thickening Agents

Thickening agents are an integral part of SMC and BMC molding technology. The earliest and still most widely used thickening system employs Group II metal oxides and hydroxides to thicken reinforced polyester molding compounds so that the compounds are essentially tack-free for handling purposes and will also have the rheology necessary for high quality molding. Sheet molding compound, when properly made, has been likened to leather in its uncured state.

The most commonly used metal oxides and hydroxides are those of magnesium and calcium. Early workers in the field explored this type of thickening extensively and also developed an extensive patent literature.[45,46,47,48,49,50,51,52]

Within recent years nonmetal oxide/hydroxide SMC thickening systems have been developed.[53] Thickening in these systems depends on a urethane reaction which by increasing the molecular weight of the polyester resin in the SMC produces the desired thickening of the molding compound.

Later work has focused on the influence of other factors on the thickening reaction and approaches for better thickening control in manufacturing.[54,55,56,57,58]

Fiber Reinforcements

Unreinforced polyesters in cast form have, like many pure cast thermoset resins, only moderate physical properties. Excellent physical properties including good toughness and impact qualities can be obtained by the use of high strength reinforcing fibers to make polyester composites. Glass fibers in the form of woven cloths, chopped glass mats, multifilament forms such as rovings and chopped rovings are the most common reinforcement used in polyester composites. Sisal, polyester, polyamide and polyaramid fibers are also used in the form of chopped fibers, filaments and woven reinforcements.

APPLICABLE MANUFACTURING PROCESSES

Overview

Most of the manufacturing processes employing unsaturated polyesters and vinyl esters depend on the liquid nature of these resins, the viscosities of which can be readily varied by increasing or decreasing the content of a monomer such as styrene and by changing the molecular weight of the unsaturated polyester alkyd. The viscosity changes as styrene content is varied is shown for a GP orthophthalic resin and a high molecular weight isophthalic resin in Figure 4-1.

Hand Layup

Hand layup was one of the first ways in which polyester resins were used. The process is very simple and consists of impregnating fiber reinforcement such as fiber glass mat or woven cloth with catalyzed resin in a female mold. Different types of reinforcement forms

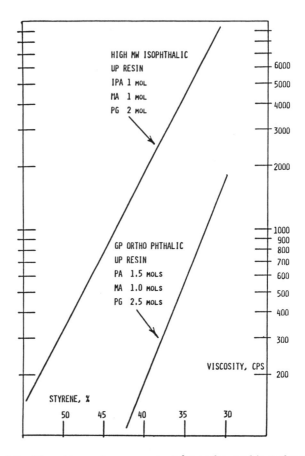

Figure 4-1: Viscosity vs styrene content for ortho- and isopolyesters.

such as woven fiber glass roving and knitted fabric may also be used alone or in conjunction with each other. This process usually employs room temperature curing resins which are prepromoted and require only the addition of catalyst, usually a ketone peroxide such as methyl ethyl ketone peroxide.

This process gives a high grade surface on the mold side. Most often, after application of mold release to the mold surface, a pigmented gel coat is applied to the mold and allowed to gel and partially cure before the lamination with reinforcement and resin.

The room temperature gel coats are usually applied by spray equipment using either small batches of catalyzed gel coat or by the use of catalyst injector guns which inject and blend the requisite amount of catalyst into the gel coat as it is sprayed on the mold. 15 to 20 mil thick (cured) gel coats are common.

The hand layup process gives molded parts with only one good surface. Precise thickness control in the molding is also difficult to achieve.

Resins for hand layup are usually thixotropic to minimize resin drainage from vertical sections of the layup. Room temperature curing systems are almost always used. All types of styrene containing polyesters can be compounded for hand layup work. Replacement of part of the styrene with methyl methacrylate to enhance weathering or paramethyl styrene to reduce monomer emissions is also possible.

Styrene containing vinyl ester resins can also be promoted and compounded for use in room temperature hand layup molding. Hand layup is often used with vinyl esters because the large pieces of chemically resistant equipment for which they are largely intended are most easily fabricated by this molding method.

Spray Layup

Spray layup, a later development than hand syrup, became possible with the development of catalyst injection spray equipment and the fitting of small roving choppers on the spray guns themselves. These developments enabled the simultaneous application of catalyzed resin and chopped fiber glass roving to molds. The chopped reinforcement is partially wet out during the spraying. Wet out and air removal is completed by "rolling out" the layup with special hand rollers.

As with hand layup, room temperature curing resins are used together with pigmented gel coats. Similarly the molding has only one good surface and precise control of thickness in the mold is difficult.

Resin Transfer Molding (RTM)

RTM is a modern version of an old molding method called the "Marco Method" developed by I. Muskat (Marco Chemicals) around 1950.

The method employs a male and a female mold. The reinforcement being used is placed between the two molds. A gel coat may be used on either one or both mold halves. After clamping the mold together with the provision of a tight seal around the mold periphery, catalyzed resin is forced into the mold by a pump until resin is seen exiting from a transparent tube placed on a vent located at the highest point in the mold. The resin "transfer" and air removal is assisted in some operations by attaching a vacuum source to the vent line.

The process generally employs room temperature curing resins. Both polyesters and vinyl esters can be used. Considerably stronger mold construction is required for RTM than for hand or spray layup because pressure and perhaps vacuum are used to force the resin into the mold and through the reinforcement.

Low viscosity resins are necessary to keep the pressure requirements moderate and to facilitate wetting out the reinforcement.

94 Handbook of Thermoset Plastics

Resins having low to medium exotherm temperatures are preferred since the curing polyester composite is totally enclosed and insulated by the mold halves which are often resin composites themselves. Excessive exotherm can cause mold damage or reduce useful mold life.

Figure 4-2 shows electrostatic precipitator collector plates made by the RTM process using a vinyl ester resin reinforced with continuous strand mat and unidirectional roving over a balsa wood core and surfaced with polyester fiber veil. These collector plates are continuously exposed to wet high voltage service.

Figure 4-2: RTM molded electrostatic collector plates. Manufacturer: Fluid-Ionic Systems, Dresser Industries. Molder: Sundt Products. Photo courtesy of: Reinforced Plastics/Composites Institute, The Society of the Plastics Industry, New York, NY.

Water Extended Polyester (WEP)

Water-in-oil unsaturated polyester emulsions on cure can result in cellular polyester composites in which all the cells are filled with water.[27] WEP resins often will have hydrophilic groups in the architecture of the alkyd backbone and employ emulsifying agents which will act in conjunction with this chemistry to allow the resin to form emulsions with water in which the resin is the continuous phase and the water is the discontinuous phase.

Cured polyester resins are not completely impermeable to water so WEP composites will have a tendency to lose water on aging which can cause shrinkage of the composite unless special stabilizers in the water such as hygroscopic agents or water soluble polymers are used. Special WEPs have been formulated so as to more readily lose water as a route to the production of reduced density polyester foams. This approach has fallen from favor as a method for making low density polyester composites since the cost of driving out the water has escalated with the rising cost of energy.

WEP is limited in strength since the water, unlike a solid filler, contributes nothing toward strength or rigidity. WEP can be a serviceable composite for articles such as flower pots and lamp bases where strength and dimensional stability are not especially critical.

Casting

Liquid polyester resins lend themselves very well to casting operations. Some typical applications include giftware and art objects, cultured marble and onyx, polymer concrete and monolithic flooring.

Giftware and art objects are usually made using special polyester resins which can cure to a clear colorless casting. This type of resin is made using highly proprietary formulations and methods. Methyl ethyl ketone peroxide catalyst is usually employed in a room temperature cure. The resin can be cast uncolored or can be colored by the use of peroxide resistant dyes or pigments. Glass, polypropylene, metal and other materials have been used for molds.

Elastomeric molds can permit the casting of artifacts with negative draft and undercuts. Silicone rubber has been the most popular elastomer system used for this purpose. Very large art pieces weighing hundreds of pounds, which have achieved international recognition, have been cast by artists such as Dwayne Valentine of Venice, California. Figure 4-3 illustrates a smaller art piece executed in clear resin by Caryl Craig.

Cultured marble and onyx, from a beginning in the early 1960s, have grown to be one of the major applications for unsaturated polyesters. These synthetic stone products are cast into either flat stock or complete bathroom items such as full size bathtubs and unilavs (bathroom counter top complete with one or more washbowls). The flat stock can be also used for tops and for wall applications such as tub surrounds. The pigmentation and appearance duplicate the appearance of the finest onyx and marble with the advantage of greater durability with respect to such properties as stain resistance. Figures 4-4, 4-5 and 4-6 illustrate three applications of cultured onyx sanitaryware: a unilav, a luxury tub installation and a toilet and bidet.

The polyester resins which dominate this application are carefully compounded to be compatible with the mineral fillers employed to give color and appearance which, with proper pigmentation, can faithfully reproduce the appearance of the natural stones. The resins must also give composites which will give the very long term service durability required of sanitaryware.

96 Handbook of Thermoset Plastics

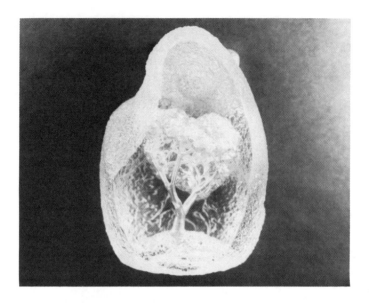

Figure 4-3: Clear unsaturated polyester art casting. Artist: Caryl Craig. Photo courtesy of: Silmar Div. Sohio Engineered Materials Co.

Figure 4-4: Polyester cultured onyx cast one piece lavatory top (unilav). Photo courtesy of: Gruber Systems, Inc., Valencia, CA.

Figure 4-5: Polyester cultured onyx cast bathtub and tub surround. Photo courtesy of: Gruber Systems, Inc., Valencia, CA.

Figure 4-6: Polyester cultured onyx commode and bidet. Photo courtesy of: Gruber Systems, Inc., Valencia, CA.

Calcium carbonate, in the form of ground limestone, is used to make cultured marble. 30 and 80 mesh are popular grades and are usually used as a mixture of two mesh sizes.

A typical resin-filler mix for cultured marble is shown in the following Table 4-21.

Table 4-21: Cultured Marble Matrix Formulation

	Pounds
Polyester marble resin	100
30 mesh calcium carbonate filler	200
80 mesh calcium carbonate filler	100
Methyl ethyl ketone peroxide catalyst	0.4-0.8

Titanium dioxide white pigment ground color to suit
Other colors to simulate marble veining to suit

A common compounding sequence will mix all of the ingredients except the veining colors. The mix, usually called the "matrix," can then have the veining pigments only partially mixed into the matrix to give, on casting and cure, the pigmented veining typical of natural marble. Large rotary mixers such as Hobart dough mixers have been used for this purpose. Other scraped wall type mixers of the Myers and Cowles types have also been successfully used. Recently continuous matrix mixing machines have been introduced for cultured marble and onyx production. Two machines of this type, the German Respecta machine and the American Venus machine have been installed in some domestic marble/onyx production plants recently. A feature common to both machines is the ability to heat the matrix above room temperature before discharge into the molds.

Cultured onyx production is very similar to the production of cultured marble. The major differences are that it is necessary to use resins which can cure without the development of any color and that the refractive indices of the filler systems are close to those of the cured resins. The most popular filler for onyx production is alumina trihydrate followed by powdered glass frits. Alumina trihydrate filler is sometimes used in conjunction with glass frit fillers.

A typical onyx mix is shown in Table 4-22.

Table 4-22: Cultured Onyx Matrix Formulation

	Pounds
Polyester onyx resin	100
Alumina trihydrate filler	200
Methyl ethyl ketone peroxide catalyst	1-2

Titania ground color to suit
Veining colors to suit

Cultured marble and onyx mixes are relatively high filled systems which must be pourable with adequate entrapped air release and correct room temperature curing characteristics. The high filler levels can exaggerate the reactivity differences of various catalysts and the effects of resin, filler and shop temperatures on the curing behavior. A number of papers and technical presentations exploring these effects have been presented by workers such as Hrouda, Husain and Zaske at the Annual and Regional Meetings of the Cultured Marble Institute.[59,60,61,62]

Polymer concrete, a more recent development, first found favor in Europe. With the correct fillers such as silica sand and chemically resistant resins, chemically resistant mortars and aggregate filled polymer concrete mixes can be formulated. Continuous mixers such as the Respecta equipment have been successfully used in this application although batch mixing equipment can also be used. Vinyl ester resins can be used for maximum chemical resistance in those environments not suitable for polyesters.

Polyester and vinyl ester resins have been very successfully applied in cast monolithic flooring. This type of flooring gives a seamless, wear resistant floor which can have considerable chemical resistance and which is easily cleaned. A monolithic cast laboratory floor is shown in Figure 4-7.

Figure 4-7: Cast monolithic polyester laboratory floor. Photo courtesy of: Silmar Div. Sohio Engineered Materials Co., Hawthorne, CA.

Acrylic Backup

Fiber glass laminate backup of vacuum formed acrylic articles such as spas and sanitaryware has become a growing application for unsaturated polyesters. The vacuum formed acrylic shape in effect becomes the service surface of the molded artifact and also serves as the mold for the application of the backup of polyester and chopped glass fiber which is usually applied by the chopper gun spray layup method. Development of good adhesion between the acrylic and the polyester fiber glass backup laminate is crucial to the success of this fabrication technique. A number of papers presented by A.H. Horner et al before the Annual Conferences of the SPI Reinforced Plastics Composites Institute have explored the performance, economics and recent developments in acrylic faced composites.[63,64,65] Thixotropic resins, prepromoted for room temperature curing, which are specially formulated to maximize adhesion, are usually used.

Matched Die Mat, Preform and Premix Molding

Both polyester and vinyl ester resins can be used in these processes which all employ matched male and female dies, usually made of metal, mounted in hydraulic molding presses. Heat cures are normally employed using such catalysts as benzoyl peroxide, tertiary butyl peroctoate and tertiary butyl perbenzoate. In production, semipositive telescoping dies are often used with flame hardened shear edges. Molding is usually to die stops. Molding temperatures can range from 225° to 250°F for mat and preform to 275° to 300°F for premix. Molding pressures are relatively low in the range of 100 to 300 psi. Press cycles can range from 2 minutes or so to longer periods depending on the thickness of the part, the mold materials and the size of the part being molded. Prototype molding has often been successfully carried out using inexpensive methods such as Kirksite dies. These, however, are not suitable for sustained production.

Mat molding, as the term implies, employs fiber glass mat for the reinforcement. Mat reinforcement, being planar, is limited to flat parts, parts with curvature in only one plane and parts which are essentially flat with only limited compound curved areas such as cafeteria trays.

Preform molding gets around the limitations of mat by the use of fiber glass preforms in which chopped glass fibers have been formed into a mat having the shape of the part to be molded. This process is applicable to small parts and also very large parts such as outboard motor boat hulls and auto body parts.

Both mat and preform molding employ ortho-phthalic and isophthalic polyesters and, to a more limited degree, vinyl ester resins. The resin molding compositions are often filled with mineral fillers such as calcium carbonates at levels of 30 to 65 PHR. Reinforcement levels are often at 35 to 40% by weight in the molded composite.

Premix molding is based on the use of a bulk molding compound made from liquid resin, filler and chopped reinforcement all combined in suitable mixers which are designed, as much as possible, to not fracture and thereby degrade the reinforcing fibers. Due to the limitations of the mixing processes, fiber lengths are usually limited to lengths shorter than can be used in mat or preform molding. For the same reasons, reinforcing fiber levels are usually lower in premix compounds. The shorter fiber lengths and lower levels of reinforcing fiber limit the physical properties of premix moldings to lower levels than can be obtained by either mat or preform molding.

Resins intended for these molding processes must have good hot strength to permit part removal from the dies without damage. The resins must also permit complete flow to the mold without pregelation and must then quickly cure for fast molding cycles. The overall combination of characteristics of isophthalic polyesters have made them favorites for press molding with mat, preform or premix.

Pultrusion

Both polyester and vinyl ester resins are successfully used in pultrusion operations. Pultruded composites will typically contain high reinforcement levels and are made by continuously pulling reinforcement through a resin impregnating bath and then through one or more heated dies the interior of which have the shape desired in the cured composite. The pultrusion process places several concurrent demands on resins. The resins must quickly and very thoroughly wet the reinforcement, even when containing filler, and must very rapidly develop a high enough green strength at die temperatures to avoid excessive pulling loads and to prevent surface cosmetic defects commonly called "sloughing." Rapid curing, high heat deflection point resins are essential for the achievement of satisfactory pulling speeds. Isophthalic polyesters are probably the most widely used resins for pultrusion. Pultruded composites, due to their generally high reinforcement levels usually have exceptionally high physical properties parallel to the direction of pultrusion. Physical properties in a direction normal to the direction of pultrusion are usually lower.

Figures 4-8, 4-9 and 4-10 illustrate a number of pultruded composite applications.

Figure 4-8 shows wind turbine blades pultruded using a vinyl ester resin reinforced with continuous strand glass mat and rovings. The blades are 9 inches wide x 11 feet 3 inches long.

Figure 4-9 illustrates an interesting application of pultrusion in molding window sash and frame components. These were molded using nonyellowing light stabilized polyester resin reinforced with glass fiber roving and continuous strand mat at a 50% reinforcement level.

Figures 4-10 and 4-11 picture an unusual pultrusion application in the manufacture of light weight, flexible poles and spreaders for the support of camouflage netting. Made for the U.S. Army, these composites are made using unsaturated polyester resin with glass fiber roving and mat reinforcement. An important plus for the composite is nondetectability by radar.

102 Handbook of Thermoset Plastics

Figure 4-8: Pultruded vinyl ester wind turbine blades. Manufacturer: Bergey Windpower Co. Molder: Morrison Molded Fiberglass Co. Photo courtesy of: Reinforced Plastics/Composite Institute, The Society of the Plastics Industry, Inc., New York, NY.

Figure 4-9: Pultruded polyester window sash and frame components. Manufacturer and molder: Omniglas Corp. Photo courtesy of: Reinforced Plastics/Composites Institute, The Society of the Plastics Industry, New York, NY.

Unsaturated Polyester and Vinyl Ester Resins 103

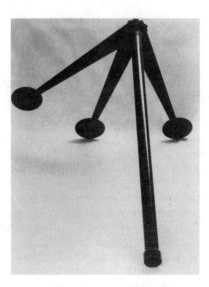

Figure 4-10: Pultruded polyester military camouflage netting support. User: U.S. Army Troop Support Command. Molder: The Pultrusion Corp. Photo courtesy of: Reinforced Plastics/Composites Institute, The Society of the Plastics Industry, New York, NY.

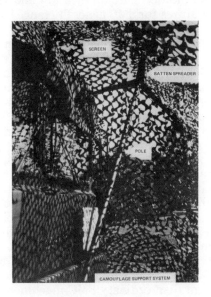

Figure 4-11: Camouflage netting supported by pultruded polyester support system. User: U.S. Army Troop Command. Molder: The Pultrusion Corp. Photo courtesy of: Reinforced Plastics/Composites Institute, The Society of the Plastics Industry, New York, NY.

Sheet and Bulk Molding Compounds (SMC and BMC)

Polyester or vinyl ester resins can be used to make SMC. Proper thickening with the usual thickening agents is an essential resin characteristic if the SMC is to be ready to mold, i.e., mature, in a reasonable time and to remain moldable long enough that it can be all used up under normal production schedules. Long moldability is especially important in SMC that is intended for use in remotely located satellite plants or SMC intended for outside sales. As previously mentioned, the most widely used thickening systems depend on the reaction of Group II metal oxides and hydroxides with carboxylic functionality on the resin. Carboxylic functionality on the shrink control or "low profile" resin used with the polyester or vinyl ester can also contribute to the thickening reaction. The thickening reaction results in the formation of a certain amount of chain extended high molecular weight fractions which can then greatly increase the viscosity of the SMC by entanglement and hydrogen bonding.

For resins to be satisfactory for SMC, they must also develop sufficient incompatibility with the shrink control additive resin on gelation and cure. This is necessary so that the shrink control additive properly precipitates as a second, monomer solvated, phase. The resultant expansion during cure can offset the curing shrinkage of the polyester or vinyl ester resin.

A good SMC resin should also have good hot strength to minimize damage to the molded part on ejection and removal from the die. Molding temperatures for SMC generally range from 275° to 300°F (135° to 149°C).

SMC is commonly made on special continuous machines designed for this purpose. Usually an SMC paste made from the molding resin, low profile resin, filler, mold release, catalyst, thickener and pigment is fed onto a continuous web of carrier film. Chopped fiber glass roving, often one inch in length, is dropped into the paste layer as it passes the glass chopping station on the machine. The film carrying the paste layer and the chopped glass is then continuously combined with a top carrier film also having a layer of SMC paste. The SMC, now encased between two layers of film, on passing between a number of rolls on the SMC machine is kneaded and squeezed to complete the impregnation of the chopped glass by the SMC paste. A take-off stand at the end of the machine rolls up the SMC which is then taken off the machine in rolls of convenient size for transfer to a "maturing" area where the rolls are stored until moldable. SMC made on equipment of this type is generally made in weights of 12 to 24 ounces per square foot. Heavier weight SMC is often more readily made on "TMC" or thick molding compound machines which are of different design than ordinary SMC machines.

Highly unsaturated isophthalic and propylene glycol maleate polyester resins are probably the msot often used resins for SMC molding. A typical low shrink SMC formulation employing a high reactivity isophthalic polyester resin is shown in Table 4-23.

The SMC paste is then combined on an SMC machine with 1 inch length chopped roving to a 25 to 30% glass content in the finished SMC.

Table 4-23: Isophthalic Resin SMC Formulation

SMC Paste	Pounds
Isophthalic polyester	100
Magnesium oxide thickener	1
Calcium carbonate filler*	150
Low profile additive**	54
Zinc stearate mold release	3
Tertiary butyl peroctoate catalyst	0.5
Tertiary butyl perbenzoate catalyst	0.5
Pigment dispersion: to suit	

*Camelwite (a trademark of Harry T. Campbell and Sons), Snowflake- (a trademark of Thompson, Weinman and Co.).
**LP-40A (a trademark of Union Carbide Corp.).

SMC is widely used now for molding automotive body parts such as grille opening panels, truck lids and hoods. Business machine housings are another growing application as are sections from which satellite reception TV antennas are assembled. There are a diversity of other applications such as for swimming pool filter tanks, seating and snowmobile body parts.

Bulk Molding Compound (BMC)

BMC can be considered to be a premix compound which incorporates shrink control additive polymer and which may also contain some thickener. As with premix, filler levels are higher and reinforcing fiber lengths and level are lower than those used in mat or preform molding and also SMC. BMC finds large application in items such as electrical parts and housings which have complicated features such as ribs and bosses and may incorporate inserts. Other applications include dinnerware and housings for small tools and appliances.

Some applications of BMC are shown in Figures 4-12 and 4-13. Figure 4-12 shows disposable frozen dinner plates made by compression molding using a glass fiber reinforced BMC formulated for microwave applications.

Figure 4-13 shows sewing machine structure and parts molded by injection molding using a flame retardant fiber glass reinforced polyester BMC. High stiffness and dimensional stability together with the flame retardance required to meet the requirements of Underwriters' Laboratories are important features of these parts.

Figure 4-12: Polyester BMC disposable dinner plates. User: Campbell Soup Co. Molder: Premix, Inc. Photo courtesy of: Reinforced Plastics/Composites Institute, The Society of the Plastics Industry, Inc., New York, NY.

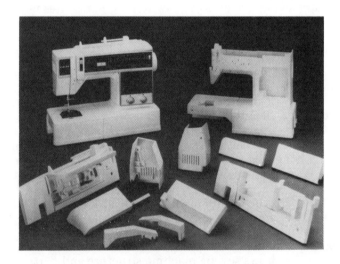

Figure 4-13: Polyester BMC sewing machine components. Manufacturer: The Singer Co. Molder: The Singer Co. and The Glastic Co. Photo courtesy of: Reinforced Plastics/Composites Institute, The Society of the Plastics Industry, Inc., New York, NY.

RECENT DEVELOPMENTS

Foamed Polyester

The preparation of cured foamed polyesters has been attempted, generally with indifferent success, since the resins have been available commercially. A recently developed foaming catalyst, Luperfoam 329 (a trademark of the Lucidol-Pennwalt Corp.), shows promise for the preparation of medium to high density polyester foams.

Foams as low as 15 pounds per cubic foot have been claimed with the use of this foaming system.[66]

Luperfoam 329 is an aqueous solution containing tertiary butylhydrazinium chloride and ferric chloride. This foaming catalyst is used at 1.25 to 1.5 PHR levels together with methyl ethyl ketone peroxide catalyst at a miminal 3 PHR level. It is reported that it is necessary to have a cobalt promoter in the polyester. The presence of even trace amounts of copper in the resin is deemed to be detrimental. 1.0 to 1.5 PHR of surfactants such as DC-193 (Dow Corning Corp.) or LK-221 (Air Products Corp.) are recommended to control cell size and promote a smooth foam rise. Auxiliary fluorocarbon blowing agents can be used to attain lower density foams.

Urethane Hybrid Resins

The preparation of urethane hybrid resins from unsaturated isophthalic polyester polyols and diisocyanates has been described by H.R. Edwards of the Amoco Chemicals Corp. The same worker has also described the use of such hybrids in casting compounds, RIM (Reaction Injection Molding) and SMC. Superior processing and physical properties were claimed in many instances.[67,68]

Reduced Styrene Emission Resins

Increased regulatory attention to styrene monomer levels in the workplace and pressures for reduction of the allowable levels have prompted development work on reduced emission resins and also directed attention to shop practices which can reduce styrene emission. The reduced emission or suppressed resins generally depend on the formation of a film on the surface of the resin. Effective film formers develop a thin surface barrier through which styrene vapor escapes more slowly than from the surface of ordinary resin. The film formers generally have limited styrene solubility and "kick out" on the surface of the resin as styrene evaporates from the surface. Paraffin waxes will function in this manner but like many effective similar materials can give problems with interlaminar bond strength on secondary lamination. P. Nylander[69] and M.J. Duffey[70] have described styrene suppressed resin systems. Most resin manufacturers now can offer reduced emission resins. Interlaminar bond strengths should be checked with such resins to ensure that no problems exist.

Table 4-24: Physical Properties Unsaturated Polyesters

All values except elongation, psi x 10^3

Castings	Rigid	Flexible
Tensile strength	6-13	0.5-3
Tensile modulus	300-640	–
Elongation at break, %	<2	40-310
Flexural strength	8.5-23	–
Flexural modulus	–	–
Compressive strength	13-30	–

Fiberglass Reinforced Composites	Preform Layup	Premix
Tensile strength	15-30	3-10
Tensile modulus	800-2,000	1,000-2,500
Elongation at break, %	1-5	<1
Flexural strength	10-40	7-20
Flexural modulus	1,000-3,000	1,000-2,000
Compressive strength	15-30	20-30

Fiberglass Reinforced Composites	Woven Cloth	SMC-Lo Shrink
Tensile strength	30-50	4.5-20
Tensile modulus	1,500-4,500	1,000-2,500
Elongation at break, %	1-2	3-5
Flexural strength	40-80	9-35
Flexural modulus	1,000-3,000	1,000-2,500
Compressive strength	25-50	15-30

Fiberglass Reinforced Composites	SMC	BMC, TMC
Tensile strength	8-25	5-10
Tensile modulus	1,400-2,500	1,500-2,500
Elongation at break, %	3	–
Flexural strength	10-36	16-24
Flexural modulus	1,000-2,200	--
Compressive strength	15-30	14-30

Reprinted with permission from *Modern Plastics Encyclopedia* 1984-1985, Copyright 1984, McGraw-Hill, Inc. All rights reserved.

MANUFACTURERS OF UNSATURATED POLYESTER AND VINYL ESTER RESINS

Trade Name	UP/VE*	Manufacturer
Advaco	UP	Advance Coatings Co.
Altek	UP	The Alpha Corp.
Aropol, Hetron	UP, VE	Ashland Chemical Co.
Cargill	UP	Cargill, Inc.
Cook	UP	Cook Paint and Varnish Co.
Derakane	VE	Dow Chemical Co.
Stypol	UP, VE	Freeman Chemical Corp.
Atlac	UP, VE	ICI Americas Inc.

(continued)

Co-Rezyn	UP, VE	Interplastic Corp.
Koplac, Dion	UP, VE	Koppers Co., Inc.
Midwest	UP	Midwest Resins Corp.
OCF	UP	Owens Corning Fiberglas Corp.
Pedigree	UP	P.D. George Co.
Pioester	UP	Pioneer Plastics Div. of LOF Plastics
Polylite	UP, VE	Reichhold Chemicals, Inc.
Epocryl	VE	Shell Chemical Co.
Silmar	UP, VE	Silmar Div., Sohio Engineered Materials Co.
USS, MR	UP, VE	USS Chemicals Div., U.S. Steel Corp.

*UP = unsaturated polyester; VE = vinyl ester

REFERENCES

1. Boenig, H.V., *Unsaturated Polyesters*, Elsevier, New York (1964).
2. Boenig, H.V., *Unsaturated Polyesters*, Elsevier, New York (1964).
3. Bjorksten, J., *Polyesters And Their Applications*, Reinhold, New York (1956).
4. Morton, M., *Rubber Technology*, Van Nostrand Reinhold, New York (1973).
5. Boenig, H.V., *Unsaturated Polyesters*, Elsevier, New York (1964).
6. Greek, B.F., *Chem. and Eng. News*, 10 (Nov. 5, 1984).
7. Kienle, R.H., *Ind. Eng. Chem.*, 22, 590 (1930).
8. Kienle, R.H., *Ind. Eng. Chem.*, 55, 229T (1936).
9. Bjorksten, J., *Polyesters And Their Applications*, Reinhold, New York (1956).
10. Parker, E.E., *Mod. Plastics*, 36, 135 (June 1959).
11. Mayo, F.R., Lewis, F.M. and Walling, C., *J. Am. Chem. Soc.*, 70, 1529 (1948).
12. Vanso-Szmercsanyi, I., Marcos, L.K. and Zahran, A.A., *J. Applied Polymer Science*, 10, 513-522 (1948).
13. Haman, K., Funke, W. and Gilch, R., *Angew. Chemie.*, 71, 596 (1959).
14. Boenig, H.V., *Unsaturated Polyesters*, Elsevier, New York (1964).
15. Hayes, B.T., Read, W.J. and Vaugham, L.V., *Chem. and Ind.*, London, 1165 (1957).
16. Cook, W.D. and Delatycki, D., *J. Macromol. Chem.*, A12(5), 769-787 (1978).
17. Demmler, K. and Schlag, J., *Kunstoffe*, 57, 566-572 (1967).
18. Office of Safety and Health Administration, U.S. Department of Welfare.
19. American Conference of Governmental and Industrial Hygienists.
20. Waters Associates, 34 Maple St., Milford, Massachusetts.
21. Varian Associates, 2700 Mitchell Dr., Walnut Creek, California.
22. Schotte, G. and Meijerink, N.L., *British Polymer J.*, 133-139 (June 1977).
23. Scheuing, D.R., *Spectra-Physics Chromatography Review*, V6, No. 1, 1-4 (January 1980) (Pub. by Spectra-Physics Corp., 2905 Stender Way, Santa Clara, California).
24. Hagnauer, G.L., *Anal. Chem.*, V54, No. 5, 265R-276R (April 1982).
25. Kerle, E.J., Connolly, W.J. and Rosenfeld, I., 29th Annual SPI RPC Conference, Paper 11G (1974).
26. Synthesis and Cure of Unsaturated Polyesters Based on TMPD Glycol for RP Applications, Public. No. N-176 (June 1974) (Pub. by Eastman Chem. Prods., Kingsport, Tennessee).
27. Pritchard, G., et al, *Thermosetting Resins for Reinforced Plastics*, Applied Science Publishers, London (1982).

28. Bowen, R.L., U.S. Patent 3,066,112 (1962).
29. Bowen, R.L., U.S. Patent 3,179,623 (1965).
30. Fekete, F., Keenan, P.J. and Plant, W.J., U.S. Patent 3,221,043 (1965).
31. Fekete, F., Keenan, P.J. and Plant, W.J., U.S. Patent 3,256,226 (1966).
32. Jackson, R.J., Ed., Epoxy Resin Technology-Developments Since 1979, Noyes Data Corp., 228 (1982).
33. May, C.A., U.S. Patent 3,345,401 (1967).
34. May, C.A., U.S. Patent 3,373,221 (1966).
35. Najvar, D.J., U.S. Patent 3,892,819 (1975).
36. Anderson, T.F. and Messick, V.B., *Developments In Reinforced Plastics-1*, Ed. by G. Pritchard, Applied Science Publishers Ltd., London (1980).
37. Brookfield Engineering Corp., 240 Cushing, Stoughton, Massachusetts.
38. Bautista, T.D., *Polymer-Plastics Technology Engineering*, 18(2), 179-207 (1982).
39. Bonsignore, P.V. and Manhart, J.H., Aluminum Hydroxide-Fire Retardant Additive and Filler for Plastics, 29th Annual SPI RPC Conference (1974).
40. Connolly, W.J. and Thornton, A.M., Alumina Trihydrate Filler In Polyester Systems, *Mod. Plastics* (October 1965).
41. Sprow, T.K., Connolly, W.J. and Kerle, E.J., Filled Polyester Spray-Up Systems Offering Improved Fire Hazard Classification, 28th Annual SPI RPC Conference (1973).
42. Wampner, F.D., Alumina Trihydrate Average Particle Size Effect on the Flammability and Physical Properties of Fiberglass Reinforced Plastics and Bulk Molding Compounds, 31st Annual SPI RPC Conference (1976).
43. Keating, J.Z., Flame and Smoke Management in Polyester Systems, 32nd Annual SPI RPC Conference (1977).
44. Zaske, O.C., Wang, M. and Wuh, J., 40th Annual SPI RPC Conference, Paper 8F (1985).
45. Frillete, V., U.S. Patent 2,568,331 (1951).
46. Fisk, C.F., U.S. Patent 2,628,209 (1953).
47. Schnell, Raichle, Prater and Bruhne, U.S. Patent 3,390,205 (1968).
48. Jernigan, J.W., U.S. Patent 3,446,259 (1969).
49. Fekete, F., Keenan, R.J. and Plant, W.J., U.S. Patent 3,256,226 (1966).
50. Fekete, F., Keenan, R.J. and Plant, W.J., U.S. Patent 3,301,743 (1967).
51. Doyle, T. and Fekete, F., U.S. Patent 3,317,465 (1967).
52. Fekete, F., 27th Annual SPI RPC Conference, Paper 12D (1972).
53. Ferrarini, J., Longnecker, D.M., Shah, N.N., Feltzin, J. and Greth, G.G., 33rd Annual SPI RPC Conference, Paper 9D (1978).
54. Burns, R., Gandhi, K.S., Hankin, A.G. and Lynskey, B.M., *Plastics and Polymers* (Brit.), 228-235 (December 1975).
55. Zaske, O.C., Fintelmann, C. and Wuh, J., 36th Annual SPI RPC Conference, Paper 23A (1981).
56. Horner, A.H., Zaske, O.C. and Brill, R., 37th Annual SPI RPC Conference, Paper 1B (1982).
57. Horner, A.H. and Brill, R., 39th Annual SPI RPC Conference, Paper 8A (1984).
58. Horner, A.H. and Brill, R., 40th Annual SPI RPC Conference, Paper 16D (1985).
59. Zaske, O.C. and Hrouda, G., The Effect of Different Commercial Ketone Peroxides on Cultured Marble Matrix Gel and Demold Times, Annual Meeting of the Cultured Marble Institute (September 22, 1975).

60. Hussain, K. and Zaske, O.C., The Effect of Ambient Temperature on the Viscosity, Gel and Demold Time of a Typical Marble Matrix, National Cultured Marble Institute Meeting (February 23, 1979).
61. Zaske, O.C., Interactions in Marble and Onyx Matrixes, Regional Cultured Marble Institute Meeting (August 17, 1979).
62. Zaske, O.C. and Husain, K., Gel Coat and Matrix Resin Chemistry Response to Water Boil Tests, National Cultured Marble Institute Meeting (February 28, 1980).
63. Horner, A.H. and Church, S.L., 28th Annual SPI RPC Conference, Paper 7C (1973).
64. Horner, A.H. and Church, S.L., 29th Annual SPI RPC Conference, Paper 7E (1974).
65. Horner, A.H., 30th Annual SPI RPC Conference, Paper 3B (1975).
66. Luperfoam 329 Technical Bulletin, Lucidol Div., Pennwalt Corp.
67. Edwards, H.R., 39th Annual SPI RPC Conference, Paper 8C (1984).
68. Hybrid Resins, Technical Bulletin IP-77, Amoco Chemicals Corp.
69. Nylander, P., 34th Annual SPI RPC Conference, Paper 6B (1979).
70. Duffy, M.J., 34th Annual SPI RPC Conference, Paper 6D (1979).

5
Allyls

Sidney H. Goodman
Developmental Products Laboratory
Technology Support Division
Hughes Aircraft Company
El Segundo, California

INTRODUCTION

The need for materials to resist the extraordinary conditions imposed by the fledgling aerospace industry in the 1950s encouraged polymer chemists to evolve a family of plastics based on diallyl phthalate (DAP). These plastics were characterized by excellent chemical resistance, low electrical loss, excellent weathering, very low mold shrinkage and good dimensional stability. For 20 years DAP compounds were among the highest priced insulating materials. In the 1970s, improvements in competitive plastics (epoxies, polyesters, polyurethanes, and many new engineering thermoplastics) significantly reduced the market for DAPs. At one point in the late 1970s there was only one supplier of monomer in the world and no more than two suppliers of compound in the U.S.

During the last few years, new lower cost DAP-based compounds were produced. More economical molding processes based on traditional compression molding and advanced injection molding techniques were also established. Thus new incentives for market growth were established and DAPs remain a commercially viable plastics family.

A thermoplastic allyl, allyl diglycol carbonate, is a colorless, optically transparent resin that is often included within the allyl family. However, discussion of this plastic is not germane to this book and the reader is referred to the bibliography (e.g., Thomas or Sare) for further detail.

CHEMISTRY

Diallyl phthalate monomers are made from propylene and phthalic anhydride as follows:

$$CH_3CH=CH_2 \xrightarrow{Cl_2} CH_3\text{-}CHCl\text{-}CH_2Cl \xrightarrow{reduc} CH_2=CHCH_2OH$$
propylene — dichloropropane — allyl alcohol

$2CH_2=CHCH_2OH$ + phthalic anhydride → diallyl phthalate

Depending on the choice of anhydride, a series of allyl monomers can be created. The two of most significant commercial importance are diallyl phthalate (diallyl orthophthalate, DAP) and diallyl isophthalate (DAIP). Diallyl fumarates and maleates have found use as highly reactive trifunctional monomers containing two kinds of polymerizable double bonds. Diallyl chlorendate is used to impart flame retardance to molding compositions. Other allyl monomers that have found application include allyl methacrylate (crosslinking agent in unsaturated polyesters and monomer intermediate) and triallyl cyanurate (crosslinking of unsaturated polyesters).

POLYMERIZATION AND PROCESSING

Polymerization and crosslinking of the allyls occur via a peroxide induced addition through the allyl unsaturation.

Typical catalysts include t-butyl perbenzoate, benzoyl peroxide, or dicumyl peroxide at levels of 2-3 phr. Dimeric peroxyethers and esters have also been used.

Allylic homopolymerization is very slow at room temperature, catalyzed formulas being stable for over a year with hardly any resin advancement. Once the temperature exceeds 150°C (300°F) the cure rate proceeds very rapidly.

Because of the vinyl-type addition reaction mechanism, no volatiles are generated during cure. Typical cure temperatures vary within the 135°-177°C (275°-350°F) range. For moldings, mold residence times are around 0.5-4.0 minutes at pressures of 500-8,000 psi for compression molding and 2,000-10,000 psi for transfer molding. Laminates made using matched-die molding or vacuum bag techniques require cure schedules of 20-35 minutes at 93°-177°C (200°-350°F). Shrinkage of allylic monomers during cure is only 12% v/v with prepolymer shrinkage less than 1% (see Figure 5-1).

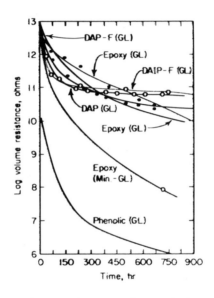

Figure 5-1: Volume resistance decrease of several thermosetting materials at 70°C and 100% RH. (GL indicates glass filler, and Min-GL indicates mineral and glass filler.) (Harper).

Allylic compounds generally fill crevices and completely surround inserts in complicated molds because of their good flow characteristics. Molds with small draft angles can be used without hampering release. Chrome-plated molds are recommended, although polished steel has been used.

The very low viscosity of the monomers is useful when admixed as a crosslinking agent in unsaturated polyesters. The overall compound viscosity is reduced and more fillers and additives can be incorporated. For many homopolymeric molding and laminating applications, the use of prepolymers have been found more useful. The prepolymers are syrupy to solid, linear, internally cyclized thermoplastic structures containing unreacted allylic groups spaced at regular intervals along the polymer chain. By using prepolymers, better handling of molding powders and laminate prepregs is obtained, along with increased control of flow and exotherm.

FORMULATION

Allyl compounds typically contain some monomer (for viscosity and reactivity control), catalyst, fillers, pigments and processing aids. Fibrous and mineral fillers comprise the bulk of fillers used. Glass fibers produce moldings with the best all around properties. Long glass fibers provide greater impact strength. Glass fibers also combine to give the highest shock and arc resistance in DAP compounds.

Acrylic fibers yield the best electrical properties particularly under high humidity exposure. Polyester fibers give impact resistance and strength in thin sections. Nylon filled systems provide high durability, i.e., the best resistance to abuse. Cellulosic and other fibrous mineral fillers generally are used to reduce cost or for some specialized property enhancement. Asbestos was a popular filler to improve impact strength at low cost (however electrical properties decreased). This filler has been effectively eliminated from use because of its carcinogenic effects. Table 5-1 shows comparative properties of various fiber-filled DAP systems.

Table 5-1: Typical Properties of Several Diallyl Phthalate Molding Compounds with Various Fillers
(Harper)

Property	Orlon	Dacron	Long glass	Asbestos	Short glass	Short glass*
Tensile strength, psi	6,000	5,000	10,000	5,500	7,000	7,000
Compressive strength, psi	25,000	25,000	25,000	25,000	25,000	28,000
Flexural strength, psi	10,000	11,500	16,000	9,600	12,000	12,000
Flexural modulus, psi $\times 10^{-6}$	0.71	0.64	1.3	1.2	1.2	1.3
Impact strength, Izod, ft-lb/in. of notch	1.2	4.5	6.0	0.4	0.6	0.5
Hardness, Rockwell M	108	108	100	100	105	110
Specific gravity at 25°C	1.31–1.45	1.39–1.62	1.55–1.70	1.55–1.65	1.6–1.8	1.65–1.75
Dielectric constant:						
At 1 kHz	3.7–4.0	3.79	4.2	....	4.4	4.1
At 1 MHz	3.3–3.6	3.4	4.2	4.5–6.0	4.4	3.4
Dissipation factor:						
At 1 kHz	0.020–0.025	0.008	0.004–0.006	0.04–0.08	0.006	0.004
At 1 MHz	0.015–0.020	0.012	0.008	0.04–0.08	0.008	0.008
Mold shrinkage, in./in.	0.009	0.010	0.002	0.006	0.003	0.003
Postmold shrinkage, in./in.	0.001	0.0006	0.0007	0.001	0.0007	0.0002
Heat-deflection temperature, °F	265	290	392	325	400	500+
Heat resistance, continuous, °F	300–500	350–400	350–400	350–400	350–400	450

* Based on diallyl isophthalate.

Particulate fillers include calcium carbonate and silicate, treated clays, and barium sulfate (barytes). These are added primarily to reduce cost and control flow.

PROPERTIES

Table 5-2 presents a general summary of allyl properties. As can be seen these plastics demonstrate high dimensional stability, excellent heat and chemical resistance, and superb electrical properties particularly under extreme temperature and humidity conditions. Since there are no corrosive volatiles released during cure, they will not attack metallic inserts nor support galvanic corrosion in the presence of moisture. Compounds based on DAIP will withstand higher temperatures, are slightly stronger, and are easier to mold than those based on DAP.

Table 5-2: General Summary of DAP Properties

Physical
    Extremely good dimensional stability.
    Almost no post-mold shrinkage.
    Chemically inert, gives off no corrosive vapors.
Mechanical
    Excellent strength in compression.
    Excellent impact resistance.
    Exceptional thin wall strength.
    Excellent for applications subjected to: sudden extreme jolts, severe stresses.
Electrical
    Retains high insulation resistance at elevated temperatures.
    Performance virtually unaffected by high ambient humidity.
Thermal
    Outstanding thermal stability at normal use temperatures.
    No decomposition below 150°C (300°F).
    Eliminates fouling of metal contacts caused by condensation of decomposition products.
Chemical
    Resistant to solvents, acids, and alkalies.
    Fungus proof.

The resistance of DAPs to humidity is demonstrated in Figures 5-2 and 5-3. Figure 5-2 shows that the insulation resistance of allyls does not deteriorate even after 4,000 hours exposure to 70°C (160°F) and 95% RH. Only polybutylene terephthalate (PBT) is better. Figure 5-3 shows that DAP and DAIP retain a much higher percentage of tensile strength after 2,880 hours exposure to 70°C (160°F) and 100% RH than either PBT or polyphenylene sulfide (PPS). Dielectric strength values for DAP are good up to 190°C (374°F) and in excess of 205°C (400°F) for DAIP. Figures 5-4 and 5-5 show the effect of frequency and temperature on the dielectric constant and dissipation factor of unfilled DAP. DAP materials are capable of withstanding extensive radiation exposure. Compounds have withstood 10^4-10^{12} rad dosages of gamma radiation without breakdown.

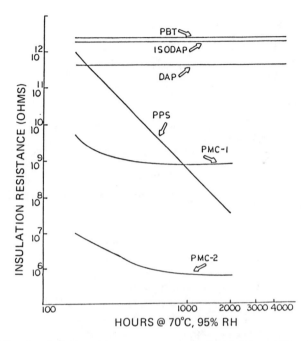

Figure 5-2: Insulation resistance vs exposure to high humidity. (Dalton and Landi).

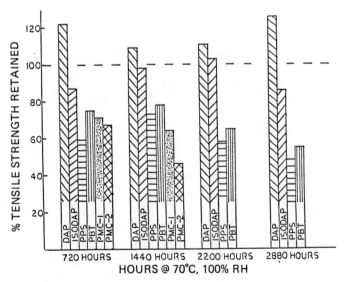

Figure 5-3: Retention of tensile strength after exposure to high humidity. (Dalton and Landi).

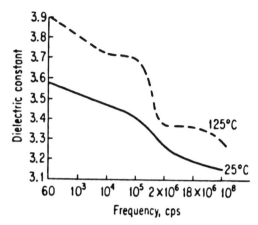

Figure 5-4: Effect of frequency and temperature on the dielectric constant of unfilled diallyl phthalate. (Harper).

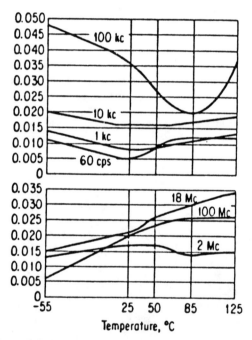

Figure 5-5: Effect of frequency and temperature on the dissipation factor of unfilled diallyl phthalate. (Harper).

Tables 5-3 through 5-7 are property listings for a variety of commercial DAP compounds. No recommendations or preference by the author is intended or implied. Rather the data is included to provide up-to-date information only. The reader is advised to contact individual suppliers for specific assistance and most recent developments.

Table 5-3: DAP Diallyl Phthalate Molding Materials (Rogers Corp.)

GRADE	1-503	1-501N	1-501AN	1-510N	2-501N	1-520	1-540	1310	2-520
Reinforcement	Orlon	Mineral	Mineral Nylon	Mineral	Mineral	Short Glass	Short Glass Mineral	Short Glass	Short Glass
Resin (DAP) Form	Ortho Granular	Ortho Granular	Ortho Granular	Ortho Granular	Iso Granular	Ortho Granular	Ortho Granular	Ortho Granular	Iso Granular
Bulk Factor	3.0	2.3	2.6	2.4	2.1	2.4	2.3	2.4	2.5
Specific Gravity (g/cm^3)	1.40	1.80	1.73	1.66	1.90	1.80	1.82	1.80	1.75
Shrinkage (Comp) in/in	.010-.012	.003-.006	.007-.010	.003-.006	.003-.006	.001-.003	.0015-.003	.001-.003	.001-.003
Izod Impact (ft. lb./in.) (C/T)	0.65	0.45	0.45	0.45	0.45	1.0	0.70	1.0	1.0
Flexural Strength (psi)	9000/ 12,000	11,000/ 14,000	10,000/ 13,000	12,000/ 15,000	10,000/ 13,000	15,000/ 18,000	10,000/ 13,000	16,000/ 20,000	15,000/ 18,000
Flexural Modulus (psi)	0.7 x 10^6	1.6 x 10^6	1.1 x 10^6	1.3 x 10^6	2.0 x 10^6	1.8 x 10^6	1.8 x 10^6	1.9 x 10^6	1.7 x 10^6
Tensile Strength (psi)	5000	6500	5500	7500	5500	8000	6500	9000	7500
Compressive Strength (psi)	29,000	24,000	25,000	21,000	27,000	28,000	23,000	28,000	30,000
Water Absorption, %	0.40	0.30	0.40	0.50	0.25	0.30	0.35	0.25	0.25
Deflection Temperature °F	290	330	300	325	525	500	425	500	525
Continuous Use Temperature °F	225	230	200	225	425	375	325	375	425
Flammability (IGN/Burn), sec.	87/440	90/373	—	110/279	115/222	70/340	107/297	100/350	98/262
UL Rating 1/8"	—	—	94 HB	—	94 HB	—	94 HB	—	94 HB
Dielectric Strength 60Hz, ST/SS, wet (vpm)	375/350	375/350	400/380	400/390	400/400	375/350	390/370	400/380	380/360
Dielectric Constant 1 khz/1 mhz, wet	3.4/3.1	4.2/3.6	4.3/3.7	4.2/3.7	4.3/3.8	4.1/3.9	4.4/4.2	4.1/3.9	3.9/3.7
Dissipation Factor 1 khz/1 mhz, wet	.016/.021	.013/.014	.015/.019	.015/.017	.015/.019	.010/.017	.013/.018	.009/.016	.011/.015
Arc Resistance (sec.)	125	140	135	135	150	135	180	140	150
MIL-M-14G Type	SDI-5	MDG	MDG	MDG	MDG	SDG	—	SDG	SDG

(continued)

Table 5-3: (continued)

GRADE	3-1-501N	3-2-520F	3-1-540	3-1-525F	1366 FR	1-530	2-530	3-1-530	3-2-530
Reinforcement	Mineral	Short Glass	Short Glass Mineral	Short Glass	Short Glass	Long Glass	Long Glass	Long Glass	Long Glass
Resin (DAP) Form	Ortho Granular	Iso Granular	Ortho Granular	Ortho Granular	Ortho Granular	Ortho Coarse Granular	Iso Coarse Granular	Ortho Coarse Granular	Iso Coarse Granular
Bulk Factor	2.0	2.4	2.2	2.3	2.4	3.5	3.5	3.5	3.5
Specific Gravity (g/cm³)	1.83	1.90	2.08	1.87	1.86	1.74	1.73	1.76	1.76
Shrinkage (Comp) in/in	.003-.006	.001-.003	.0015-.003	.0015-.003	.0015-.003	.0015-.004	.0015-.004	.0015-.004	.0015-.004
Izod Impact (ft. lb./in.) (C/T)	0.60	1.0	0.70	1.0	1.1	3.0-7.0	3.0-7.0	3.0-7.0	3.0-7.0
Flexural Strength (psi)	11,000/ 14,000	14,000/ 17,000	13,000/ 16,000	16,000/ 19,000	16,000/ 20,000	17,000/ 20,000	17,000/ 20,000	18,000/ 21,000	18,000/ 21,000
Flexural Modulus (psi)	1.7×10^6	1.4×10^6	2.4×10^6	1.8×10^6	1.7×10^6	1.8×10^6	1.4×10^6	1.6×10^6	1.7×10^6
Tensile Strength (psi)	6500	7500	7500	8000	9000	8500	7500	8500	8500
Compressive Strength (psi)	25,000	26,000	28,000	28,000	26,000	27,000	25,000	27,000	28,000
Water Absorption, %	0.30	0.25	0.25	0.25	0.25	0.35	0.35	0.35	0.35
Deflection Temperature °F	325	525	400	450	400	500	550	500	550
Continuous Use Temperature °F	225	425	300	350	350	400	450	400	450
Flammability (IGN/Burn), sec.	106/23	119/40	159/32	110/40	120/20	95/275	95/339	110/40	110/40
UL Rating 1/8"	—	94V-0	94V-0	94V-0	94V-0	—	—	94V-0	—
Dielectric Strength 60Hz, ST/SS, wet (vpm)	400/375	375/350	370/350	400/375	400/375	400/350	400/350	400/350	400/350
Dielectric Constant 1 khz/1 mhz, wet	4.3/3.7	4.1/3.9	4.4/4.3	4.3/4.1	4.4/3.9	4.1/4.0	3.8/3.6	4.2/3.9	4.1/3.9
Dissipation Factor 1 khz/1 mhz, wet	.013/.016	.010/.013	.011/.015	.010/.016	.009/.017	.011/.019	.010/.019	.010/.018	.010/.019
Arc Resistance (sec.)	135	150	180	130	140	135	135	130	135
MIL-M-14G Type	MDG	SDG-F	—	SDG-F	SDG-F	GDI-30	GDI-30	GDI-30F	GDI-30F

Test methods used are according to A.S.T.M., Fed. Std. 406, or Military Specifications.

Table 5-4: DAP Molding Materials Properties (Cosmic Plastics, Inc.)

Compound Number		DIALLYL ORTHO PHTHALATE RESIN (Regular)		
		D69	D62	D63
Filler		Long Glass Fiber		
Forms		Flakes	Flakes	Flakes
Standard(s) or Flame Retardant (F)		S	F	S
Specific Gravity		1.62	1.76	1.77
Bulk Factor		6	6	6
Molding Pressure PSI		500-8000	500 - 8000	500 -8000
Molding Temperature °F		275 - 350	275 - 350	275 - 350
Molding Shrinkage in./in.		0.001 - 0.004	0.001 - 0.004	0.001 - 0.004
Heat Distortion Temperatures °F		500	500	500
Continuous Heat Resistance °F		350 - 400	350 - 400	350 - 400
Dimensional Stability % max.		0.01	0.01	0.01
Thermal Expansion $10^{-5}/°C$		2.4	2.4	2.4
Flame Resistance: Ignition Time sec. (min)		90	155	90
Burning Time sec. (max)		300	30	285
Water Absorption: %48 hrs. @50°C %		0.2	0.2	0.2
Impact Strength: Ft - lb./in. notch (Izod)		4 - 10	4 - 10	4 - 10
Flexural Strength: PSI		17,000	17,000	17,000
Compressive Strength: PSI		29,000	29,000	29,000
Tensile Strength: PSI		7000 - 10,000	7000 - 10,000	7000 - 10,000
Arc Resistance: Seconds		180	180	180
Dielectric Strength:volts/mil. Step by Step	DRY	400	400	374
	WET	400	400	364
Dielectric Breakdown: Kv Step by Step	DRY	62	62	62
	WET	60	60	60
Dielectric Constant: 1KC/1MC	DRY	4.0/3.9	4.2/4.0	4.0/4.3
	WET	4.2/4.1	4.4/4.2	4.2/4.4
Dissipation Factor: 1KC/1MC	DRY	.006/.013	.006/.014	.007/.013
	WET	.008/.015	.009/.016	.009/.015
Surface Resistance: megohms	AS IS	10^{10} plus	10^{10} plus	10^{10} plus
30 Days @100% R.H. @158°F		200,000	26,000	10.000
Volume Resistance: megohms	AS IS	10^{10} plus	10^{10} plus	10^{10} plus
30 Days @100% R.H. @158°F		200,000	20,000	10,000
Water Abstract Conductances 10^{-6} MHOS/CM		30	30	35
Certifiable to Military Specifications		MIL-M-14G MIL-P-19833	MIL-M-14G MIL-P-19833	MIL-M-14G MIL-P-19833
and Type Design		GDI-30	GDI-30F	GDI-30
Flammability Rating: UL-94⅛"		HB	VO	HB

(continued)

Table 5-4: (continued)

Compound Number		DIALLYL ORTHO PHTHALATE RESIN (Regular)	
		D33	D72
Filler		Short Glass Fiber	
Forms		Granules	Granules
Standard(s) or Flame Retardant (F)		S	F
Specific Gravity		1.70	1.76
Bulk Factor		2.5	2.5
Molding Pressure PSI		500 - 8000	500 - 8000
Molding Temperature °F		275 - 350	275 - 350
Molding Shrinkage in./in.		0.001 - 0.004	0.001 - 0.004
Heat Distortion Temperatures °F		500	500
Continuous Heat Resistance °F		350 - 400	350 - 400
Dimensional Stability % max.		0.01	0.01
Thermal Expansion 10^{-5}/°C		1.2	1.6
Flame Resistance: Ignition Time sec. (min)		90	110
Burning Time sec. (max)		300	60
Water Absorption: %48 hrs. @50°C %		0.2	0.2
Impact Strength: Ft - lb./in. notch (Izod)		0.5 - 1.2	0.5 - 1.2
Flexural Strength: PSI		13,000	13,000
Compressive Strength: PSI		25,000	25,000
Tensile Strength: PSI		7000 - 10,000	7000 - 10,000
Arc Resistance: Seconds		145	145
Dielectric Strength: volts/mil. Step by Step	DRY	385	380
	WET	340	360
Dielectric Breakdown: Kv Step by Step	DRY	62	62
	WET	60	60
Dielectric Constant: 1KC/1MC	DRY	4.2/4.0	4.2/4.0
	WET	4.3/4.2	4.3/4.2
Dissipation Factor: 1KC/1MC	DRY	.006/.013	.007/.013
	WET	.012/.015	.011/.015
Surface Resistance: megohms	AS IS	10^{10} plus	10^{10} plus
30 Days @100% R.H. @158°F		10,000	10,000
Volume Resistance: megohms	AS IS	10^{10} plus	10^{10} plus
30 Days @100% R.H. @158°F		10,000	10,000
Water Abstract Conductances 10^{-6} MHOS/CM			
Certifiable to Military Specifications		MIL-M-14G	MIL-M-14G
and Type Design		SDG	SDG-F
Flammability Rating: UL-94⅛"		HB	VO

(continued)

Table 5-4: (continued)

Compound Number			DIALLYL ORTHO PHTHALATE RESIN (Regular)		
			224	306	D45(D44)
Filler			Dacron	Orlon	Mineral
Forms			Flakes	Flakes	Powder
Standard(s) or Flame Retardant (F)			S	S	S
Specific Gravity			1.42	1.39	1.67
Bulk Factor			6	6	2.3
Molding Pressure PSI			500 - 8000	500 - 8000	500-8000
Molding Temperature °F			275 - 350	275 - 350	275 - 350
Molding Shrinkage in./in.			.007 - .009	0.007 - 0.009	0.003 - 0.007
Heat Distortion Temperatures °F			290	280	350
Continuous Heat Resistance °F			350 - 400	350-400	350
Dimensional Stability % max.			0.06	0.07	0.08
Thermal Expansion 10^{-5}/°C			4.0	4.2	4.5
Flame Resistance:	Ignition Time sec. (min)		80	80	105
	Burning Time sec. (max)		300	300	280
Water Absorption: %48 hrs. @50°C %			0.2	0.2	0.4
Impact Strength: Ft - lb./in. notch (Izod)			6 - 12	1.5 - 3.0	0.50
Flexural Strength: PSI			12,500	12,500	12,000
Compressive Strength: PSI			25,000	30,000	28,000
Tensile Strength: PSI			5200	5000	6000
Arc Resistance: Seconds			125	125	180
Dielectric Strength: volts/mil. Step by Step		DRY	360	360	380
		WET	350	340	360
Dielectric Breakdown: Kv Step by Step		DRY	62	63	54
		WET	58	60	48
Dielectric Constant: 1KC/1MC		DRY	3.4/3.5	3.2/3.1	5.8/4.2
		WET	3.5/3.6	3.4/3.3	6.5/4.4
Dissipation Factor: 1KC/1MC		DRY	.005/.010	.020/.013	.039/.038
		WET	.008/.012	.022/.015	.12/.041
Surface Resistance: megohms		AS IS	10^{10} plus	10^{10} plus	10^{10} plus
30 Days @100% R.H. @158°F			10^6	10^5	10,000
Volume Resistance: megohms		AS IS	10^{10} plus	10^{10} plus	10^{10} plus
30 Days @100% R.H. @158°F			10^6	10^5	6000
Water Abstract Conductances 10^{-6} MHOS/CM					
Certifiable to Military Specifications			MIL-M-14G	MIL-M-14G	MIL-M-14G
and Type Design			SDI-30	SDI-5	MDG
Flammability Rating: UL-94⅛"					HB

(continued)

Table 5-4: (continued)

Compound Number		DIALLYL META PHTHALATE RESIN (High Heat Resistant)	
		K66	K61
Filler		Long Glass Fiber	
Forms		Flakes	Flakes
Standard(s) or Flame Retardant (F)		F	S
Specific Gravity		1.78	1.60
Bulk Factor		6	6
Molding Pressure PSI		500 - 8000	500 - 8000
Molding Temperature °F		275 - 350	275 - 350
Molding Shrinkage in./in.		0.001 - 0.004	0.001 - 0.004
Heat Distortion Temperatures °F		500	500
Continuous Heat Resistance °F		425 - 500	425 - 500
Dimensional Stability % max.		0.01	0.01
Thermal Expansion 10^{-5}/°C		2.4	2.4
Flame Resistance: Ignition Time sec. (min)		127	90
Burning Time sec. (max)		23	300
Water Absorption: %48 hrs. @50°C %		0.2	0.2
Impact Strength: Ft - lb./in. notch (Izod)		4 - 10	4 - 10
Flexural Strength: PSI		17,000	17,000
Compressive Strength: PSI		30,000	29,000
Tensile Strength: PSI		7000 - 10,000	7000 - 10,000
Arc Resistance: Seconds		180	180
Dielectric Strength: volts/mil. Step by Step	DRY	475	410
	WET	420	406
Dielectric Breakdown: Kv Step by Step	DRY	64	64
	WET	60	60
Dielectric Constant: 1KC/1MC	DRY	4.0/3.8	4.0/3.8
	WET	4.1/3.9	4.1/3.9
Dissipation Factor: 1KC/1MC	DRY	.008/.012	.008/.012
	WET	.009/.013	.009/.013
Surface Resistance: megohms	AS IS	10^{10} plus	10^{10} plus
30 Days @100% R.H. @158°F		50,000	50,000
Volume Resistance: megohms	AS IS	10^{10} plus	10^{10} plus
30 Days @100% R.H. @158°F		30,000	30,000
Water Abstract Conductances 10^{-6} MHOS/CM		30	30
Certifiable to Military Specifications		MIL-M-14G MIL-P-19833	MIL-M-14G MIL-P-19833
and Type Design		GDI-30F	GDI-30
Flammability Rating: UL-94⅛"		VO	HB

(continued)

Table 5-4: (continued)

Compound Number		DIALLYL META PHTHALATE RESIN (High Heat Resistant)		
		K77	K31	K43
Filler		Short Glass Fiber	Short Glass Fiber	Mineral
Forms		Granules	Granules	Powder
Standard(s) or Flame Retardant (F)		F	S	S
Specific Gravity		1.78	1.70	1.67
Bulk Factor		2.5	2.5	2.3
Molding Pressure PSI		500 - 8000	500 - 8000	500 - 8000
Molding Temperature °F		275 - 350	275 - 350	275 - 350
Molding Shrinkage in./in.		0.001 - 0.004	0.001 - 0.004	0.003 - 0.007
Heat Distortion Temperatures °F		500	500	500
Continuous Heat Resistance °F		425 - 500	425 - 500	425 - 500
Dimensional Stability % max.		0.01	0.01	0.08
Thermal Expansion 10^{-5}/°C		1.5	2.5	4.6
Flame Resistance: Ignition Time sec. (min)		110	90	105
Burning Time sec. (max)		59	300	280
Water Absorption: %48 hrs. @50°C %		0.2	0.2	0.4
Impact Strength: Ft - lb./in. notch (Izod)		0.7 - 1.2	0.7 - 1.2	0.5 - 1.0
Flexural Strength: PSI		14,000	14,000	13,000
Compressive Strength: PSI		25,000	25,000	32,000
Tensile Strength: PSI		7000 - 10,000	7000 - 10,000	8000
Arc Resistance: Seconds		180	180	180
Dielectric Strength: volts/mil. Step by Step	DRY	400	400	440
	WET	380	380	420
Dielectric Breakdown: Kv Step by Step	DRY	64	64	55
	WET	60	60	50
Dielectric Constant: 1KC/1MC	DRY	4.2/3.9	4.1/4.0	4.8/4.0
	WET	4.3/4.2	4.3/4.1	5.2/4.0
Dissipation Factor: 1KC/1MC	DRY	.007/.014	.006/.012	.035/.031
	WET	.010/.015	.010/.015	.063/.039
Surface Resistance: megohms	AS IS	10^{10} plus	10^{10} plus	10^{10} plus
30 Days @100% R.H. @158°F		10,000	10,000	10,000
Volume Resistance: megohms	AS IS	10^{10} plus	10^{10} plus	10^{10} plus
30 Days @100% R.H. @158°F		10,000	10,000	6000
Water Abstract Conductances 10^{-6} MHOS/CM				
Certifiable to Military Specifications		MIL-M-14G	MIL-M-14G	MIL-M-14G
and Type Design		SDG-F	SDG	MDG
Flammability Rating: UL-94 1/8"		VO	HB	HB

Table 5-5: Plaskon® DAP Long-Glass-Filled Molding Compounds—
Typical Molded Properties and Molding Parameters
(Plaskon Electronic Materials Inc., subsidiary of Rohm and Haas Co.)

PROPERTY	52-20-30	52-40-40	FS-4	FS-80	Test method
Raw material					
Resin type	ortho	ortho	meta	meta	
Physical form	flake	flake	flake	flake	
Bulk factor	6	6	6	6	ASTM D1895
Mechanical					
Flexural strength, psi	15,000	15,000	15,000	15,000	FED-STD-406
Flexural modulus, psi x 10^6	1.3	1.3	1.4	1.4	ASTM D 790
Impact strength, ft-lb/in. (notch)	4.0	3.9	3.6	3.4	FED-STD-406
Barcol hardness	63	63	65	65	ASTM D 2583
Electrical					
Arc resistance, sec	130	133	138	145	FED-STD-406
Dielectric strength, v/mil (step-by-step)	350	350	350	340	FED-STD-406
Dielectric constant, at 1 line MHz	4.2	4.3	4.0	4.1	FED-STD-406
Dissipation factor, at 1 line MHz	0.016	0.016	0.014	0.015	FED-STD-406
Flammability					
Rating	STD	FR	STD	FR	
Ignition coil test, ignition time/ burning time, sec	—	>90/<90	—	>90/<90	MIL-M-14
UL flammability	no	no	no	no	UL 94
Oxygen index, %	26	30	29	34	ASTM D 2863
Comparative tracking index, sec	—	—	—	—	UL 746A
Other properties					
Specific gravity	1.72	1.79	1.64	1.74	ASTM D 792
Water absorption, %	0.35	0.32	0.35	0.35	FED-STD-406
Heat distortion temperature, (HDT), °F	500	500	500	500	
°C	260	260	260	260	
MIL-M-14 certification*					
type GDI-30	yes	yes	yes	yes	
type GDI-30F	no	yes	no	yes	MIL-M-14
PROCESSING PARAMETERS					
Molding pressure, psi					
compression	2500	2500	2500	2500	
transfer	5000	5000	5000	5000	
Preforming procedure	semiauto**	semiauto	semiauto	semiauto	
Mold shrinkage					
-compression, in./in.	0.0025	0.0025	0.0025	0.0025	
Post mold shrinkage, in./in.	0.0007	0.0007	0.0005	0.0005	

*Military certification requires individual batch testing.
**auto = automatic

Table 5-6: Plaskon® DAP Short-Glass-Filled Molding Compounds—
Typical Molded Properties and Molding Parameters
(Plaskon Electronic Materials Inc., subsidiary of Rohm and Haas Co.)

PROPERTY	52-01	FS-5	52-70-70*	73-70-70*	FS-10*	Test Method
Raw material						
Resin type	ortho	meta	ortho	ortho	meta	
Physical form	granular	granular	granular	granular	granular	
Bulk factor	2.3	2.3	2.3	2.4	2.3	ASTM D1895
Mechanical						
Flexural strength, psi	12,000	11,000	14,000	13,000	13,000	FED-STD-406
Flexural modulus, psi x 10^6	1.3	1.4	1.3	1.4	1.4	ASTM D970
Impact strength, ft-lb/in. of notch	.6	.6	.6	.6	.6	FED-STD-406
Barcol hardness	63	65	63	63	65	
Electrical						
Arc resistance, sec.	150	175	130	130	175	FED-STD-406
Dielectric strength, v/mil (step-by-step)	350	360	350	350	350	FED-STD-406
Dielectric constant, at 1 MHz	4.5	4.6	4.4	4.4	4.6	FED-STD-406
Dissipation factor, at 1 MHz	0.016	0.014	0.014	0.016	0.013	FED-STD-406
Flammability						
Rating	STD	STD	FR	FR	FR	
Ignition coil test, ignition time/ burning time, sec	—	—	>90/<90	>90/<90	>90/<90	
UL flammability (VO grade available)	no	no	yes ($1/16$ in.)	yes ($1/16$ in.)	yes ($1/16$ in.)	UL 94
Oxygen index, %	26	29	43	45	> 50	ASTM D 2863
Comparative tracking index, sec	—	—	600+	600+	360+	UL 746-A
Other properties						
Specific gravity	1.93	1.93	1.91	1.91	1.91	ASTM D792
Water absorption, %	0.25	0.30	0.25	0.25	0.30	FED-STD-406
Heat distortion temperature·						
(HDT), °F	> 500	> 500	> 500	> 375	> 460	
°C	> 260	> 260	> 260	> 190	> 240	
MIL-M-14 certification*						
Type SDG	yes	yes	yes	yes	yes	
Type SDG-F	no	no	yes	yes	yes	
PROCESSING PARAMETERS						
Molding pressure, psi						
compression	2500	2500	2500	2500	2500	
transfer	5000	5000	5000	5000	5000	
injection	4000	4000	4000	4000	4000	
Preforming procedure	auto**	auto	auto	auto	auto	
Mold shrinkage						
-compression, in./in.	0.0025	0.0025	0.0025	0.0025	0.0025	
Post mold shrinkage, in./in.	0.0007	0.0005	0.0007	0.0007	0.0005	

*Various qualified grades available having essentially same molded properties.
*Military certification requires individual batch testing.
**Automatic

128 Handbook of Thermoset Plastics

Table 5-7: Plaskon® DAP Mineral-Filled Molding Compounds—
Typical Molded Properties and Molding Parameters
(Plaskon Electronic Materials Inc., subsidiary of Rohm and Haas Co.)

PROPERTY	51-01 CAFR	775 CAF	FS-6 CAF	Test method
Raw material				
Resin type	ortho	ortho	meta	
Physical form	granular	granular	granular	
Mechanical				
Bulk factor	2.3	2.8	2.3	ASTM D 1985
Flexural strength, psi	10,000	10,000	10,500	FED-STD-406
Flexural modulus, psi x 10^6	1.2	1.0	1.3	ASTM D 790
Impact strength, ft-lb/in. notch	0.4	0.4	0.4	FED-STD-406
Barcol hardness	55	50	57	ASTM D 2583
Electrical				
Arc resistance, sec	135	138	175	FED-STD-406
Dielectric strength, v/mil step-by-step	375	400	350	FED-STD-406
Dielectric constant, at 1 MHz	4.3	4.5	4.1	FED-STD-406
Dissipation factor, at 1 MHz	0.037	0.028	0.035	FED-STD-406
Flammability				
Rating	FR	Std	Std	MIL-M-14
Ignition coil test, ignition time/burning time, sec	>90/<90	—	—	
UL flammability	*	no	no	UL 94
Oxygen index, %	> 30	24	23	ASTM D 2863
Comparative tracking index, sec	—	—	—	UL 746A
Other properties				
Specific gravity	1.75	1.67	1.86	ASTM D 782
Water absorption, %	0.55	0.65	0.50	FED-STD-406
Heat distortion temperature				
HDT, °F	300-350 °F	300-350 °F	300-350 °F	ASTM D 648
°C	150-170 °C	150-170 °C	150-170 °C	ASTM D 648
MIL-M-14 certification**	MDG	MDG	MDG	MIL-M-14
PROCESSING PARAMETERS				
Molding pressure, psi				
compression	2500	2500	2500	
transfer	5000	5000	5000	
injection	4000	4000		
Preforming procedure***	auto	auto	auto	
Mold shrinkage				
-compression, in./in.	0.0050	0.0065	0.0035	
Post mold shrinkage, in./in.	0.0007	0.0005	0.0007	

*V-O grade available in black
**Military certification requires individual batch testing
***auto = automatic
Note: All Plaskon mineral filled DAP molding compounds are asbestos-free.

APPLICATIONS

The major use of DAP compounds is in the electrical/electronic industry. Connectors for electronic communications, computers and aerospace systems consume a large volume of molding compounds. Insulators, potentiometers, circuit boards, potting vessels, trim pots, coil forms, switches, and TV components represent other end-uses of allylics (see Figure 5-6).

Figure 5-6: Electronic connectors, switches and other devices molded from DAP. (Photo courtesy of Rogers Corp.).

Sealants have been made based on allyl resins. They are used in the vacuum impregnation of metal castings and in ceramic compositions.

DAP prepolymers are used for improved surface laminates, plywood, hardboard and particle board. They are usually applied as an overlay by means of a resin-treated nonwoven acrylic fabric. In addition tubing, ducting, radomes, junction boxes, aircraft and missile parts find wide use of reinforced DAP.

Allylic monomers, principally DAP, are used in the crosslinking of unsaturated polyesters and alkyds. They are found in preform or mat binders, laminating prepregs or in wet lay-up formulas as well as in rope, granular and premix gunk molding compounds. The DAP content of these systems usually varies between 10–15%.

DAP is preferred over styrene, especially in large moldings, because of its low vapor pressure at molding temperatures [around 2.4 mm Hg @ 150°C (300°F)]. This low volatility allows for higher molding temperatures which translates into faster molding cycles.

In addition, mold shrinkage is lowered. The DAP, when mixed in styrenic-based formulations, extends the shelf life also. DAP polyesters have good mechanical and electrical properties in the 70°C (160°F) range.

In formulating a DAP-based polyester, the proportion of DAP and glycol strongly influences the high temperature properties of the compound. DAP polyesters have been reported with temperature resistance values approaching those of triallyl cyanurate polyesters. Laminates of DAP polyesters have been prepared with flexural strengths of 66,600 psi at room temperature and 27,500 psi at 260°C (500°F). The use of DAP increases the cost of a polyester system and trade-offs must be made against the value of the improved properties obtained.

TRADE NAMES

Cosmic	Monomers, polymers and compounds	Cosmic Plastics
Dapex	Compounds	Rogers Corp.
Plaskon	Compounds	Plaskon Electronic Materials, Inc.

REFERENCES AND BIBLIOGRAPHY

Allied Chemical Co., Plaskon Polyester Resins Premix Molding, Bulletin 851-36. Diallyl Phthalate Mineral Filled, Bulletin 612-100 (October 1978). High Performance Molding Compounds -DAP, Bulletins 612-101 and 2 (July 1978).

Beacham, H.H., Diallyl Phthalate Resin and Monomer, *Plastics Design & Processing*, pp 20-23 (April 1967).

Beacham, H.H. and Johnston, C.W., How to Formulate Heat Resistant DAP Polyesters, *Plastics Technology*, pp 44-46 (May 1963).

Cosmic Plastics, Inc., Data & Property Sheet (1984).

Dalton, J.L. and Landi, V.R., Resistance of Diallyl Phthalate and Other Engineering Plastics to Demanding End Use Conditions, Private communication, Rogers Corp. (1983).

DuBois, J.H. and John, F.W., *Plastics*, 5th Ed., Van Nostrand Reinhold, New York, pp 38-39 (1974).

Harper, C.A., Ed., *Handbook of Materials & Processes for Electronics*, McGraw-Hill Book Co., Inc., New York, pp 1-18 and 1-19 (1970).

Hayes, W.A. Jr., A Case for Thermosets vs. Thermoplastics, Private communication, Rogers Corp., (September 1983).

Landi, V.R., Long Term Test Data Helps Connector Material Choice, Reprint from *Electronic Packaging & Production* (May 1983).

Luh, C.H., A New Look at DAP for Electronics Insulation, *Insulation/Circuits* (October 1981).

Pixley, D. and Richards, P., Thermoset or Thermoplastic? Reprint from *Plastics Design Forum* Focus Issue (April 1981).

Powers, P.O. and Brother, G.H., The Chemistry of Plastics, in *Handbook of Plastics*, by Simonds, H.R. et al., 2nd Ed., D. Van Nostrand Co., Inc., Princeton, NJ, p 1054 (1955).

Rogers Corp., Diallyl Phthalate Molding Materials, Bulletin J4208 (1981).
Sare, E.J., Allyl, in *Modern Plastics Encyclopedia*, Vol 60, No. 10A, McGraw-Hill Inc., New York, p 18 (1983-4).
Schwartz, S.S. and Goodman, S.H., *Plastics Materials & Processes*, Van Nostrand Reinhold, New York, pp 339-345 (1982).
Thomas, J.L., Allyl, in *Modern Plastics Encyclopedia*, Vol 58, No. 10A, McGraw-Hill Inc., New York, pp 10-12 (1981-2).

6
Epoxy Resins

Sidney H. Goodman

*Developmental Products Laboratory
Technology Support Division
Hughes Aircraft Company
El Segundo, California*

INTRODUCTION

In the late 1930s, Dr. Pierre Castan in Switzerland and Dr. S.O. Greenlee in the United States synthesized the first resinous reaction products of bisphenol A and epichlorohydrin. These materials were characterized by terminal epoxide groups and were the semination of the epoxy family of plastics. The commercial production and introduction of this family occurred in 1947. New types of epoxies proliferated from the 1950s through the 1970s with at least 25 distinct types available by the late 1960s.

The generic term *epoxy*, epoxide in Europe, is now understood to mean the base (thermoplastic, uncured) resins as well as the resultant crosslinked (thermoset, cured) plastic. Chemically, an epoxy resin contains more than one a-epoxy group situated terminally, cyclicly, or internally in a molecule which can be converted to a solid through a thermosetting reaction.

The a-epoxy, or 1,2-epoxy, is the most common type of functional moiety. Ethylene oxide,

$$CH_2-CH_2 \atop \diagdown O \diagup$$

is the simplest type of 1,2-epoxy. This ring is also referred to as the oxirane ring. Another common group in this resin class is the glycidyl group,

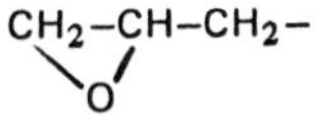

$$CH_2-CH-CH_2-$$
$$\underset{O}{\diagdown\diagup}$$

RESIN TYPES

Diglycidyl Ether of Bisphenol A

The *diglycidyl ether of bisphenol A (DGEBA)* continues to this day to represent the most common type of epoxy resin. It is the product of the following reaction:

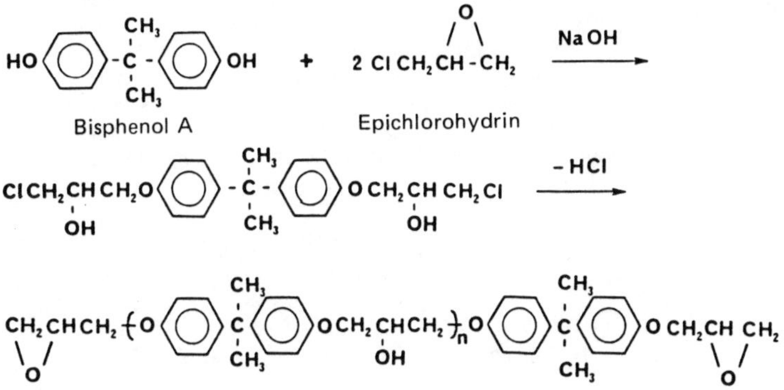

$$HO-C_6H_4-C(CH_3)_2-C_6H_4-OH + 2\,ClCH_2CH-CH_2 \xrightarrow{NaOH}$$

Bisphenol A   Epichlorohydrin

$$ClCH_2CHCH_2O-C_6H_4-C(CH_3)_2-C_6H_4-OCH_2CHCH_2Cl \xrightarrow{-HCl}$$
$$OHOH$$

$$CH_2CHCH_2\!-\!\!\left[O\text{-}C_6H_4\text{-}C(CH_3)_2\text{-}C_6H_4\text{-}OCH_2CHCH_2\right]_{\!n}\!\!-\!O\text{-}C_6H_4\text{-}C(CH_3)_2\text{-}C_6H_4\text{-}OCH_2CHCH_2$$

The basic commercial version of this resin is the one having a molecular weight of 380. Purified versions (n = essentially 0) have molecular weights as low as 344. Higher molecular weight versions (n = 1-10) have been produced by reducing the amount of epichlorohydrin and reacting under more alkaline conditions. Tables 6-1 through 6-5 list some commercial grades of these resins.

Changes in the base resin structure have been made in order to adjust final plastics properties. Higher reactivity, greater crosslink density, higher temperature and chemical resistance are obtained through the use of novolac and some types of peracid epoxies.

Novolacs

Novolacs are epoxidized phenol-formaldehyde novolacs.

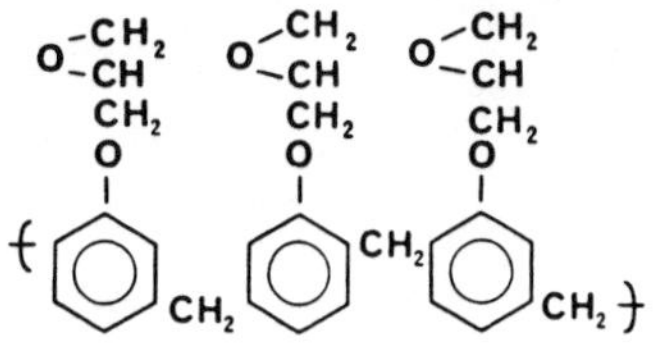

Table 6-1: Standard Undiluted BIS Resins
(Courtesy of: Dow Chemical Co.)

	EEW	Viscosity ..(cp @ 25°C)..	Color*
D.E.R. 331	182-190	11,000-14,000	3
Epon 8280**	185-192	11,000-15,000	3 max.
Epon 828	185-192	11,000-15,000	3 max.
Araldite 6010	185-196	12,000-16,000	2
Epi-Rez 5101	180-200	13,000-15,000	3
Epi-Rez 510	180-200	10,000-16,000	3
Epi-Rez WD-510	190-216	7,500-12,500	3
Epotuf 37-140	180-195	11,000-14,000	3
Gen Epoxy 190	180-195	11,000-14,000	3

*Gardner-Holdt.
**Special vacuum-casting resin characterized by rapid foam breakdown under vacuum.

Standard undiluted resins for all general purposes requiring performance up to 400°F. Aliphatic polyamines or polyamides satisfactory up to approximately 230°F. Anhydrides such as phthalic satisfactory to approximately 200°F. Aromatic amines and anhydrides satisfactory to 400°F. The anhydrides are effective viscosity reducers to permit higher filler loading.

Table 6-2: Lowest-Viscosity Resins
(Courtesy: Dow Chemical Co.)

Products	EEW	Viscosity .(cp @ 25°C).	Color*
D.E.R. 332	172-176	4,000-6,000	75 APHA max.
Epon 825	172-178	4,000-6,000	1 max.
Epi-Rez 508	171-177	3,600-5,500	1
Araldite 6004	187 max.	5,000-6,400	2 max.

*Gardner-Holdt.

The low equivalent weight resins are virtually pure diglycidyl ethers of Bisphenol A. They are the lowest viscosity undiluted Bisphenol A resins available. They are so pure, however, that they crystallize on storage. The crystals melt on warming above 125°F. They fit all the general uses of D.E.R. 330 or D.E.R. 331 resins with the following advantages:

    a. Increased HDT
    b. Lower viscosity
    c. More chemical uniformity
    d. Longer pot life with most curing agents
    e. Better wetting of glass reinforcements
    f. Very pale color
    g. Better electrical properties

Table 6-3: High-Viscosity Resins
(Courtesy: Dow Chemical Co.)

Products	EEW	Viscosity ..(cp @ 25°C)..	Color*
D.E.R. 317	192-203	16,000-25,000	5
D.E.R. 337	230-250	400-800**	3**
Araldite 6020	196-208	16,000-20,000	3
Epon 830	190-198	17,000-22,500	9
Epon 834	230-280	410-970**	5**

*Gardner-Holdt.
**@ 70% NV in DOWANOL DB glycol ether solvent.

The lower EEW resins in this series have the same general properties as D.E.R. 331 except viscosity.

The higher EEW resins are very viscous liquids finding main use in coatings or adhesive systems where solvents may be used to reduce viscosity.

As EEW increases, pot life is shorter, HDT decreases, exotherms decrease; impact, elongation, and adhesion improve.

Table 6-4: Low-Melting Solid Resins
(Courtesy: Dow Chemical Co.)

Products	EEW	Viscosity*	Color**	Durran's SP °C
D.E.R. 661	475-575	G-J	1	70-80
Gen Epoxy 525	475-525	G-J	1	70-80
Epon 1001	450-550	D-G	4 max.	
Epi-Rez 520 C	450-525	C-G	3	65-75
Epotuf 37-301	450-525	E-J	3	65-75
Araldite 7065	455-500	G-J	2	68-78
Araldite 7071	450-550	D-G	2	65-75

*Gardner-Holdt at 40% NV in DOWANOL DB @ 25°C.
**Gardner @ 40% NV in DOWANOL DB @ 25°C.

Primary uses in amine cured protective coatings and for prepreg glass cloth for electrical laminates. D.E.R. 661 resin modified with polyamines or polyamides is used where high chemically resistant performance is required coupled with a room temperature or low bake application.

Blends of ketone solvents (MEK or MIBK) with aromatics (Xylene or Toluene) are generally suitable for thinning these systems. Higher boiling solvents such as glycol ethers can be used in amounts of 5-15% to improve flow and film surface properties.

Systems of D.E.R. 661 resin can be used on all substrates—metal, wood, glass, masonry by all applications—brushing, spraying, dipping, etc.

Coatings end uses include pipe and drum linings, maintenance finishes, marine finishes.

Table 6-5: High Molecular Weight Solid Resins
(Courtesy: Dow Chemical Co.)

Products	EEW	Viscosity*	Color**	Durran's SP°C
D.E.R. 667	1,600-2,000	$Y-Z_1$	3	115-130
Gen Epoxy 1800	1,600-2,000	$Y-Z_1$	1	113-123
Epon 1007	2,000-2,500	$Y-Z_1$	5 max.	
Araldite 6097	1,000-2,500	$Z-Z_2$	3	125-135
Araldite 7097	1,650-2,000	X-Z	3	113-123
Epi-Rez 540-C	1,400-1,800	X-Z	3	120-130
Epi-Rez 540-F	1,600-2,000	$X-Z_1$	4	127-133
Epotuf 37-307	1,550-2,000	$V-Z_1$	3	116-124
Epotuf 37-306	2,000-2,500	Y-Z	3	NS

*Gardner-Holdt @ 40% NV in DOWANOL DB @ 25°C.
**Gardner @ 40% NV in DOWANOL DB @ 25°C.

Optimum epoxy coating can be obtained by modifying D.E.R. 667 resin with urea or melamine formaldehyde or phenolic resins. These systems in the blended solution form have excellent pot life and can be stored for several months without noticeable viscosity change. To cure the coating, high bakes of 300°-400°F from 15-30 minutes are required. Phenolic modified systems require the maximum bake schedule for complete cure. The addition of 1% phosphoric acid will catalyze the cure at somewhat lower temperatures. Ketones and aromatic solvents are used to thin D.E.R. 667.

End-use applications include tank and drum linings, wire enamels, collapsible tube coatings, metal furniture finishes.

The number of glycidyl groups per molecule per resin is a function of the number of available phenolic hydroxyls in the precursor novolac, the extent of reaction, and the extent of chain extension of the lowest molecular species during synthesis. Table 6-6 describes a number of commercial novolac resins.

Table 6-6: Epoxy Novolac Resins
(Courtesy: Dow Chemical Co.)

Products	EEW	Viscosity (cp @ 25°C)	Color*	Durran's SP°C	Solvent/% NV
D.E.N. 431	172-179†	1,100-1,700**	3		100
D.E.N. 438-EK85	176-181†	600-1,600	2		MEK/84-86
D.E.N. 438	176-181	20,000-50,000	2		100
D.E.N. 438-A85	176-181	500-1,200	2		Acetone/84-86
D.E.N. 438-MK75	176-181†	200-600	2		MIBK 74-76
D.E.N. 439	191-210	4,000-10,000***	3	45-58	100
D.E.N. 439-EK85	191-210***,†	4,000-10,000	3		MEK/84-86
Ciba EPN 1138	176-181	20,000-50,000**	2		
Ciba EPN 1138 A-85	176-181				Acetone/85
Ciba EPN 1139	172-179	1,400-2,000**	3		100
Epi-Rez 5155	175-190	30,000-90,000**	10		100
Epi-Rez 5109	180-200	6,000-16,000	10		100

*Gardner.
**At 125°F.
***85% in MEK.
†On solids.

The multi-functional epoxy novolacs have greater heat and chemical resistance than Bisphenol A-derived resins when cured with appropriate hardeners.

Peracid Resins

Of the peracid resins the cyclic types contribute to higher cross-link densities. These resins have lower viscosities and color compared to novolac and DGEBA types.

$$\text{C}_6\text{H}_5-\text{COOH} + \text{C=C} \rightarrow \text{C}_6\text{H}_5-\text{COH} + \text{C}-\text{O}-\text{C}$$

perbenzoic acid   olefin   benzoic acid   epoxy

A typical such resin is vinylcyclohexane dioxide,

$$\text{O}\triangleleft\bigcirc-\text{CH}-\text{CH}_2 \text{ (O)}$$

A series of peracid based resins are also made for modification of standard resin systems. They alter such properties as cure rate, flexibility, and heat deflection temperature. These resins are acyclic aliphatic resins such as epoxidized soya or linseed oils, polyglycols and/or polybutadiene.

$$\text{O}\triangleleft\bigcirc-\text{CH}_2\text{OC}-\bigcirc\triangleright\text{O}$$

3,4-epoxycyclohexylmethyl-3,4-epoxycyclohexane carboxylate

Table 6-7 lists commercial types of peracid epoxies.

Table 6-7: Peracid Epoxies
(Courtesy: Dow Chemical Co.)

Products	EEW	Viscosity ...(cp @ 25°C)...	Color	Composition
ERL 4206	70-74	<5	1*	Vinyl cyclohexane dioxide
ERL 4221	131-143	350-450	1*	Epoxy cyclohexyl methyl epoxy cyclohexane carboxylate
ERL 4289	205-216	500-1,000	1*	Cycloaliphatic epoxide
Araldite				
CY-178	213	900	1	Cycloaliphatic epoxies
CY-179	140	350	1	Cycloaliphatic epoxies
RD-4	77	20	1	Cycloaliphatic epoxies
CY-182	160	P-U**	2-3	Cycloaliphatic epoxies
CY-183	154	L-Q**	1-2	Cycloaliphatic epoxies
ERL 4234	133-154	7,000-17,000††	2*	Epoxy cyclohexyl-spiro-epoxy cyclohexane dioxide
ERL 4090	390-430	450***	2*	Modified cycloaliphatic epoxies
ERL 4205	91-102	<100†	2*	Cycloaliphatic epoxide

*Gardner, max.   †At 45°C.
**Gardner-Holdt.   ††At 100°F.
***At 40°C.

Hydantoin Resins

In recent years, the hydantoin resins have shown increased popularity for increasing temperature resistance and improving mechanical properties, particularly in structural composites. Numerous structural modifications are feasible with hydantoin resins as shown in Figure 6-1.

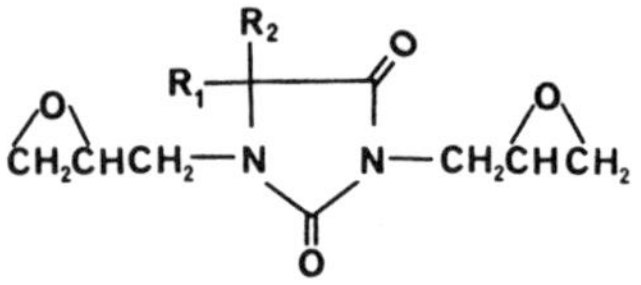

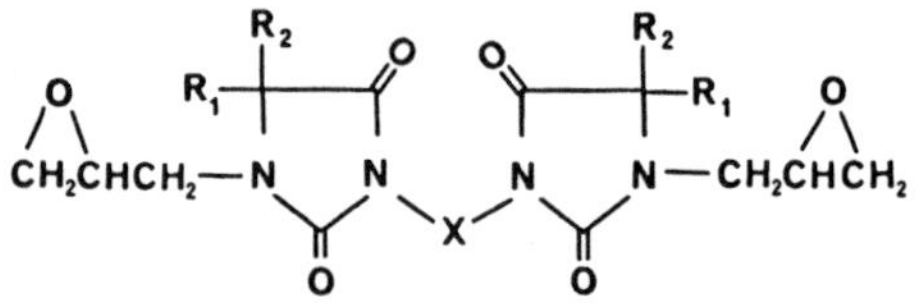

Figure 6-1: Hydantoin epoxy resin structures. R_1 and R_2 can be alkyl groups such as methyl, ethyl and pentamethylene; X can be methylene, bis-hydroxyethyl esters of various chain lengths, or urethane or urea groups.

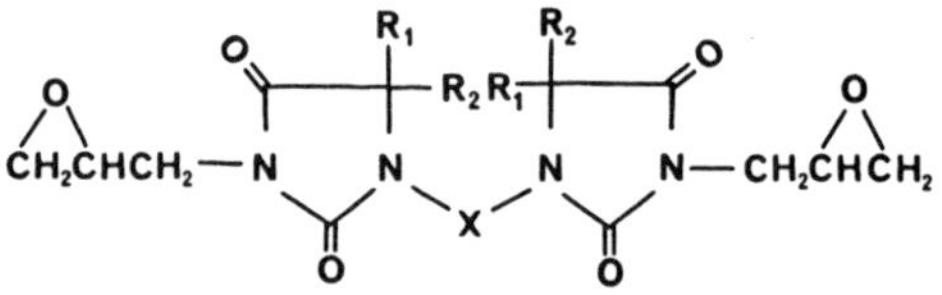

Other Types

Other new types of polyfunctional resins include tetraglycidyl methylene dianiline,

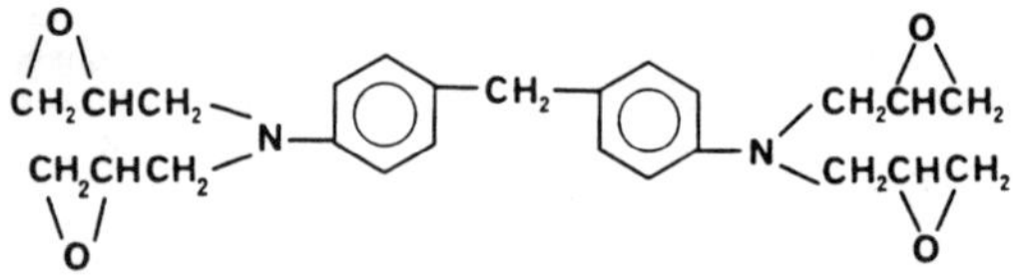

and a novolac synthesized from bisphenol A.

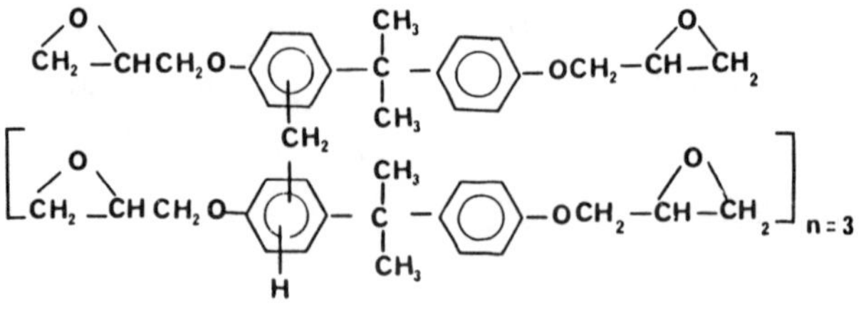

In order to build in flexibility, a series of epoxy resins are based on glycerol,

$$\text{CH}_2-\text{CH}-\text{CH}_2$$
$$\quad|\qquad|\qquad|$$
$$\text{OH}\quad\text{OH}\quad\text{OH}$$

and other polyglycols (polyethylene and polypropylene, most often); oils, such as cashew nut oil,

organic acids such as dimerized fatty acids;

aliphatic diacid glycidyl esters where n = 0, 2, 4, 5 and 8

and modified bisphenols such as bisphenol F.

There is also a whole series of resins based on elastomeric modification. The first of this series used carboxy-terminated polybutadiene/acrylonitrile (CTBN) liquid elastomers. These telechelic polymers are macromolecular diacids which coreact easily to form epoxy terminated adduct resins,

Amine terminated elastomers of this same type were introduced in the mid-1970s. Although they can also be adducted to DGEBA type resins, they are more frequently combined with curing agents. Table 6-8 lists some commercial versions of flexible epoxies.

Table 6-8: Flexible Epoxy Resins
(Courtesy: Dow Chemical Co.)

Products	EEW	Viscosity (cp @ 25°C)	Color*	Type
D.E.R. 736	175-205	30-60	1	Polyglycol
D.E.R. 732	305-335	55-100	1	Polyglycol
D.E.R. 741	364-380	2,500-4,500	125 APHA	
Epon 812	150-170	120-200	3 max.	Glycerine
Epon 872 X-75	650-750	1,500-3,000**	10 max.	
Epon 872	650-750	1,500-2,500	10	75% NV Xylene
Epon 871	390-470	400-900	12 max.	
Araldite 508	400-455	2,000-5,000	5	Polyglycol
Epotuf 37-151	450-550	30,000-70,000	4	
Epi-Rez 505	550-650	300-500	8	
Epi-Rez 5060	270-320	220-520	8	
Epi-Rez 5132	400-450	40,000-60,000	Dark	
Epi-Rez 50821	245-270	900-1,300	4	
Epi-Rez 50833	220-240	1,750-2,250	5	

*Gardner.
**75% NV in Xylene.

Flame retardancy is often introduced into epoxy systems with halogen or phosphorus based additives. However, some resins have been provided which have these constituents prereacted in the resin. Chlorinated and brominated versions of DGEBA are the most common.

New developments in resin synthesis continues beyond the more traditional types. Among them is a polyfunctional resin based on triphenyl methane designed for high and low temperature (-195°C to 200°C, -319°F to 392°F) resistance.

A resin having a low-temperature cure with high temperature properties is triglycidyl p-aminophenol.

Resins having a high resistance to weather degradation and attack by biological organisms have substantial amounts of fluorine in the backbone structure,

$$\overset{O}{\underset{2}{CH_2CHCH}}-O-\underset{CF_3}{\overset{CF_3}{\underset{|}{C}}}-\underset{R_f}{\bigcirc}-\underset{CF_3}{\overset{CF_3}{\underset{|}{C}}}-O-CH_2\overset{O}{CHCH_2}$$

where $R_f = C_3F_7$ or C_7F_{15}.

Epoxy-silicone hybrid resins have been developed for use in the molding of microelectronic packages. A typical structure is

$$\overset{O}{CH_2CHCH_2}O(CH_2)_3 \left[\underset{CH_3}{\overset{CH_3}{\underset{|}{Si}}}-O \right] \underset{CH_3}{\overset{CH_3}{\underset{|}{Si}}} (CH_2)_3 OCH_2 \overset{O}{CHCH_2}$$

Since 1979, a number of new epoxy resins with unusual characteristics have been patented. However, widespread commercialization has not yet occurred. Included in this group are epoxy-lignin resins, ethynyl-terminated epoxies and water-soluble glycidyl ethers of glycerin.

CURATIVES AND CROSSLINKING REACTIONS

The conversion of epoxy resins from the thermoplastic state to tough, hard, thermoset solids can occur via a variety of crosslinking mechanisms. Epoxies can catalytically homopolymerize or form a heteropolymer by coreacting through their functional epoxide groups with different curatives. In epoxy technology, curatives are most frequently called *curing agents*. Often the terms *hardener, activator*, or *catalyst* are applied to specific types of curing agents. It is advisable to clearly distinguish between true catalytic curing agents, which participate in the crosslinking via the traditional chemical concept of catalysis, and multifunctional crosslinking agents which become chemically bound in the final three-dimensional structure. The latter, therefore, can strongly influence the properties of the end plastic. Too often this fact is overlooked or not understood, causing nonoptimized formulations to be used under inappropriate circumstances. Consultation with established epoxy formulating chemists is rigorously advised before indiscriminate changes in formula or cure conditions are made in order to effect property changes.

Stoichiometry

In the same vein, attention must be paid to the stoichiometric relationships between curing agents and resins. Catalytic curatives are added at relatively low levels (0-5 parts per hundred of resin, phr). Since their behavior during cure is truly catalytic, the application principles that apply to other catalytic polymerizations (e.g., with polyester resins) are the same with epoxies.

On the other hand, multifunctional coreactants require that the user address the stoichiometric balance between the reacting species. An epoxy formulator will often establish the correct reactive ratio and supply the system accordingly. Many users, however, develop individualistic recipes and must therefore be able to calculate and optimize the proportions of curatives and resins. An example of a simple stoichiometric calculation for DGEBA and a typical polyamine is shown in Table 6-9.

Table 6-9: Example of a Stoichiometric Calculation

Resin: DGEBA

Amine Curative: Triethylene Tetramine (TETA)

$$\underline{H_2N}(CH_2)_2\underline{NH}(CH_2)_2\underline{NH}(CH_2)_2\underline{NH_2}$$

Molecular weight of amine;

6 carbons =	6 × 12 =	72
4 nitrogens =	4 × 14 =	56
18 hydrogens =	18 × 1 =	18
Molecular weight =		146

There are 6 amine hydrogens functionally reactive (underlined) with an epoxy group. Therefore,

$$\frac{146 \text{ grams/mol}}{6 \text{ equivalents/mol}} = 24.3 \text{ grams/equivalent}$$

Thus, 24.3 grams of TETA are used per equivalent of epoxy. If the DGEBA has an equivalent weight of 190 (380 g/mol/2 eq./mol), then 24.3 grams of TETA are used with 190 grams of DGEBA, or

24.3/190 = 12.8 grams of TETA per hundred grams of DGEBA.

Often a commercial curing agent's chemical structure is kept proprietary or the amount of reactive functional group is ambiguous. In such cases, an amine or active hydrogen equivalent will be provided by the vendor from which an appropriate mix ratio can be calculated. It is also important when performing stoichiometric balances to be aware of reactive groups that may be bifunctional (e.g., anhydride, olefin, etc.).

Experience has determined that a precise stoichiometric balance does not always produce a cured resin system having optimized properties. Consequently a formulator will run experiments to establish the variance of properties of interest with mix ratio. Figure 6-2 shows such a variance. Note that the optimum level of TETA is about 12.5 phr, almost exactly the theoretical value calculated in Table 6-9. It is not uncommon, however, for the mix ratio to depart by 80-110% of theoretical. The selected and optimized ratio will subsequently be published in data sheets or on package labels for the user's convenience or the vendor will prepackage the resin and curing agent in an appropriate volumetric or weight proportion.

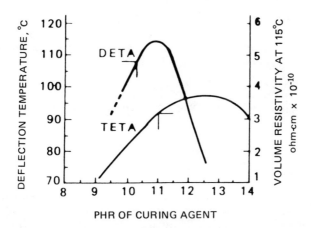

Figure 6-2: Effect of concentration of DETA and TETA on deflection temperatures of DGEBA, (From *Handbook of Epoxy Resins* by Lee and Neville. Copyright 1967 by McGraw-Hill, Inc. Used with permission of McGraw-Hill Book Co.)

Earlier the ambiguousness of establishing an accurate equivalency was mentioned. Such a situation arises, for example, with aminopolyamides whose structure is too complex to determine just how many hydrogens will coreact with oxirane rings. Table 6-10 shows the variance of hardness obtained with a wide spread of curing agent levels. The final use ratio will be selected, in these cases, based on the best combination of desired cured-resin properties.

Table 6-10: Variation of Hardness with Mix-Ratio
(Skiest)

Weight Ratio, Versamid 125/ERL-2795	Barcol Hardness, Impressor Model GYZJ-935	Shore Durometer A Hardness
40/60	60-65	–
50/50	20-25	–
60/40	–	90
65/35	–	85
70/30	–	50
75/25	–	30
80/20	–	5

144 Handbook of Thermoset Plastics

Epoxy curing agents can be divided into two major classes: alkaline and acidic. The alkaline class includes Lewis bases, primary and secondary amines and amides, and other nitrogen containing compounds.

ALKALINE CURING AGENTS

Lewis Bases

Lewis bases contain an atom with an unshared electron in its outer orbital. The main types of Lewis bases used in epoxy resin chemistry are tertiary amines. They catalyze epoxy polymerization if some hydroxyl containing molecules are present.

$$R_3N + CH_2{-}CH{-} \longrightarrow R_3N^+ + -CH_2-CH{-}\underset{2}{CH\,CH} \longrightarrow -CH_2CH{-}\!\!-CH_2-CH{-}$$

Primarily used in adhesives, laminating, and coatings, the tertiary amines are widely used as accelerators for acid anhydride and aromatic amine curing agents. They are rarely used at more than 1.5 phr unless they are used for low temperature curing of epoxy adhesive formulas, in which case they may be used as high as 15 phr. Excess tertiary amine does degrade cured-resin properties.

Many of the popular tertiary amines contain hydroxyl groups for enhanced reactivity. Examples of tertiary amines include tris-dimethylaminomethylphenol (DMP-30), dimethylaminoethanol (S-1) and diethylaminoethanol (S-2), benzyldimethylamine, a-methylbenzyldimethylamine, and triethyl- and trimethylamine.

Cured systems behave similarly to aliphatic amines in large masses. The more steric hindrance of the nitrogen, the less reactive the catalyst.

A tertiary amine salt (DMP-30 tri-2-ethyl hexoate) has been used in electrical applications because of improved resistance to humidity and good metal adhesion. Chemical resistance, however, is poor and all properties drop rapidly with increasing temperature.

Primary and Secondary Aliphatic Amines

The epoxy-primary amine reaction is:

$$RNH_2 + CH_2{-}CH{-}R' \longrightarrow R\overset{H}{N}CH_2\overset{OH}{C}HR'$$

The epoxy-secondary amine reaction is:

$$\text{R}\underset{\text{H}}{\text{N}}\text{CH}_2\underset{\text{OH}}{\text{CHR}'} + \text{CH}_2\overset{\text{O}}{-}\text{CH}-\text{R}' \longrightarrow \text{R}\,\text{N}\begin{matrix}\text{CH}_2\text{CH}-\text{R}'\\|\\ \text{OH}\end{matrix}\begin{matrix}\text{OH}\\|\\ \end{matrix}$$

$$\text{R N}\begin{matrix}\nearrow \text{CH}_2\text{CH}-\text{R}' \\ \searrow \text{CH}_2\text{CH}-\text{R}' \\ |\\ \text{OH}\end{matrix}$$

The hydroxyls formed are further reactive, however the tertiary amine is generally too sterically hindered to contribute much to cure.

Aliphatic amines constitute the largest group of epoxy curing agents. They can be used as is or adducted to modify volatility, toxicity, reactivity, and stoichiometry. They are characterized by short pot lives and high exotherms. They are skin sensitizers and some can cause respiratory difficulties.

The most common aliphatic polyamines are those that belong to the following homologous series: diethylene triamine (DETA), triethylenetetramine (TETA), and tetraethylene pentamine (TEPA). Typical properties are shown in Table 6-11. Systems cured with these three curing agents generally have similar properties, particularly electrical and chemical resistance. Another common aliphatic amine is diethylaminopropylamine (DEAPA). It provides more pot life than the DETA types and even requires some heat to complete the cure. DEAPA cured resins are softer with lower heat deflection temperatures.

Table 6-11: Comparative Mechanical Properties of DETA and TETA Cured Epoxy Castings at 25°C (Bruins)

Property	DETA 10-11 phr	TETA 13-14 phr
Heat deflection temperature, °C	95-124	98-124
Flexural strength, psi	14,500-17,000	13,900-17,700
Flexural modulus, X 10^{-5} psi	5.0-5.4	4.4-4.9
Compressive strength, psi	16,500	16,300
Tensile strength, psi	11,400	11,400
Ultimate elongation, %	5.5	4.4
Izod impact strength, ft-lb/in. of notch	0.4	0.4
Hardness, Rockwell M	99-108	106

Amine Adducts

Adduction of DETA type curing agents is done to reduce their volatility, alter the reaction rate, and/or increase mix ratios. The oldest adduct is the product of DETA and DGEBA. It provides a shorter cure time because the adduct is a partially reacted substance with added hydroxyls. Thus less reactant in a more accelerated reaction is required to reach the gel point.

Other types of adducts are based on ethylene and propylene oxides and alkylene polyamines. Examples are N-(2-hydroxypropyl)-

ethylenediamine and ethylene oxide/DETA. These adducts tend to be more hygroscopic, calling for careful storage. They are not recommended for casting applications but do perform well in laminating and patching kits.

Amine terminated polyglycols appeared in the early 1970s in order to introduce flexibility into the three-dimensional crosslinked structure of epoxies. A series of bi-, tri-, and tetrafunctional polyoxypropylene amines were commercialized. While providing for increased flexibility, they also increased pot life and decreased viscosity of formulations without the attendant age hardening that accompanies the use of nonreactive flexibilizers.

Cyclic Amines

Some applications require only intermediate cured-resin properties between the aliphatic and aromatic amines. The cycloaliphatic amines fill this gap. The four major cycloaliphatics are piperidine, N-aminoethylpiperazine (AEP), menthanediamine, and m-xylylenediamine (MXDA).

Piperidine,

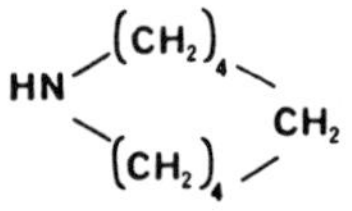

$$HN \underset{(CH_2)_4}{\overset{(CH_2)_4}{<}} CH_2$$

has one active hydrogen for reaction, however the resultant tertiary amine has sufficient catalytic strength to promote continued polymerization of epoxy. It provides long pot life and lower exotherm with other properties equivalent to the aliphatic polyamines. Its toxicity has, however, made it the subject of governmental restrictions which have significantly limited its current use.

N-aminoethylpiperazine,

$$H_2N(CH_2)_2 N \underset{(CH_2)_2}{\overset{(CH_2)_2}{<}} NH$$

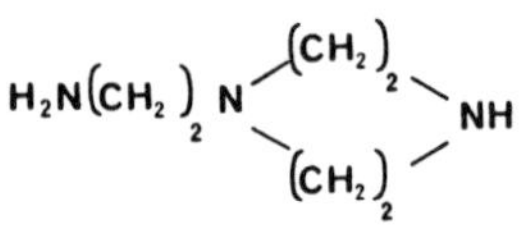

contributes improved impact strength when compared to the DETA series of hardeners. Although gel time and exotherm are also comparable, a post cure of 38°-66°C (100°-150°F) is required for complete cure.

Menthanediamine,

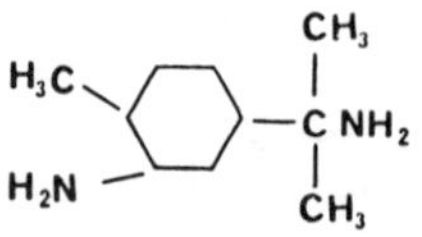

$$H_3C-\underset{H_2N}{\bigcirc}-\underset{CH_3}{\overset{CH_3}{\underset{|}{C}}}NH_2$$

makes processing easier through reduced viscosity of resin mixtures. It improves the temperature resistance as compared to the aliphatics. Its properties are not as good, however, as the aromatics.

m-Xylylenediamine,

yields the same properties as menthanediamine but contributes hardly any color to formulations. It is popularly used in so-called "water-white" castings. Table 6-12 shows some general properties of the cycloaliphatic amines.

Table 6-12: Mechanical Properties of Cycloaliphatic Amine Cured Epoxy Resins
(Bruins)

	Piperidine	Aminoethyl-piperazne	Menthane-diamine	Meta-xylylene-dimaine
Heat deflection temperature, °C	75-110	100-120	148-158	130-150
Flexural modulus, psi X 10^{-5}				4.36
Flexural strength, psi	13,500-15,500		15,500-17,500	
Compressive strength, psi	13,000-16,000	8,700	10,500	15,200
Tensile strength, psi	7,000-9,500	9,000	9,000	10,600
Ultimate elongation, %	6.0-8.5	8.8	2.9	6.7
Izod impact strength, ft-lb/in. of notch	0.3-0.5	1.0-1.2	0.3-0.4	
Rockwell M hardness	90-96	95-105	105	

Aromatic Amines

Aromatic amines generally contribute the best properties of the amine cured epoxies. Specifically they increase temperature and chemical resistance, extend pot life (although exotherms remain high) and always require heat for cure. In addition, in recent years many aromatics have become increasingly scrutinized and regulated by government agencies because of their potential health hazards. In some instances, it has been guilt by association because of their structural resemblance to aniline-based suspect carcinogens. One formulator has marketed an adducted aromatic amine that is reported to minimize toxicity with minimal loss in properties. In any case, the reader is advised to consult with vendors on the current hazard status and proper handling procedures before use.

The three major aromatics are m-phenylenediamine (MPDA),

$$\text{C}_6\text{H}_4(\text{NH}_2)_2$$

4,4'-methylenedianiline (MDA),

$$\text{H}_2\text{N}-\text{C}_6\text{H}_4-\text{CH}_2-\text{C}_6\text{H}_4-\text{NH}_2$$

and 4,4'-diaminodiphenyl sulfone (DADS),

$$\text{H}_2\text{N}-\text{C}_6\text{H}_4-\text{SO}_2-\text{C}_6\text{H}_4-\text{NH}_2$$

Because of their reduced reactivity, these materials lend themselves to formulations which are easily B-staged. Typically, these B-stage systems are stable at room temperature for months. Molding compounds, tape adhesives, and laminating prepregs are key applications for such formulas.

MPDA contributes cured resin temperature resistance of 150°-177°C (300°-350°F) as compared to 93°-107°C (200°-225°F) for aliphatics. It is a skin and clothing stainer and must be melted into resins. MDA's properties are somewhat less than MPDA, however, because its polarity is less, it contributes improved dielectric constant and loss factor. DADS yields the highest heat deflection temperature of the aromatic amines. With proper selection of resin, systems resistant to 204°C (400°F) are common.

A series of eutectic blends of MPDA and MDA have been commercialized, many with proprietary additives to retard crystallization. This tends to maintain hardener liquidity and lets one mix resin and hardener at lower temperatures. This then extends pot life and improves handling. Table 6-13 lists some typical properties of aromatic amine cured epoxy.

Table 6-13: Properties of Aromatic Amine Cured Epoxy Resins (Bruins)

Property	MPDA	MDA	DDS	Eutectic
Heat deflection temp., °C	150	144	190	145
Flexural modulus, psi X 10^{-5}	4.0	3.9		4.4
Flexural strength, psi	15,500	17,500	17,900	16,400
Compressive strength, psi	10,500	10,500		10,500
Tensile strength, psi	8,000	8,100	8,550	8,000
Ultimate elongation, %	3.0	4.4	3.3	4.8
Izod impact strength, ft-lb/in. of notch	0.2-0.3	0.3-0.5		0.5
Rockwell M hardness	108	106	110	105-110

Aromatic amine adducts have also been prepared for the coatings industry. MPDA, MDA, and DADS have been adducted with styrene oxide, phenyl glycidyl ether, cresyl glycidyl ether, and low molecular weight DGEBA. Accelerators such as phenols or organic acids, (e.g., salicylic) are added to the adduct to promote ambient temperature cures. The adducts are usually dark in color and high in viscosity. They tend to produce brittle films unless nonreactive diluents such as dibutyl phthalate or benzoyl alcohol are added to reduce viscosity and flexibilize the formulation.

A series of composites have been formulated using new novel bisimide amines as the epoxy curative. These curing agents have a structure of

$$H_2N\text{-}Ar\text{-}N\begin{smallmatrix}C=O\\C=O\end{smallmatrix}\text{-}\bigcirc\text{-}C(CF_3)_2\text{-}\bigcirc\text{-}\begin{smallmatrix}C=O\\C=O\end{smallmatrix}N\text{-}Ar\text{-}NH_2$$

Typical amines used include DADS and MDA in the Ar structure. The increased aromaticity creates higher temperature and improved moisture resistance over current epoxy/graphite systems.

Polyamides

The polyamides used to cure epoxies are in fact aminopolyamides. The literature also refers to them as amidopolyamines. They are fundamentally dimerized or polymerized fatty acids that have been coreacted with various aliphatic amines such as ethylenediamine, DETA, TETA, and TEPA,

$$\begin{array}{l}\text{cyclohexene ring with substituents:}\\ -(CH_2)_7\overset{O}{C}\overset{H}{N}(CH_2)_2\overset{H}{N}R\\ -(CH_2)_7\overset{O}{C}\overset{H}{N}(CH_2)_2\overset{H}{N}R\\ -(CH_2)_2CH=CH(CH_2)_4CH_3\\ -(CH_2)_5CH_3\end{array}$$

where R = other dimer units and amine units.

The resultant molecules are very large and contain varying levels of primary and secondary amine hydrogens, reactive amide, and carboxyl groups—all of which can contribute to epoxy curing. Establishment of mix ratios is thus more a function of property selection rather than stoichiometric balance (see earlier discussion on Stoichiometry).

Formulations made from these polyamides are the bases of "user-friendly" systems, since the tolerances on mix ratio are very broad. Although the resultant properties do vary, these systems find application in uses that do not require optimized, highly specific properties; for example, in two-tube household glues where ease of mixing volumetrically is more important than maximum shear strength.

The aminopolyamides introduce considerably reduced volatility and dermatitic potential, increased flexibility and impact strength, and water resistance (even to the point of effecting underwater cure). They have poor chemical resistance and low heat deflection temperature. Principle applications are coatings and adhesives with lesser use in laminates and castings.

Other Amines

Other amine-containing curatives fall into the catalytic class of curing agents. Long popular for use in stable one-can systems is dicydiandiamide,

$$H_2NCNHC \equiv N$$
$$\overset{\overset{NH}{\|}}{}$$

This material is used in catalytic quantities even though its breakdown products have been shown to participate in coreactive crosslinking. Various imidazoles have been used for similar applications. Typical are 2-ethyl, 3-methyl, and 2-ethyl-4-methyl imidazoles.

$$CH_3CH_2C \overset{N-CHCH_3}{\underset{\underset{H}{N}}{\overset{\|}{\diagdown}}} \overset{|}{\underset{CH_2}{\diagup}}$$

Heat is required for full cure of these systems and the final resin exhibits high temperature, electrical, and chemical resistance.

ACID CURING AGENTS

The acidic class of epoxy curing agents includes Lewis acids, phenols, organic acids, carboxylic acid anhydrides, and thiols.

Lewis Acids

Lewis acids contain empty orbitals in the atomic outer shell. Metal halides, like zinc, aluminum and ferric chlorides, and adducted BF_3 compounds, e.g., BF_3-monoethylamine or BF_3-etherate, are the most commonly used to cure epoxies. Most Lewis acids are latent catalysts used in heat-curing stable one-can systems with room temperature shelf lives up to one year. Although the electrical properties are good, metallic corrosion can occur from decomposition by-products, thus precluding them from numerous insulation and encapsulation applications.

Trifluoromethane sulfonic acid (triflic acid) and its salts have recently been used to catalyze hydroxyl/epoxy reactions for coatings applications. Blocking these acids with selected amines provides extended shelf stability. Heating unblocks the compound and allows the catalysis of the polymerization to proceed. Diethylammonium triflate has cocured epoxies with polyols, phenolics, and aminoplasts as well as homopolymerized DGEBA. As a result new one-component very high solids epoxy coatings have been commercialized.

Phenols

Phenols will react with epoxies, however they are seldom used as sole curing agents. They perform much better as reactive accelerators for other curing agents. They can etherify epoxies as follows:

$$\text{Ph-OH} + \text{R} + \underset{}{\text{C}\overset{O}{-}\text{C}} \xrightarrow{\text{ROH}} \underset{}{\overset{OH}{\text{C}} - \overset{OR}{\text{C}}}$$

Typical phenols are phenol-formaldehyde resoles and novolacs and substituted phenol.

Organic Acids

Organic acids are infrequently used alone as curing agents. The reaction mechanism is the key to the utility of the acid anhydrides. The esterification proceeds as follows:

$$-\overset{O}{\underset{}{\text{C}}}\text{OH} + \underset{}{\overset{O}{\text{C}} - \text{C}} \longrightarrow -\overset{O}{\underset{}{\text{C}}}\text{O}\overset{}{\underset{}{\text{C}}}\overset{OH}{\underset{}{\text{C}}}-$$

The alcoholic hydroxyl which is formed can etherify as described for the phenols. Both the esterification and etherification reaction are temperature dependent. High temperatures promote ester formation; low temperatures promote ether formation. Steric considerations such as position of the oxirane ring and nature of the carboxyl group will also influence the course of these reactions. Tertiary amines tend to enhance esterification and retard etherification.

When organic acids are used, they act as accelerators like the phenols. Typical acids include dimerized and trimerized fatty acids, phthalic, oxalic and maleic acid, and carboxy-terminated polyesters.

Cyclic Anhydrides

The cyclic anhydrides have been used most successfully with epoxy resins. Ring opening is effected by the presence of active hydrogens present as hydroxyls or water, or by a Lewis base.

$$\text{(anhydride)} \xrightarrow[\text{ROH}]{H_2O} \text{(diacid / monoester)} \quad + \quad \underset{\text{epoxide}}{\diagdown\!C\!-\!C\!\diagup} \longrightarrow -\overset{O}{\underset{\|}{C}}-O-\overset{OH}{\underset{|}{C}}-\overset{}{\underset{|}{C}}-$$

Anhydrides are the second largest curatives for epoxies and are especially suited for electrical insulation applications. They are not skin-sensitizing, however their vapors can be irritating. The liquid anhydrides are easily blended into epoxy resins. The solid anhydrides, on the other hand, need heat and extremely good mixing for proper blending. Formulations have low viscosity, long pot life and low exotherm. They have higher temperature resistance than the aliphatic amines although not as good as some of the aromatic amines. Elevated temperature cure and postcuring are generally required.

Because of the competing esterification and etherification reactions, cure schedules are often detailed and relatively complex. Consideration must be given to gel time/temperature, postcure time/temperature, presence or absence and type of accelerator, amount of hydroxyl groups present, anhydride/epoxy (A/E) ratio and stoichiometry, and amount of free acid.

Tertiary amines are the most favored accelerators, typical ones being BDMA and DMP-30. These promote esterification as mentioned earlier. Acid accelerators, like BF_3 complexes, phenols, and dibasic acids promote etherification. These considerations strongly influence the optimization of the A/E ratio and thus the cured resin properties.

Phthalic anhydride (PA),

is the least expensive anhydride and is used where formulation cost is of primary importance and overall performance is secondary. It is used in laminates, castings, and pottings and provides medium range heat deflection temperatures. PA sublimes readily and must be reacted quickly with resin. It yields low exotherms in the production of large castings.

Hexahydrophthalic anhydride (HHPA),

is a low melting solid providing good general purpose properties. In electrical encapsulation and filament winding it adds resilience without significant loss in mechanical properties. It does not sublime like PA and epoxy mixtures have lower viscosities combined with long pot life, low exotherm and very light color.

Nadic methyl anhydride (NMA),

is also used in electrical laminating and filament winding. It is a liquid easily blended into resins. Cured products have light color, excellent arc resistance and high heat deflection temperature.

Dodecenylsuccinic anhydride (DDSA),

is another easily mixed liquid anhydride. The dodecenyl group contributes added flexibility and impact resistance to systems. In addition this anhydride yields the most outstanding electrical resistance properties of this class of curatives. It has a high equivalent weight, so in order to optimize cost vs properties, it is frequently admixed with other anhydrides.

Tetrahydrophthalic and maleic anhydrides (THPA and MA) are primarily used in anhydride blends. The THPA can cause darkening of cured resins, but contributes to lower cost while yielding properties similar to HHPA. MA by itself produces very brittle systems. In blends, however, it contributes to compressive strength with some loss in tensile and flexural strength.

Pyromellitic dianhydride (PMDA),

is a high melting solid of limited solubility in epoxies. Blending with other anhydrides is common in order to facilitate incorporation into formulations. PMDA/MA blends have generated heat deflection temperatures of 250°C (480°F). This curing agent is one of the earliest dianhydrides developed to maximize temperature resistance by significantly increasing crosslink density. Tensile and flexural strengths are reduced as a result but electrical properties are maintained.

Two other high melting solid anhydrides which provide high temperature resistant epoxy systems are trimellitic anhydride (TMA),

and benzophenonetetracarboxylic dianhydride (BTDA),

They have found wide use in molding powders and prepreg for laminating. Heat deflection temperatures of 200°–300°C (392°-572°F) are common.

Chlorendic anhydride,

is the major halogenated anhydride for incorporation of flame resistance into cured systems. It contains 57% chlorine yet holds good electrical and mechanical properties to its heat deflection temperature [200°C (392°F)].

Many of the solid anhydrides can be blended into eutectic liquids for improved mixing into resin formulas. The eutectics may be liquid at room or near-room temperature. Examples include a 70/30 mix of chlorendic and HHPA, a 75/25 NMA/THPA, and a 50/50 DDSA/HHPA, all of which have melting points below 25°C (77°F). Table 6-14 describes some generalized properties of various anhydride-cured epoxies. Figure 6-3 compares the relative reactivity of anhydride curing agents to those of the other curing agents described here.

Table 6-14: Mechanical and Electrical Properties
of Anhydride Hardened Epoxy Resins
(Bruins)

Property	DDSA (130 phr)	PA (75 phr)	HHPA (80 phr)	NMA (80 phr)	PMDA/MA 17/23	Chlorendic (110 phr)
Heat deflection temp., °C	66–70	110–152	110–130	150–175	225	145–190
Flexural modulus, psi $\times 10^{-5}$	3.8	4.02	4.01	4.4	4.2	5.2
Flexural strength, psi	13,500	16,000	18,800	14,000	10,940	17,000
Compressive strength, psi	10,600	22,000	16,800	18,300	46,000	20,500
Tensile strength, psi	8,100	11,800	11,400	10,000	3,670	12,000
Ultimate elongation, %	4.5	4.8	7.4	2.5	0.9	2.6
Izod impact strength, ft-lb/in. of notch	0.3–0.4	0.46	0.3–0.4	0.48	0.34	
Rockwell M hardness		100	105	111	109	111
Dielectric constant, 60/10^6 cps	3.1/2.8	4.0/3.5	4.0/3.5	3.15/3.0	3.73/334	3.4/3.0
Dissipation factor, 60/10^6 cps	0.001/0.01	0.001/0.02	0.007/0.02	0.002/0.02	0.007/0.026	0.003/0.02

Polysulfides and Mercaptans

Several liquid polysulfide polymers have been available for curing and modifying epoxies for many years. They have the general structure,

$$HS-(-C_2H_4OCH_2OC_2H_4S-S-)_n-C_2H_4OCH_2OC_2H_4-SH$$

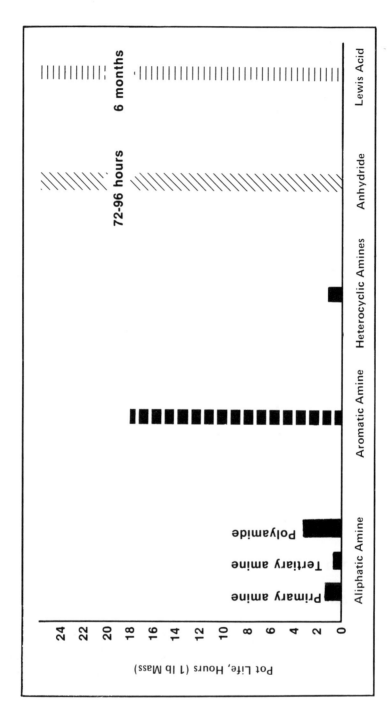

Figure 6-3: Pot life of curing agent/diglycidyl ether of bisphenol-A (DGEBPA). (Nielsen)

They are mercaptan terminated and these end groups are sufficiently acidic to create a gel but generally not strong enough to complete cure. Consequently they are added as reactive modifiers to other curing agent formulas. They impart impact resistance and toughness, increased flexibility and reduced shrinkage.

In the last ten years, new accelerated polymercaptans have been commercialized which are extremely fast (seconds to minutes) room temperature curatives. In the past some BF_3 adducts could gel epoxies in seconds at 25°C (77°F). However, they were not active at lower temperatures and the properties were not very good. These new polymercaptans can cure at temperatures as low as -40°C (-40°F). This has allowed them to be used in adhesive formulations requiring rapid cure such as bonding of highway markers (where traffic must be allowed to resume in a matter of hours) and repair and patch kits (so-called 5-minute systems).

FORMULATION PRINCIPLES

Epoxy resins have achieved their commercial success due in no small way to their amenability to a variety of formulating techniques. Not only can different resins and curatives be brought together to achieve property goals, but many additives, reactive and nonreactive, can be included for further optimization. The epoxy resin formulator's skill lies in his/her ability to effect the appropriate handling, curing, processing, and property tradeoffs needed by deftly manipulating the plethora of potential recipe ingredients.

Table 6-15 describes a typical epoxy formulation. By adjusting the type and quantity of ingredients in Table 6-15 the formulator can create systems which vary from stable free-flowing molding powders to highly viscous caulks and adhesives to clear water thin castable liquids. With the conclusion of our summary of resins, curatives and catalysts, we will now briefly review the remaining ingredients.

Table 6-15: A Typical Epoxy Formula

Resin Side	Curative Side*
Epoxy Resin(s)	Curing Agent(s)
Epoxide-Containing Reactive Diluents	Catalysts and Accelerators

Non-reactive diluents and Resinous Modifiers**
Fillers (reinforcing and/or non-reinforcing)**
Colorants (pigments and dyes)**
Rheological Additives (thixotropes, viscosity suppressants)**
Property Promoters (wetting agents, adhesion promoters, flame retardant additives)**
Processing Aids (deaerating agents, mold release agents)**

*In one-can systems, both sides are combined together and are stored under shelf-stable conditions until ready for use.

**May be commonly added to either or both sides provided no interfering reactions can take place.

Epoxy-Containing Reactive Diluents

Epoxy-containing reactive diluents are either low viscosity epoxy resins or monoepoxides. They are called reactive because the epoxide moieties will react with the curing agents and their presence must be accounted for in the stoichiometric analysis. Table 6-16 lists some typical monoepoxides. Tables 6-17 and 6-18 list some commercial DGEBA resins that have been prediluted with reactive diluents.

Table 6-16: Commercial Epoxy-Containing Monofunctional Reactive Diluents*

$CH_3(CH_2)_n CH-CH_2$ (epoxide)
Olefin oxides, b.p. 168°C at 37 mm

$CH_2=C(CH_3)-COOCH_2CH-CH_2$ (epoxide)
Glycidyl methacrylate (mol. wt. 142) b.p. 75°C at 10 mm

$CH_3(CH_2)_4 CH-CHCH_3$ (epoxide)
Octylene oxide (mol. wt. 128) 85%, 15% 1,2-isomer b.p. 77°C at 45 mm

$CH_2=CHCH_2OCH_2CH-CH_2$ (epoxide)
Allyl glycidyl ether (mol. wt. 114)

$CH_3(CH_2)_3OCH_2CH-CH_2$ (epoxide)
Butyl glycidyl ether (mol. wt. 130)

Cyclohexene vinyl monoxide (mol. wt. 124), b.p. 169°C at 760 mm

Styrene oxide (mol. wt. 120)

Dipentene monoxide (mol. wt. 152), b.p. 75°C at 10 mm

Phenyl glycidyl ether (mol. wt. 150)

p-Butyl phenol glycidyl ether

α-Pinene oxide (mol. wt. 152), b.p. 62°C at 10 mm

Cresyl glycidyl ether (mol. wt. 165)

3 (Pentadecyl) phenol glycidyl ether

($C_{12-14}H_{22-26}O_3$) Glycidyl ester of *tert*-carboxylic acid (mol. wt. 240-250), b.p. 135°C at 760 mm

*From *Handbook of Epoxy Resins* by Lee and Neville. Copyright 1967 by McGraw-Hill, Inc. Used with permission of McGraw-Hill Book Co.

Table 6-17: Lowest-Viscosity Diluted Resins
(Courtesy: Dow Chemical Co.)

Products	EEW	Viscosity (cp @ 25°C)	Color*	Diluent
Araldite 506	172-185	500-700	3	BGE-12%
Araldite 507	185-192	500-700	7	CGE
Araldite 509	189-200	500-700	NS	Epoxide 7
Epi-Rez 504	170-180	150-210	3	BGE-20%
Epi-Rez 5071	180-195	500-900	3	BGE-12%
Epi-Rez 5077	185-200	500-700	7	CGE
Epon 815	175-195	500-700	5	BGE-12%
Epotuf 37-130	175-195	500-900	3	BGE-12%
Epotuf 37-137	185-200	500-700	3	CGE

*Gardner.

These resins are similar to D.E.R. 331 resin diluted with a reactive diluent to reduce viscosity. This permits easier laminating and/or higher filler loading for cost reduction. The resins are suitable for most general purposes where use temperatures will not exceed 200°F and are generally used with polyamine or polyamide hardeners.

Table 6-18: Medium-Viscosity Diluted Resins
(Courtesy: Dow Chemical Co.)

Products	EEW	Viscosity (cp @ 25°C)	Color*	Diluent
Araldite 502	232-250	2,100-3,600	3	DBP
Epi-Rez 50840	190-210	2,000-5,000	5	Water Dispersible
Epon 820	180-194	4,000-10,000	5	PGE

*Gardner.

General comments on utility of PGE-containing resins same as for D.E.R. 330 and D.E.R. 331 resins, but the PGE introduces greater toxicity and dermatitis problems. The DBP modified resins are generally suitable for uses only at ambient temperatures, are cured with polyamines and polyamides, and are generally softer, less brittle, and less solvent resistant than 100% reactive resins.

Monoepoxides are often skin sensitizers yet provide very effective viscosity reduction at low concentrations. They can adversely affect final physical properties. Figure 6-4 shows viscosity reductions obtainable from typical monoepoxides. Table 6-19 shows the effect of diluents on the properties of a DGEBA/TETA system.

The low viscosity epoxies (e.g., butadiene dioxide, resorcinol diglycidyl ether) are used at higher concentrations to get equivalent viscosity reduction. These diluents do not degrade properties at normal use levels and can, in some cases, even improve selected final properties. In recent years toxicological problems have been uncovered with some of the traditional resins (e.g., vinylcyclohexane dioxide) and they have been discontinued.

160 Handbook of Thermoset Plastics

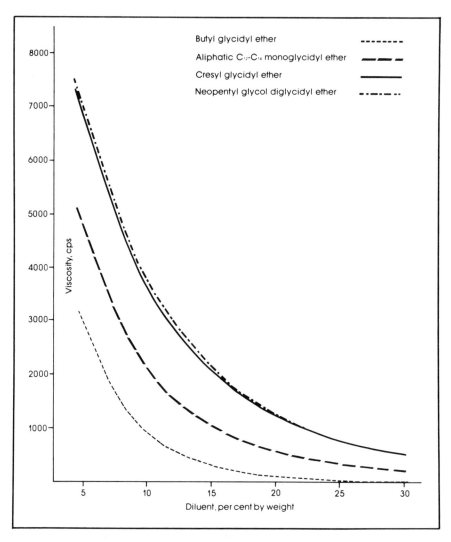

Figure 6-4: Comparison of diluent efficiencies. (DiBenedetto, 1980).

Resinous Modifiers

Resinous modifiers include PVC, polyesters, polyurethanes, silicones, furfurals, acrylics, and butadiene-acrylonitrile resins. Coal tar and phenolic modifiers are also used. These materials are added in order to impart or enhance specialized properties. Often the property enhancement of the modified epoxy is a direct result of the contribution of a characteristic property of the added modifier resin. In some instances, the modifying resin is sufficiently low cost and can act as an extender in the formula.

Table 6-19: Effect of Diluent on Properties of TETA-Cured DGEBA*

Diluent and Mixed Viscosity, Centipoises	Pot Life at 23°C	Exotherm, °C	Water Absorption, Weight Increase, %	Flexural Strength, psi	DT, °C	Weight Loss in 48 hr at 200°C, %
None	40	200	0.73	20,300	120	1.89
Octylene oxide:						
1,500	55	205	1.10	14,185	74	5.20
500	72	180	1.18	11,440	65	Unstable
200	100	164	1.04	7,797	58	Unstable
Styrene oxide:						
1,500	48	200	0.71	19,403	93	2.39
500	69	193	0.80	18,854	82	4.29
200	84	178	1.06	16,303	70	6.20
3(Pentadecyl) phenol glycidyl ether:						
1,500	54	164	1.19	14,010	76	2.84
500	68	148	1.48	6,443	55	4.15
Butyl glycidyl ether:						
1,500	55	204	1.01	18,433	101	2.12
500	63	207	1.32	17,237	85	2.72
200	76	193	1.63	16,400	75	3.77
Allyl glycidyl ether:						
1,500	48	204	1.11	18,817	95	1.86
500	59	194	1.51	18,047	76	2.37
200	66	189	1.72	15,827	74	3.14
Phenyl glycidyl ether:						
1,500	47	198	1.04	19,717	–	2.37
500	49	209	1.26	19,523	–	3.28
200	54	198	1.35	19,600	–	4.93
Cresyl glycidyl ether:						
1,500	46	202	0.98	22,067	92	2.98
500	46	204	1.24	20,000	78	4.40
200	51	201	1.35	15,230	63	6.58
Glycidyl methacrylate:						
1,500	41	208	1.30	18,507	105	2.93
500	43	224	1.64	17,510	95	3.78
200	46	224	2.02	18,400	87	4.43

*From *Handbook of Epoxy Resins* by Lee and Neville. Copyright 1967 by McGraw-Hill, Inc. Used with permission of McGraw-Hill Book Co.

Among the many properties affected by resinous modifiers are flexibility, toughness and impact strength; peel strength, adhesion to substrates and chalk resistance of coatings.

Nonreactive Diluents

Nonreactive diluents are the equivalent of secondary plasticizers used with vinyls. They do not participate in the crosslinking reaction and in fact can be sufficiently mobile that they can ultimately migrate out of the polymerized mass. Such migration is of long duration and controllable to some degree. The benefits of adding nonreactives may often offset the migration problem. Primary benefits include viscosity and cost reduction, extension of pot life and decrease in exotherm. Consistent with plasticization in other plastics, the nonreactives tend to degrade mechanical, electrical and resistance properties as their concentration increases. Typical nonreactive diluents include monomeric styrene, bisphenols, hydrocarbon oils, and phthalate ester plasticizers like dioctyl and dibutyl phthalate.

Fillers

Fillers play a primary role in epoxy resin formulation. Reinforcing fibers such as glass, graphite and polyaramid improve mechanical properties to such an extent that epoxies can be used in many structural applications. Nonreinforcing fillers include powdered metals (electrical and thermal conductivity), alumina (thermal conductivity), silica (cost reduction, minor strength enhancement), mica (electrical resistance), talc and calcium carbonate (cost reduction), barium sulfate (barytes, density control), and carbon and graphite powders (lubricity). Increasing filler content generally increases viscosity and makes processing more difficult. Specific gravity usually increases although some fillers like hollow glass or phenolic microballoons create syntactic foams of significantly reduced density. Table 6-20 shows general effects of some commonly used fillers in epoxies.

Some interesting developments have occurred in the last five years bearing on the use of fillers in epoxies. Silver coated glass spheres have been used to replace very expensive powdered silver and gold to create electrical conductivity in epoxy systems. On a volume basis, the use of these spheres can provide equivalent conductivity at one-twentieth the cost of silver. Table 6-21 compares the electrical conductivity of various metals and filled conductive epoxies. Table 6-22 compares the thermal conductivity of metals, oxides, and filled conductive epoxies.

Polyvinylidene chloride (PVDC) microspheres are used as a low cost syntactic foam filler. However, they react with amines and melt if the exotherm gets too high. Recent work has produced a series of PVDC spheres which are stable if blended in the resin first and castings are limited to 1½ inches of thickness. Comparative costs, calculated as of early 1984, of microspheres vs other fillers is shown in Table 6-23.

The use of synthetic sodium aluminum silicate in epoxy coatings provides improved opacity over titanium dioxide. These new silicates are prepared by the reaction of aluminum sulfate and sodium silicate as opposed to being mined naturally. Better control of structure, particle size and reinforcement characteristics is obtained.

Table 6-20: General Effects on Properties of Some Commonly Used Fillers
(Courtesy: Shell Chemical Co./EPON® Resins)

Filler	Increase							Decrease						
	Thermal conductivity	Thermal-shock resistance	Impact resistance	Compressive strength	Arc resistance	Machinability	Electrical conductivity	Cost	Cracking	Exotherm	Coefficient of expansion	Density	Settling	Shrinkage
Bulk fillers:														
Sand	X			X				•	X	X				X
Silica	X			X				•	X	X				X
Talc						•		X						X
Clay		X		X		•		X	X			X		X
Calcium carbonate						•		X	X					X
Calcium sulfate (anhydrous)	X			X	•			X		X				X
Reinforcing:														
Mica		X	•					X	X			X		
Wollastonite	X	X	•					X	X					
Chopped glass		X	•							X				
Wood flour		X	X		X			X					•	
Sawdust		X	X		X			X					•	
Specialty:														
Quartz	X		X	X					X		•			X
Alumina	X	X		X						X	•			X
Hydrate alumina					•							X		
Li Al silicate	X			X					X	X	•			X
Beryl	X			X					X	X	•			X
Silica aerogel													•	
Bentonite													•	
Graphite						X	•							
Powder metals	X	X	X	X		X	X	X	X	X	X			X
Ceramic spheres												•		

● Major use
X Minor use

Source: Harper, *Electronic Packaging with Resins*, McGraw-Hill Book Co., New York, 1961

Colorants and Dyes

A wide variety of colorants can be used with epoxies. Dyes are less frequently used because of the natural tendency of clear epoxies to yellow when exposed to UV light. Some blue dyes are effective in

Table 6-21: Electrical Conductivity of Metals, Conductive Plastics and Various Insulation Materials at 25°C
(Bolger and Morano)

	Specific Gravity (gms/cm^3)	ρ = Volume Resistivity (ohm cm)
Silver	10.5	1.6×10^{-6}
Copper	8.9	1.8×10^{-6}
Gold	19.3	2.3×10^{-6}
Aluminum	2.7	2.9×10^{-6}
Nickel	8.9	10×10^{-6}
Platinum	21.5	21.5×10^{-6}
Eutectic solders	—	$20\text{-}30 \times 10^{-6}$
Best silver-filled inks and coatings	—	1×10^{-4}
Best silver-filled epoxy adhesives	—	1×10^{-3}
Graphite	—	1.3×10^{-3}
Low cost silver-filled epoxy adhesives	—	1×10^{-2}
Graphite or carbon-filled coatings	—	10^2 to 10
Oxide-filled epoxy adhesives	1.5-2.5	$10^{14} - 10^{15}$
Unfilled epoxy adhesives	1.1	$10^{14} - 10^{15}$
Mica, polystyrene & other best dielectrics	—	10^{16}

prolonging the perception of yellowing. Table 6-24 lists many of the pigments found acceptable for use in epoxy systems. Most inorganics except chrome greens, natural siennas and ochers, and zinc sulfide white are used. Organic pigments are generally limited to carbon blacks and phthalates.

Other Additives

Rheological additives include viscosity depressants (usually solvents, surface activators, or diluents) and thixotropic agents. Pyrolitic silicas, bentonite clays, and castor oil derivatives are the most common thixotropes. New hydrophobic fumed silicas have received much attention as stable thickening agents (see Figure 6-5).

New developments in property modification have been primarily in the areas of improved adhesion via nonsilane adhesion promoters (organometallics), improved strength-to-weight reinforcing fibers (polyaramids), and improved flame retardants. One study assessed the affect of tris(dibromopropyl)phosphate and dechlorane on epoxy flammability. Table 6-25 shows a number of flame retardant additives used with or without halogenated epoxies. Organophosphorous polyols can be coreacted into epoxies to enhance flame retardance of epoxy prepreg material used to make printed circuit boards (see Table 6-26).

Table 6-22: Thermal Conductivity of Metals, Oxides and Conductive Adhesives
(Bolger and Morano)

	Thermal Conductivity at 25°C (Btu/hr°F ft²/ft)
Silver	240
Copper	220
Beryllium oxide	130
Aluminum	110
Steel	40
Eutectic solders	20-30
Aluminum oxide	20
Best silver-filled epoxy adhesives	1 to 4
Aluminum-filled (50%) epoxy	1 to 2
Epoxy filled with 75% by wt. Al_2O_3	0.8 to 1
Epoxy filled with 50% by wt. Al_2O_3	0.3 to 0.4
Epoxy filled with 25% by wt. Al_2O_3	0.2 to 0.3
Unfilled epoxies	0.1 to 0.15
Foamed plastics	0.01 to 0.03
Air	0.015

Table for Conversion of Thermal Conductivity Units

g cal/cm² sec °C/cm	w/cm² °C/cm	Btu/ft²hr°F/ft	Btu/ft²hr°F/in
1.0	4.19	242	2900
0.23	1.0	58	690
4.13×10^{-3}	0.0173	1.0	12.0
3.44×10^{-4}	1.44×10^{-3}	0.083	1.0

Heat transfer formula:

$$q = \frac{k \Delta T}{x}$$

k = thermal conductivity
ΔT = temperature drop across material
q = heat flow/unit area
x = material thickness

Table 6-23: Comparative Costs of Fillers and Extenders
(Melber, et al.)

Name	Specific gravity (g/ml)	Cost, $/lb	Volume cost, $/ft³
Typical resin	1.2	0.70	52.40
Glass fiber	2.5	0.75	117.00
Expanded PVDC microspheres	0.032	6.62	13.22
Hollow glass spheres	0.15	1.64	15.35
Solid glass spheres	2.48	0.33	51.07
Calcium carbonate	2.7	0.04	6.74
Aluminum trihydrate	2.4	0.25	37.44
Phenolic microballoons	0.17	3.75	39.78

Table 6-24: Applicability of Colorants to Epoxy Resins
(Courtesy: Shell Chemical Co./EPON® Resins)

Applicability key
4 - Recommended
3 - Applicable
2 - Limited conditions
1 - Economy, low quality

Pigment type	Shades	Applicability rating
Violets, maroons, and reds		
Cadmium sulfoselenide[1]	Maroon to light red	4
Quinacridone	Maroon to medium red	4
Bon (2B-Ca salt)	Maroon to light red	3
Bon (2B-Mn salt)	Maroon to light red	4
Lithol rubine	Bluish red	2
Ba and Ca lithols	Maroon to light red	2
Pigment scarlet	Bluish red	4
Thioindigoid[2]	Maroon	3
PTA toners	Violet to medium red	2
Red lake C	Light red	1
Pyrazolone	Light red	3
Naphthol	Light red to dark red	3
Iron oxide	Maroons and brick reds	4
High molecular weight desazo red	Medium to light red	2
Vat reds	Medium reds	2
Oranges and yellows		
Cadmium sulfoselenide[3]	Orange to very light yellow	4
Chrome yellow[4]	Medium to very light yellow	4
Chrome orange[5]	Yellowish to red orange	3
Molybdate orange[6]	Orange and orange red	3
Vat colors	Oranges and yellows	2
Benzidine yellow	Light yellow	4
Benzidine yellow, xylidide	Light to deep yellow	4
Nickel-azo ("greengold")	Greenish yellow	4
Strontium yellow[7]	Very light yellow	4
Zinc chromate	Light yellow	4
Ni-Ti yellow[8]	Very light yellow	4
Iron oxide	Reddish to yellow tan	4
Greens and blues		
Phthalocyanine[9]	Blue and green	4
PTA/PMA toners	Blue and green	2
Chromium oxide	Dull green	4
Hydrated chromium oxide	Bluish green	4
Chrome green	Dark bluish to light yellow-green	4
Pigment green B	Dark green	3
Iron blue[10]	Dark blue	4
Ultramarine	Blues and violets	2
Indanthrone	Blue	4
Cobalt blue	Blue	4
White		
Titanium dioxide, rutile	White	4
Titanium dioxide, anatase	White	4
Zinc oxide	White	4
Antimony oxide[11]	White	4

(continued)

Table 6-24: (continued)

Pigment type	Shades	Applicability rating
Blacks and browns		
Channel black	Jet black	4
Furnace black	Black	4
Lamp black	Bluish black	4
Iron oxide	Brown and black	4
Bone black	Black	4
Soluble dyes		
Oil soluble		
Azo	Yellow to red (red) green,	2
Anthraquinone	blue, black, brown	
Acetate dyes	Wide range	2
Basic dye bases	Wide range	1

[1]Discolors with S or Cu metal contact.
[2]Varies within class.
[3]Discolors in contact with copper.
[4]Stains with sulfide.
[5]Stains with sulfide.
[6]Stains with sulfide.
[7]Sulfide-stable.
[8]Sulfide-stable.
[9]The very red shades of blue crystallize in aromatics.
[10]Color destroyed in reducing atmosphere.
[11]Mostly used as flameproofer with Cl and Br.
Source: Oleesky and Mohr, Handbook of Reinforced Plastics (SPI), Reinhold Publishing Corp., New York, 1964.

Table 6-25: Flame-Retardant Additives for Epoxy Resins (Davis)

Additive	Level Needed, %	Manufacturer
Phosphorus compounds		
Triphenylphosphine	5-10	BASF Wyandotte, M&T Chemical
Tris-β-chloroethyl phosphate	5-10	Stauffer Chemical
Hydrates		
Alumina trihydrate	20-25	Alcoa, Kaiser, Reynolds, others
Halogenated compounds		
Octabromobiphenyl	20-25	White Chemical
Decabromobiphenyl	20-25	Velsicol
Decabromodiphenyl oxide	20-25	Dow Chemical, Great Lakes Chemical, Saytech
Dechlorane	20-30	Hooker Chemical
Synergists*		
Antimony oxide	2-10	Chemetron, Harshaw, M&T Chemical, McGean, NL Industries, Nyacol, Samincorp
Zinc borate	2-10	U.S. Borax & Chemical Corp.
Molybdic oxide	5-7	Climax Molybdenum Co.

*Used in conjunction with other flame-retardant additives.

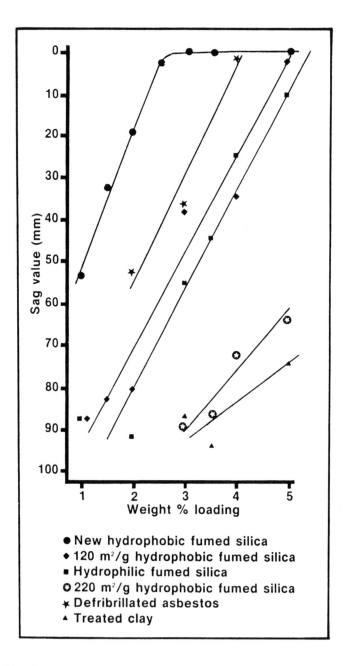

Figure 6-5: Sag values of epoxy sealants after aging four weeks as a functon of loading. (Cochrane and Miller).

Table 6-26: Flame Retardance of FR-T In Epoxy Resin Castings*
(Reprinted from "New Flame Retardant for Epoxy Glass Laminates," by E.R. Fretz and J. Green, *Printed Circuit Fabrication*, Vol. 6, No. 5, p 57 (May 1983).

Flame Retardant (wt %)		UL-94	Average Burn	
FR-T	TBBP-A	Rating	Times, sec.	LOI**
None	None	Burn	–	19.8
0	26	V-O	1.8	32.7
7	9	V-O	1.4	34.3
7	5	V-O	2.5	31.0
14	0	V-O	3.8	30.4

*Castings cured with Nadic® methyl anhydride/phthalic anhydride (2/1). Trademark of Allied Corporation.
**ASTM D2863-74.

PROPERTIES

It should be patently obvious that the properties of epoxy resins can vary over a very wide range, depending on the selection of a formulation's ingredients, their relative proportions, the processing of the formula and the configuration and environment of the final part.

As with any versatile class of resins (e.g., vinyls, polyesters, etc.) it becomes difficult to present an all-inclusive tabulation of the properties for every possible formulation. Table 6-27 presents a reasonable summary of some of the most important properties of some basic epoxy configurations. The data allows the reader to make some broad stroke comparisons with similar data of other plastics. Ultimately, however, consultation with epoxy formulators and review of vendor data sheets/literature combined with effective and application-specific testing will provide the most appropriate data base from which design decisions can be made.

Some generalizations about epoxy resin properties are possible. Liquid resins and curatives can form low viscosity, easily modified systems. They can cure at temperatures from -40°C (-40°F) to 200°C (392°F), depending on the curing agents used. They exhibit very low shrinkage and do not evolve volatile by-products during cure. Cure schedules can be varied within wide boundaries to accommodate different processing methods and applications.

Because of the presence of significant polarity, epoxies wet and adhere exceptionally well to many surfaces. Mechanical properties of cast epoxy exceed most other castable plastics. Epoxies are excellent electrical and thermal insulators. They can be formulated to resist temperatures as high as 290°C (550°F). They are selectively resistant to a broad range of environments and chemicals. They are highly resistant to caustics, oils, and many solvents with fair acid resistance. Chlorinated hydrocarbons and some organic acids will

Table 6-27: General Properties of Epoxies

				Casting resins and compounds		
	Properties		ASTM test method	Unfilled	Silica-filled	Aluminum-filled
Processing	1. Melting temperature, °C. T_m (crystalline)			Thermoset	Thermoset	Thermoset
	T_g (amorphous)					
	2. Processing temperature range, °F. (C = compression; T = transfer; I = injection; E = extrusion)					
	3. Molding pressure range, 10^3 p.s.i.					
	4. Compression ratio					
	5. Mold (linear) shrinkage, in./in.		D955	0.001-0.010	0.0005-0.008	0.001-0.005
Mechanical	6. Tensile strength at break, p.s.i.		D638	4000-13,000	7000-13,000	7000-12,000
	7. Elongation at break, %		D638	3-6	1-3	0.5-3
	8. Tensile yield strength, p.s.i.		D638			
	9. Compressive strength (rupture or yield), p.s.i.		D695	15,000-25,000	15,000-35,000	15,000-33,000
	10. Flexural strength (rupture or yield), p.s.i.		D790	13,000-21,000	8000-14,000	8500-24,000
	11. Tensile modulus, 10^3 p.s.i.		D638	350		
	12. Compressive modulus, 10^3 p.s.i.		D695			
	13. Flexural modulus, 10^3 p.s.i.	73° F.	D790			
		200° F.	D790			
		250° F.	D790			
		300° F.	D790			
	14. Izod impact, ft.-lb./in. of notch (⅛-in. thick specimen)		D256A	0.2-1.0	0.3-0.45	0.4-1.6
	15. Hardness	Rockwell	D785	M80-110	M85-120	M55-85
		Shore/Barcol	D2240/D2583			
Thermal	16. Coef. of linear thermal expansion, 10^{-6} in./in./°C.		D696	45-65	20-40	5.5
	17. Deflection temperature under flexural load, °F.	264 p.s.i.	D648	115-550	160-550	190-600
		66 p.s.i.	D648			
	18. Thermal conductivity, 10^{-4} cal.-cm./sec.-cm.^{2}-°C.		C177	4.5	10-20	15-25
Physical	19. Specific gravity		D792	1.11-1.40	1.6-2.0	1.4-1.8
	20. Water absorption (⅛-in. thick specimen), %	24 hr.	D570	0.08-0.15	0.04-0.1	0.1-4.0
		Saturation	D570			
	21. Dielectric strength (⅛-in. thick specimen), short time, v./mil		D149	300-500	300-550	

(continued)

Table 6-27: (continued)

			ASTM test method	Casting resins and compounds	
		Properties		Flexibilized	Cyclo-aliphatic
Processing	1.	Melting temperature, °C. T_m (crystalline)		Thermoset	Thermoset
		T_g (amorphous)			
	2.	Processing temperature range, °F. (C = compression; T = transfer; I = injection; E = extrusion)			
	3.	Molding pressure range, 10^3 p.s.i.			
	4.	Compression ratio			
	5.	Mold (linear) shrinkage, in./in.	D955	0.001-0.010	
Mechanical	6.	Tensile strength at break, p.s.i.	D638	2000-10,000	8000-12,000
	7.	Elongation at break, %	D638	20-70	2-10
	8.	Tensile yield strength, p.s.i.	D638		
	9.	Compressive strength (rupture or yield), p.s.i.	D695	1000-14,000	15,000-20,000
	10.	Flexural strength (rupture or yield), p.s.i.	D790	1000-13,000	10,000-13,000
	11.	Tensile modulus, 10^3 p.s.i.	D638	1-350	495
	12.	Compressive modulus, 10^3 p.s.i.	D695		
	13.	Flexural modulus, 10^3 p.s.i. 73° F.	D790		
		200°F.	D790		
		250°F.	D790		
		300°F.	D790		
	14.	Izod impact, ft.-lb./in. of notch (⅛-in. thick specimen)	D256A	3.5-5.0	
	15.	Hardness Rockwell	D785		
		Shore/Barcol	D2240/ D2583	Shore D70-89	
Thermal	16.	Coef. of linear thermal expansion, 10^{-6} in./in./°C.	D696	20-100	
	17.	Deflection temperature under flexural load, °F. 264 p.s.i.	D648	73-250	200-450
		66 p.s.i.	D648		
	18.	Thermal conductivity, 10^{-4} cal.-cm./ sec.-cm.^{2}-°C.	C177		
Physical	19.	Specific gravity	D792	1.05-1.35	1.16-1.21
	20.	Water absorption (⅛-in. thick specimen), % 24 hr.	D570	0.27-0.5	
		Saturation	D570		
	21.	Dielectric strength (⅛-in. thick specimen), short time, v./mil	D149	235-400	

(continued)

Table 6-27: (continued)

		Properties	ASTM test method	Bisphenol molding compounds		
				Glass fiber-reinforced	Mineral-filled	Low density glass sphere-filled
Processing	1.	Melting temperature, °C. T_m (crystalline) T_g (amorphous)		Thermoset	Thermoset	Thermoset
	2.	Processing temperature range, °F. (C = compression; T = transfer; I = injection; E = extrusion)		C: 300-330 T: 280-380	C: 250-330 T: 250-380	C: 250-300 I: 250-300
	3.	Molding pressure range, 10^3 p.s.i.		1-5	0.1-3	0.1-2
	4.	Compression ratio		3.0-7.0	2.0-3.0	3.0-7.0
	5.	Mold (linear) shrinkage, in./in.	D955	0.001-0.008	0.002-0.010	0.006-0.010
Mechanical	6.	Tensile strength at break, p.s.i.	D638	5000-20,000	4000-10,000	2500-4000
	7.	Elongation at break, %	D638	4		
	8.	Tensile yield strength, p.s.i.	D638			
	9.	Compressive strength (rupture or yield), p.s.i.	D695	18,000-40,000	18,000-40,000	10,000-15,000
	10.	Flexural strength (rupture or yield), p.s.i.	D790	8000-30,000	6000-18,000	5000-7000
	11.	Tensile modulus, 10^3 p.s.i.	D638	3000		
	12.	Compressive modulus, 10^3 p.s.i.	D695			
	13.	Flexural modulus, 10^3 p.s.i. 73° F.	D790	2000-4500	1400-2000	500-750
		200°F.	D790			
		250°F.	D790			
		300°F.	D790			
	14.	Izod impact, ft.-lb./in. of notch (1/8-in. thick specimen)	D256A	0.3-10.0	0.3-0.5	0.15-0.25
	15.	Hardness Rockwell	D785	M100-112	M100-M112	
		Shore/Barcol	D2240/ D2583			
Thermal	16.	Coef. of linear thermal expansion, 10^{-6} in./in./°C.	D696	11-50	20-60	
	17.	Deflection temperature under flexural load, °F. 264 p.s.i.	D648	225-500	225-500	200-250
		66 p.s.i.	D648			
	18.	Thermal conductivity, 10^{-4} cal.-cm./sec.-cm.^{2}-°C.	C177	4.0-10.0	4-35	4.0-6.0
Physical	19.	Specific gravity	D792	1.6-2.0	1.6-2.1	0.75-1.0
	20.	Water absorption (1/8-in. thick specimen), % 24 hr.	D570	0.04-0.20	0.03-0.20	0.2-1.0
		Saturation	D570			
	21.	Dielectric strength (1/8-in. thick specimen), short time, v./mil	D149	250-400	250-400	380-420

(continued)

Table 6-27: (continued)

	Properties	ASTM test method	Novolak molding compounds		
			Mineral- and glass- filled, encapsulation	Mineral- and glass- filled, high temperature	Glass- filled, high strength
Processing	1. Melting temperature, °C. T_m (crystalline)				
	T_g (amorphous)		145-155	195	
	2. Processing temperature range, °F. (C = compression; T = transfer; I = injection; E = extrusion)		C: 280-360 T: 250-380 I: 290-350	T: 375-385	C: 290-330 T: 290-330
	3. Molding pressure range, 10^3 p.s.i.		0.25-3.0	0.3-2.5	2.5-5.0
	4. Compression ratio			1.5-2.5	6-7
	5. Mold (linear) shrinkage, in./in.	D955	0.004-0.008	0.003-0.006	0.0002
Mechanical	6. Tensile strength at break, p.s.i.	D638	5000-12,500	6000-9000	18,000-27,000
	7. Elongation at break, %	D638			
	8. Tensile yield strength, p.s.i.	D638			
	9. Compressive strength (rupture or yield), p.s.i.	D695	24,000-48,000	30,000-44,500	30,000-38,000
	10. Flexural strength (rupture or yield), p.s.i.	D790	10,000-21,800	10,000-17,000	50,000-70,000
	11. Tensile modulus, 10^3 p.s.i.	D638	2100		
	12. Compressive modulus, 10^3 p.s.i.	D695			
	13. Flexural modulus, 10^3 p.s.i. 73° F.	D790	1400-2400		2.8-4.2
	200°F.	D790			
	250°F.	D790			
	300°F.	D790			
	14. Izod impact, ft.-lb./in. of notch (⅛-in. thick specimen)	D256A	0.3-0.5	0.4	25-34
	15. Hardness Rockwell	D785	M115		
	Shore/Barcol	D2240/ D2583	Barcol 70-74	Barcol 78	Barcol 60-74
Thermal	16. Coef. of linear thermal expansion, 10^{-6} in./in./°C.	D696	18-43	35	
	17. Deflection temperature under flexural load, °F. 264 p.s.i.	D648	310-446	500	
	66 p.s.i.	D648			
	18. Thermal conductivity, 10^{-4} cal.-cm./ sec.-cm.^{2}-°C.	C177	8-24.8		
Physical	19. Specific gravity	D792	1.6-2.0	1.90-1.94	1.84
	20. Water absorption (⅛-in. thick specimen), % 24 hr.	D570	0.04-0.29	0.17	
	Saturation	D570	0.15-0.3		
	21. Dielectric strength (⅛-in. thick specimen), short time, v./mil	D149	325-425	410-450	380-400

(Reprinted by permission from *Modern Plastics Encyclopedia* for 1983-1984, copyright McGraw-Hill, Inc. All rights reserved.)

attack epoxy systems. Epoxies will discolor when exposed to ultraviolet energy. They tend towards brittleness but can be toughened at lower use temperatures (<104°C, 200°F). Many epoxies and curing agents are skin sensitizers. Although they are not the most expensive of thermosets, they are not the least expensive either. Under some conditions of high heat and humidity (>120°C, 250°F, 95% RH) significant loss of properties has been recorded. However, polymer breakdown (reversion) as occurs with some silicones and polyurethanes is extremely rare.

APPLICATIONS

Epoxies find application in five major areas: coatings, electrical and electronic insulation, adhesives, composites, and construction. The total epoxy market in 1980 was about 300 million pounds, of which half went towards coatings. The other half was distributed within the structural markets consisting of the other four areas listed. Figure 6-6 shows the breakdown of the structural market as of 1980. The growth of epoxies has continued steadily upwards. The 18 million pounds of epoxy adhesive produced in 1980, for example, increased to 33 million pounds by 1982 and are projected to increase to 83 million pounds by 1995.

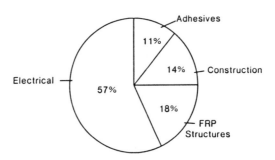

Figure 6-6: 1983 market for epoxy structural resins. (Brown and McCrea, courtesy of Shell Chemical Co./EPON® Resin).

These categorizations define the nature of the epoxy systems and span all major commercial enterprises. For example, the medical and dental field uses epoxy castings, encapsulants and adhesives; space exploration uses epoxy composites, adhesives and electrical insulation;

the automotive industry uses protective coatings and adhesives, etc.

As noted, coatings consume fully 50% or more of epoxy resin production. Epoxy's chemical resistance, toughness and durability, and adhesion are the prime features for this arena. They are used in appliance and automotive primers, can coatings, industrial maintenance paints, and product and marine finishes. Figure 6-7 shows an example of an epoxy coating.

Figure 6-7: Epoxy coated pipe. (Courtesy of Shell Chemical Co./EPON® Resin).

New pollution control constraints have prompted developments in water-based, high solids and solventless coating systems. Powder coatings for such areas as thick-film pipe coatings continue to consume large volumes of resin. Two-component, air-dried, solventless systems are adaptable to new spray application processes in maintenance coatings. Two-component, water-based emulsion paints are being used in architectural applications. Traditional coal-tar epoxies and zinc-rich wash coat primers remain staples for maintenance and marine protection coatings. Ultraviolet curable epoxy coatings have been developed for environmental protection of printed circuit boards.

The high resistivity and relatively low dissipation factor, combined with high mechanical properties, are the characteristics that allow for the widespread use of epoxies in electrical and electronic insulation. Encapsulation and coating of transistors, switches, coils, insulators and integrated circuits are routine. New casting processes are providing dimensional stability, eliminating stress build-up and surface defects and significantly reducing demold time. In Europe, epoxies continue to dominate porcelain in large outdoor transformers, switching gear and high voltage insulators. Figure 6-8 shows an example of an epoxy used in such applications.

176 Handbook of Thermoset Plastics

Figure 6-8: Electrical/electronic devices encapsulated with epoxy resin. (Courtesy of Shell Chemical Co./EPON® Resin).

New encapsulants are being developed based on the concept of simultaneous interpenetrating networks (SIN). In this situation two different monomers are polymerized simultaneously to form interpenetrating three-dimensional networks. An example of one such system is an SIN based on epoxy and poly(n-butyl acrylate). The major advantage of this approach for epoxy castables is improved resistance to crack growth.

Many epoxies are cast for nonelectrical applications. Recent novel applications for such structural castings include large bearings for an oceanic oil rig swivel buoy (see Figure 6-9), acid-resistant pump impellers, and sleeves for ship stern-tube assemblies. Plans are in progress to build a new deep-diving submersible from acrylic and epoxy resins which will provide a one-person, one-atmosphere diving capability to depths of 6,500 feet.

Resin transfer molding (RTM) is a new process that has been found very useful in rapid molding of liquid epoxies. Typical parts made via RTM include propeller blades, industrial fan blades and support beams. New epoxy systems are being examined for use in the reaction injection molding (RIM) process. RIM has been dominated by urethanes, however these new epoxies, particularly reinforced

Epoxy Resins 177

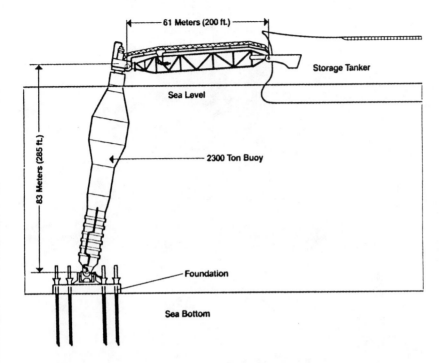

Figure 6-9: To assure continuous flow of North Sea Oil, liquid epoxy resin was pumped into 12 mounting areas of a swivel buoy that weighs in at 4.6 million pounds. (Wilson, in *Materials Engineering*, April 1983).

versions, have higher tensile and flexural moduli, are more versatile to formulate, and provide higher service temperatures. Successful applications include skateboards and snow skis.

Since their introduction, epoxies have been a dominant force in adhesives and bonding. Volatile-free curing and minimal shrinkage combined with excellent lap-shear strength make epoxies the premier adhesive. Major recent developments have focused on new latent curatives for one-can systems that are room temperature stable for over a year, yet will cure in minutes at temperatures as low as 100°C, 212°F. New epoxy systems have successfully bonded to and filled enamel, dentin and cementum in the dental field (see Figure 6-10).

In 1978 the United States Air Force began a major program to determine whether adhesive bonding with epoxy could replace rivets that are traditionally used in aircraft assemblies. The largest adhesively bonded primary structure ever assembled, a 42' long by 18' wide fuselage section was thoroughly tested (see Figure 6-11). The program validated the technology to the point that new aircraft designs will begin to use as much adhesive on primary structures as are currently used for secondary and nonstructural aircraft elements.

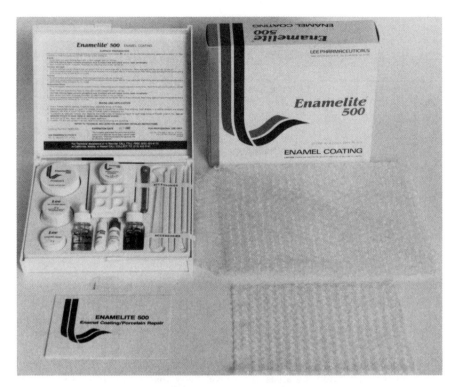

Figure 6-10: Epoxy enamel coating for teeth. (Courtesy of Lee Pharmaceuticals).

Figure 6-11: PABST program fuselage. (Courtesy of McDonnell Douglas Corp.).

Several factors in the automotive industry have promoted the replacement of welding, riveting and other traditional metal joining processes with epoxy adhesive bonding. In car, bus and truck plants adhesive processes reduce noise and eliminate hazardous materials (e.g., lead) and processes. The adhesives help meet crash, rollover and other safety regulations and reduce weight for improvement of fuel efficiency.

Glass, graphite and polyaramid reinforced epoxy composites continue to find major use in industries such as space, printed circuitry, tanks and pressure vessels, and pipe. Epoxy composites provide high strength-to-weight ratios, good thermal, electrical and chemical resistance and are compatible with every reinforced plastics process.

A novel use of graphite/epoxy composite has been reported in the music field. A violin (see Figure 6-12) was constructed with the composite replacing traditional woods. Cost and fabrication time were substantially reduced. The overall tone balance was rated good with excellent high notes by the concert artists who played the instrument.

Figure 6-12: Graphite/epoxy violin. (Courtesy of L.K. John, inventor).

Industrial chemically-resistant flooring remains a major use of epoxy resins in the construction trade. Sand filled compositions having excellent oil, water, solvent, and caustic resistance and superb adhesion to concrete are the primary epoxy systems used. Some decorative "pour-a-floor" systems are still popular because of the ease of application and excellent adhesion to glass, quartz, marble chips and other attractive inclusion materials. Other construction uses are coal-tar based paving materials, grouts, and adhesives for segmental bridge construction and airport runway repair.

TRADE NAMES

Able-	Formulations	Ablestik Laboratories
Ajicure	Curing Agents	Ajinomoto Co., Inc.
Anca-	Curing Agents	Pacific Anchor Chemical
APCO	Formulations and Curing Agents	Applied Plastics Co.
Araldite	Resins and Curing Agents	Ciba-Geigy Corp.
Bakelite	Resins and Curing Agents	Union Carbide Corp.
DEN, DER, DOW	Resins and Curing Agents	Dow Chemical Co.
Epi-Cure, Epi-Rez	Resins and Curing Agents	Celanese Resins Div.
Epon	Resins and Curing Agents	Shell Chemical Co.
Eposet	Formulations	Hardman Co., Inc.
Epotek	Formulations	Epoxy Technology Inc.
Epotuf	Resins and Curing Agents	Reichhold Chemicals Inc.
Hysol	Formulations	Dexter Corp.
Scotch-	Formulations	3M Co.
Stycast	Formulations	Emerson & Cuming, Inc.
Uniset	Formulations	Amicon Corp.
Versamid	Curing Agents	Henkel Corp.

REFERENCES AND BIBLIOGRAPHY

Allied Corporation, Various Technical Data Bulletins on Boron Trifluoride Complexes, 1984.

Alm, R., Formulation Techniques Using Triflic Acid Salts, *Modern Paint and Coatings*, Vol. 70, No. 10, p 88 (October 1980).

Anon, Graphite/Epoxy Composite Violins Have Excellent Tone Compared To Wood, *Materials Engineering*, Vol. 93, No. 1, p 12 (January 1981).

Barker, A., Adhesive Consumption May Rise 60% by Volume by 1995, *Adhesives Age*, Vol. 27, No. 1, p 32 (January 1984).

Bolger, J.C., Epoxies for Manufacturing Cars, Buses, & Trucks, *Adhesives Age*, Vol. 23, No. 12, p 14 (December 1980).

Bolger, J.C. and Morana, S.L., Conductive Adhesives: How and Where They Work, *Adhesives Age*, Vol. 27, No. 7, p 17 (June 1984).

Braasch, H., New Adhesive Withstands Temperature Extremes, NASA Tech. Briefs, New Technology Report, p 1 (Spring 1978).

Brown, R.E. and McCrea, R.E., Competition, Chances for Growth for Epoxy Adhesive Markets, *Adhesives Age*, Vol. 25, No. 2, p 21 (February 1982).

Bruins, P.F., *Epoxy Resin Technology*, Interscience Publishers, New York (1968).

Burns, P., Recent Developments in Epoxy Resins, Term paper submitted to fulfill requirements of Ch.E. 478, University of Southern California (April 1984).

Catsiff, E.H., Dee, H.B. and Seltzer, R., Hydantoin Epoxy Resins, *Modern Plastics*, Vol. 55, No. 7, p 54 (July 1978).

Cochrane, H. and Miller, D., Hydrophobic Fumed Silica as a Rheology Control Agent for Epoxy Adhesives, Sealants, *Adhesives Age*, Vol. 25, No. 11, p 22 (November 1982).

Crozier, D., Morse, G. and Tajima, Y., The Development of Improved Chemical Analysis Methods for Epoxy Resins, *SAMPE Journal*, Vol. 18, No. 5, p 17, (September/October 1982).

Davis, W., Flame Retardants for Thermosets, Part II: Epoxies, *Plastics Compounding*, Vol. 2, No. 4, p 53 (July/August 1979).

Denoms, S.D., Coloring the Tough Ones: Thermosets, *Plastics Compounding*, Vol. 4, No. 3, p 45 (May/June 1981).

DiBenedetto, M., Using Solvents and Reactive Diluents in Epoxy Systems, *Modern Paint and Coatings*, Vol. 70, No. 7, p 39 (July 1980). Aromatic Amine Adducts for High-Performance Coatings, *Modern Paint and Coatings*, Vol. 71, No. 7, p 36 (July 1981).

DiStasio, J.L., Ed., *Epoxy Resin Technology, Developments Since 1979*, Noyes Data Corp., Park Ridge, New Jersey (1982).

Dow Chemical Co., Dow Liquid Epoxy Resins, Bulletin No. 190-224-76 (1976).

Dow Corning Corp., Dow Corning 631 Semiconductor Grade Molding Compound, Bulletin (1980).

Driver, W.E., *Plastics Chemistry and Technology*, Van Nostrand Reinhold Co., New York (1979).

Fritz, E.R. and Green, J., New Flame Retardant for Epoxy Glass Laminates, *Printed Circuit Fabrication*, Vol. 6, No. 5, p 55 (May 1983).

Graham, J.A. and O'Connor, J.E., Epoxy With Low-Temperature Cure and High Temperature Properties Developed, *Adhesives Age*, Vol. 21, No. 7, p 20 (July 1978).

Heinze, R.E. and Ritter, J.R., Unique Spheres Impart Electrical Conductivity in Reinforced Plastics, Presentation to the 31st Annual Technical Conference, Reinforced Plastics/Composites Institute, The Society of the Plastics Industry, Inc., Section 8-A, p 1 (1976).

Kubiak, R.S. and Harper, R.C., The Development of Non-Urethane Materials for the RIM Process, Presentation to the 35th Annual Technical Conference, Reinforced Plastics/Composites Institute, The Society of the Plastics Industry, Inc., Section 22-C, p 1 (1980).

Lee, H., Advances in Biomedical Adhesives and Sealants, *SAMPE Journal*, Vol. 20, No. 4, p 13 (July/August 1984).

Lee, H. and Neville, K., *Handbook of Epoxy Resins*, McGraw-Hill Book Co., New York (1967).

Lee, S.M., Encapsulation, State-of-the-Art (Part I), *SAMPE Journal*, Vol. 14, No. 6, p 5 (November/December 1978).

Melber, G.E., Gibbons, K.M. and Anderson, T.F., Organic Microspheres for Supertough Syntactic Foams, *Plastics Compounding*, Vol. 7, No. 2, p 19 (March/April 1984).

Misra, S.C., Manson, J.A. and Van Der Hoff, J.W., Coatings From Epoxy Latexes, *Modern Paint and Coatings*, Vol. 68, No. 12, p 27 (December 1978).

Naitove, M.H. and Colangelo, M., At RP Meeting: An Upbeat Mood, Modest Advances in Technology, *Plastics Technology*, Vol. 29, No. 3, p 48 (March 1983).

Nielsen, P.O., Properties of Epoxy Resins, Hardeners, and Modifiers, *Adhesives Age*, Vol. 25, No. 4, p 42 (April 1982).

Plastics Engineering Staff, Plastics Gain in Stature as Use in Construction Reaches 7 Billion Pounds, *Plastics Engineering*, Vol. XXXVI, No. 7, p 17 (July 1980).

Schwartz, S.S. and Goodman, S.H., *Plastics Materials and Processes*, Van Nostrand Reinhold Co., New York (1982).
Scola, D.A. and Pater, R.H., The Properties of Novel Bisimide Amine Cured Epoxy/Celion 6000 Graphite Fiber Composites, *SAMPE Journal*, Vol. 18, No. 1, p 16 (January/February 1982).
Shell Chemical Co., EPON Resins for Fiberglass Reinforced Plastics, Bulletin No. SC:72-25 (January 1972). EPON Resins for Electrical & Electronic Embedment, Technical Bulletin SC:226-78 (December 1978).
Skiest, I., *Epoxy Resins*, Reinhold Publishing Co., New York (1958).
Sperling, L.H., Interpenetrating Polymer Networks: A New Class of Materials, *Materials Engineering*, Vol. 92, No. 3, p 67 (September 1980).
Thrall, E.W. Jr., PABST Program Test Results, *Adhesives Age*, Vol. 22, No. 10, p 22 (October 1979).
Union Carbide Corp., Bakelite Liquid Epoxy Resins and Hardeners, Bulletin F42461 (October 1969). Cycloaliphatic Epoxide Systems, Bulletin F-42953A (June 1976).
Van Dover, L.K., Berg, C.J. and Foshay, R.W., UV Curable Epoxy Resins for Printed Circuit Board Coatings, Presentations to the Electrical/Electronics Insulation Conference, Boston (November 1975), and NEPCON WEST, Anaheim, 3M Co. Electronic Products Div., Bulletin E-LUVER (36.3) BPH (1976).
Vazirani, H.N., Flexible Epoxy Resins, *Adhesives Age*, Vol. 23, No. 10, p 31, (October 1980).
Villani, T., Epoxy, in *Modern Plastics Encyclopedia*, Vol. 58, No. 10A, McGraw-Hill Inc., New York (1981).
Waddill, H.G., Reaction Injection Molding (RIM) with Epoxy Resin Systems, Presentation to the 35th Annual Technical Conference, Reinforced Plastics/Composites Institute, The Society of the Plastics Industry, Inc., Section 22-B, p 1 (1980).
Wason, S.K., Synthetic Sodium Aluminum Silicates, *Plastics Compounding*, Vol. 6, No. 5, p 29 (September/October 1983).
Wehrenberg, R.H., Epoxies: Versatile, New Materials for RIM, *Materials Engineering*, Vol. 91, No. 6, p 42 (June 1980).
Weiss, J., Epoxy Hydantoins as Matrix Resins, NASA Contractor Report 166024, Contract NAS1-16551, NASA Langley Research Center, Hampton, Virginia (March 1983).
Wilson, J.M., Cast Epoxies Replace Metals in Mechanical Designs, *Materials Engineering*, Vol. 97, No. 4, p 28 (April 1983).
Wittenwyler, C.V., Achieving Performance Balance in Fire-Retardant Epoxy Systems, *Modern Plastics*, Vol. 55, No. 12, p 67 (December 1978).

7
Thermoset Polyurethanes

Isao Shimoyama
DFC Company
Los Angeles, California

INTRODUCTION

Thermoset polyurethane products have a multitude of applications such as: flexible foams for cushions, mattresses, and insulations; rigid foams for insulations, structural and decorative applications; solid or micro foam products used for tooling dies, bumpers, printing rollers, hydraulic seals and other oil resistant elastomer uses. There are myriads of other uses including: vibration dampers, torque transmission couplings, drive belts, transfer belts, lining of rock crusher equipment, forklift tires, roller skate wheels, caster wheels, swim fins (Figure 7-1), ski boots, caulking, sealants, coatings, and premium grade paints, to name a few. Lately there has been a significant increase in the use of polyurethanes by the automobile industry from the bumpers to the paint and to the sealants. The plywood industry has begun to use polyurethane compositions for filling in cut knot holes instead of matching cut plywood skins into cut knot holes. The use of polyurethane for shoe soles is a natural application for the scuff and wear resistant elastomer.

Much has been written on all aspects of polyurethanes and the types and varieties available. This chapter discusses the fundamentals of polyurethanes. It is written in an attempt to enlighten and to orient the curious engineers and uninformed chemists, students, and

184 *Handbook of Thermoset Plastics*

Figure 7-1: Sporting goods like these Force Fin® swim fins represent a major market for polyurethanes. (Photo courtesy of Bob Evans Designs.)

others, and to provide them with enough basic and starting information which can be used as thinking tools to cut through all the mumbo-jumbo of polyurethane information available. Readers should be able to correctly differentiate between "apples and oranges" and categorize the wet samples and ask intelligent "know how" questions such as:

What is the free isocyanate content of the prepolymer?

What is the isocyanate molecule used?

What is the base polyol used in this system?

What is the overall hydroxyl number or the isocyanate equivalent of the curative?

What kind of curative system is used? or Is the curative an aromatic or an aliphatic system?

And even, What type of catalyst system is used in this compound?

And finally, What is the isocyanate index used to produce the final product?

The above questions are more profound than the usual and normal uninitiated's question, What is the mix ratio? If enough clear and positive information is obtained, one should be able to successfully predict the properties that are to be expected from casting the wet samples. If any need arises, he can correctly crossbreed different suppliers' prepolymers and curatives for investigational purposes and even calculate the equivalent weights of the products. He can even prepare his own prepolymer or curative to be used with suppliers' products to produce a desired product with desired properties and may save himself the time required to produce the more difficult components.

If the information on a product is suspect or is not forth-coming the reader should be able to verify for himself the mix ratio by testing the products using test procedures outlined in this chapter. Also the reader can discover by running these tests, the designs and the thinking behind the product which the supplier had used to produce his submitted products.

When one becomes accustomed to thinking and using the polyurethane fundamentals, the chemistry and compounding of the polyurethanes become simple and challenging. Interestingly, polyurethanes are one product, due to their ability to be tested, that will clearly point out weaknesses and faults to the compounder and allow him to learn as he recompounds the shortcomings out of his formulations. Polyurethanes, being an elastic product, will display faults by either not setting up into the predicted elastomer or being softer or harder than predicted, etc. In comparison, epoxy compounding seems much simpler because the epoxy compounds can accommodate a wide margin of error in their mix ratios and still produce hard rigid materials. Polyurethane compounding is not that forgiving.

This chapter is written in two parts. The first part is the general description of polyurethanes: What Are Polyurethanes?; followed by special handling requirements of the polyurethane raw materials and compounded products; the description of the ingredients that make up polyurethanes; and finally, compounding of the polyurethanes. The end of the first part concludes with ideas contributing to more recent and advanced products that are being produced with newly available commercial raw materials aided by new engineering gimmicks.

The second part consists of: the development of the industrial math used by the polyurethane industry proving that the confusing and seemingly artificial means of calculations are indeed based on fundamental chemical calculation methods; the technical dissertation on key diisocyanates used by the polyurethane industry; the test methods used to test polyurethane ingredients; procedures used to actually "cook" prepolymers using the various types of diisocyanates; the reference to ASTM test methods used to test the physical properties of produced polyurethanes; helpful hints are included where applicable.

POLYURETHANE CHEMISTRY

The basic polyurethane chemistry can simply be explained by the following six equations.[1]

1. R-N=C=O + H-O-R' → R-N(H)-C(O)-O-R'
    isocyanate alcohol   urethane

2. R-N=C=O + R'-NH$_2$ → R-N(H)-C(O)-N(H)-R'
    isocyanate amine   urea

3. R-N=C=O + H$_2$O → R-N(H)-C(O)-O-H
    isocyanate water   carbamic acid

    R-N(H)-C(O)-O-H   → R-NH$_2$ + CO$_2$
    carbamic acid   amine   carbon dioxide
                                      gas

    R-N=C=O + R-NH$_2$ → R-N(H)-C(O)-N(H)-R
    isocyanate amine   urea

    Summary of reaction
    product of isocyanate → urea + CO$_2$ (gas)
    and water

4. R-N=C=O + R-N(H)-C(O)-N(H)-R" → R-N(-C(O)-N(H)-R")
    isocyanate urea |
                                          C=O
                                          |
                                          R-N-H
                                          biuret

5. R-N=C=O + R'-N(H)-C(O)-O-R" → R'-N(-C(O)-O-R")
    isocyanate urethane |
                                         C=O
                                         |
                                         R-N-H
                                         allophanate

6. R-N=C=O + R'-C(O)-O-H → R-N-C-O-R' + CO$_2$
    isocyanate carboxylic   amide   carbon
                acid   dioxide

WHAT ARE POLYURETHANES?

The isocyanate is the basis for the whole polyurethane industry.

$$-N=C=O$$

isocyanate radical

Technically, polyurethanes are the reaction products of molecules containing two or more isocyanate groups (polyisocyanates) with molecules containing two or more hydroxyl groups (glycols, polyols, glycerine, etc.) and water and phenols, to form the chain, making a series of interconnected isocyanates and hydroxy-containing molecules. In general they include the reaction product of polyisocyanate molecules with molecules containing two or more active hydrogen compounds such as those on the nitrogen (amines), and compounds with single active hydrogen (ureas, allophanates), and sulfur (mercaptans) groups. If the stoichiometry is correct, when one mixes urethane ingredients in any amount, the product produced can be a single gigantic interconnected macro-molecule polymer of *polyurethane*.

Since the configurations of the ingredients can be varied to produce these macro-molecules, many, many different types of polyurethane chains and spatial configurations can be designed for any specific purposes, such as hardness, toughness, elongation, scuff resistance, chemical resistance, higher heat resistance, etc., with one's curiosity limited only by one's imagination. However, the more practical limit of freely designing viable molecular structures for commercial exploitation has been the commercial availability of raw materials.

Another limiting factor in producing a commercially viable product is the hazard factor. Unfortunately, polyurethane raw materials have been the target for investigations by various organizations with resulting negative effects on existing polymer formulations. Designed products have had to be altered to meet the requirements put forth by ruling bodies to eliminate potential suspected hazards posed to the public by the ingredients used to form the original products. Many formulations with ingredients that provide the "best" design product have been sacrificed and deleted and the substitutions resulted in products lower in quality. Toluene diisocyanate (TDI), the original and most common of the polyisocyanates, has been classified as a poison. Products containing TDI are being avoided by customers' industrial hygienists and these customers are requiring that any products containing TDI either be eliminated from their purchasing list or be substituted with products containing methylene diphenyl diisocyanate (MDI). Unfortunately, the initial substitution of MDI for TDI produced products which did not simulate the fast gelation of the TDI products, and required gross changes in production procedures to overcome these disadvantages. However, these disadvantages have recently been slowly eliminated with innovative formulations, engineering, equipment, and application of heat.

Another widely used polyurethane chemical which produces products with exceptionally high physical properties, 4,4'-methylenebis(o-chloroaniline), commonly referred to as MOCA, has been attacked as a potential suspect carcinogen. The polyurethane industry has spent millions of dollars and has expended many, many multilaboratory efforts to develop a product equal to or better than MOCA, but the substitutes have been unable to duplicate *all* the excellent properties provided when MOCA is used. At this writing, all the domestic producers of MOCA have discontinued their production of this chemical and the source is now overseas.

POLYURETHANE RAW MATERIALS AND MOISTURE

A basic problem associated with polyurethanes, and not found in working with most other thermoset polymers, is that *the reaction between the isocyanate, a vital ingredient in polyurethanes, and water forms carbon dioxide gas.* Water, being a very small molecule, can react with significant amounts of polyisocyanates and can precipitateureas, or if the polyisocyanate radicals are of very high molecular weights, the water reaction can cause premature gelation of the product. Today's raw materials have been carefully prepared and shipped in predried, sealed containers to maintain the integrity of the system. The polyurethane raw materials (other than the isocyanates) prior to shipping from various sources have undergone vacuuming to remove entrapped bubbles. Their water content has been lowered to a point that any reactions with isocyanates will not generate carbon dioxide gas. Since the natural moisture content equilibrium has been shifted so far to the dry side, the polyurethane raw material products are now like a dry blotter and will readily absorb moisture from the atmosphere (the products are much drier than the moisture in the ambient surroundings) in an attempt to restore the moisture content equilibrium with the environment. If unlimited amounts of air are allowed to freely come into contact with the dried materials, the moisture content can reach as high as 2% by weight. The ideal maximum moisture content for the ingredients to produce bubble-free solid elastomers is in the vicinity of 0.005% by weight. In general, upon mixing the poorly handled ingredients, carbon dioxide is generated, and the gas is trapped by the polyurethane mass as the products begin to gel and the designed product is altered since it may contain unwanted ureas and foam. Furthermore, if a considerable amount of moisture is involved, the product can be off-ratio, because the unaccounted for moisture consumed some of the reaction-required isocyanates.

The importance of protecting the polyurethane ingredients and compounded materials from exposure to the atmosphere is very difficult to instill upon novices since one cannot see the moisture being absorbed by the super dry polyurethane systems and ingredients. Only when one attempts to use the polyurethanes does one begin to see the evidence of this carelessness because the normally solid product will now show significant foaming.

What are other means that water can enter the urethane systems? The water may come from the ambient air, as stated above, but it may be present as an impurity in the other ingredients used (i.e., moisture in powders, moisture in other liquid ingredients, etc.) or it may come from naturally condensed moisture on surfaces of the mixing equipment and containers. This condensed surface moisture is a significant factor in the calculations of polyurethane chemistry. When one puts a cast iron skillet onto a hot stove one can actually see the moisture leaving the surface, starting from the center, as the skillet begins to heat up. This moisture on the surfaces of the production equipment and packaging containers *must* be eliminated before polyurethane materials are allowed to come into contact with those surfaces. Processing equipment can be heat dried and vacuumed prior to introduction of polyurethane ingredients. Containers can be stored in heated dry boxes. A quick and popular method for drying containers is to heat them to 180+°F (82°C) with hot air from a hot air blower. Many prepolymers have had their shelf life shortened due to packagers not taking the proper precautions to eliminate the moisture from the inside surfaces of the containers prior to packaging.

HANDLING OF POLYURETHANE COMPONENTS

Further precautions in handling compounded polyurethanes must be taken to assure that the designed product can be produced. It is a cardinal sin to submit to the inclination to shake containers containing polyurethane ingredients prior to opening them. All one succeeds in doing is putting back into the mass the bubbles that were carefully vacuumed out of the system by the supplier before shipping the product.

Containers with compounded polyurethanes must *not* be opened unless one intends to use the product immediately. Once the container is opened, the degradation process of the contents will begin at that time. Most suppliers will have a warning on the label stating that once the container is opened, it must be used in a specified short period or one will be confronted with the loss of the contents due to premature gelation.

High speed mixers for open can mixing must *not* be used. They will uncontrollably whip air into the prepared urethane ingredients and will entrap the air with its moisture content into the mass. A slow speed mixer or paddle should be used to carefully stir the ingredients so that bubbles are not introduced into the mix. The sides and bottom of the mixing container *and* the surface of the mixing blade must be scraped to ensure that all ingredients are mixed thoroughly. If time allows, the ingredients should be carefully mixed for at least three minutes. In more advanced use stages, it is best to very quickly vacuum the ingredients at the end of the mixing to remove any bubbles from the system before casting.

A few "do's and don'ts" for handling polyurethane producing chemicals are:

Do not shake the can.

Do not open the container unless it is to be used immediately.

Keep the cover tightly on the container and only open it long enough to quickly remove samples, and close the system again immediately. If dry nitrogen is available, blanket the space in the container with a shot of dry nitrogen each time before resealing the container.

Occasionally vacuum the container to remove bubbles.

Mix the ingredients carefully to avoid incorporating air into the mixing mass. Scrape the side and bottom surfaces to thoroughly mix the entire contents. Mix for a minimum of three minutes, if time will allow.

Vacuum the mix prior to casting.

Equipment for mixing polyurethanes for in-line manufacturing of solid elastomers are available on the market. Generally, the pumping process includes separate chambers for each ingredient for vacuuming and heating prior to mixing.

TYPES OF POLYURETHANE SYSTEMS

Compounded polyurethane elastomers are available as: (1) castable liquids, (2) thermoplastic pellets or millable gums.

Castable Liquids

Compounded polyurethane products generally start off as liquids. The available liquid polyurethane systems can be divided into the following categories:

(1) A one shot system.

(2) A two component system.

(3) A moisture cured system.

A *one shot system* is one in which all the ingredients are mixed and poured into cavities or onto surfaces. The production of flexible and rigid polyurethane foam on the production line is a good example of the one shot system.

The more popular system is the *two component system* (the type generally bought by end users). One component consists of the isocyanate (prepolymer) ingredients, and the other component contains the hydroxyls, amines, catalysts, emulsifiers, antifoams, leveling aids, and all the other ingredients that will react and modify the isocyanate materials to produce the desired polyurethane end product.

The *moisture cured system* is made up of large polymers with a small percentage of free isocyanates as end groups extended in a solvent system. The single component product is applied onto surfaces and the solvent is allowed to evaporate and the moisture in the air reacts with the small amount of free isocyanate and completes the polymerization process for film formation. Technically, one can consider the moisture cured system as a two component system—the basic isocyanate-containing product (the prepolymer) being the first component and the moisture in the air being the second component.

What are the significances of the systems?

Why is not everything made in the one shot system where the pure isocyanate is the second component and everything else is in the other component?

The reason lies in critical handling characteristics.

In the *one shot system*, the pure diisocyanate is directly reacted with the polyols to form the final product. The polyol may be viscous and the diisocyanates may be very low in viscosity, or the polyol may be a pasty solid and the isocyanate may be low in viscosity, or the polyol may be low in viscosity and the isocyanates may be in a crystalline solidlike form. The thorough mixing of the two extremely different viscosity products can be difficult. The weight ratios required may be difficult to measure. In the *one shot system* the ratios can be as far apart as 100 parts polyol to 4-8 parts by weight of diisocyanates. These diverse ratios can be handled in house, but will not be practical out in the field.

Heat is usually generated by the reaction between the diisocyanate and the polyols. The exothermic reaction temperatures developed may be such that excessive heat is generated and entrapment of this heat can result in shrinkage, scorched centers, and distortions as well as other stresses.

Furthermore, the reaction rates when the ingredients are mixed may be too fast and difficult to control, or the induction period for the reaction may be too long and be unacceptable. The pure diisocyanate can and will crystallize upon being subjected to colder temperatures.

The answer which tends to eliminate the above problems is through the *adduct* or *prepolymer* system (a basic terminology in polyurethanes). The diisocyanate portion of the component is modified by reacting a predetermined excess of the diisocyanate with a polyhydroxyl-containing product (a process called adduction or capping to form an adduct or a prepolymer).

$$2OCN-R-NCO + HO-R'-OH \longrightarrow OCN-R-\underset{H}{\underset{|}{N}}-\underset{O}{\underset{\|}{C}}-O-R'-O-\underset{O}{\underset{\|}{C}}-\underset{H}{\underset{|}{N}}-R-NCO$$

An example of a simple adduct (or prepolymer) is shown above where two mols of diisocyanate are reacted with one mol of a glycol to produce a molecule with unreacted isocyanates attached to both ends of the glycol. Essentially the molecule is still an active diisocyanate but is a longer chain diisocyanate molecule usually with higher viscosity. During this prepolymer-producing reaction, the exothermic reaction calories are generated but in the process of cooling, the additional exothermic heat that would have been generated in producing the final product is avoided. The prepolymer system thus reduces the potential heat generation that usually results in shrinkage and other stresses in the formation of the final polymer.

Again, this adduction can take many different approaches to achieve desired compounding goals. Instead of a simple adduct, one can go a step further and react three mols of diisocyanate with two mols of glycols, or four mols of a diisocyanate with three mols of glycols, etc. The assumption is that the larger the prepolymer molecule, the more quickly the end product can be produced, since much of the polymerization has been accomplished previously. Each larger molecule results in a smaller percentage by weight of free NCO per molecule along with an increase in viscosity.

As an example:

(1) Gram molecular weight of TDI is 174.2. Gram molecular weights of the isocyanates are 84. The percent free NCO is 48.22.

(2) Gram molecular weight of two mols of TDI reacted with one mol of propylene glycol = 424 grams. Result is a larger diisocyanate product. Gram molecular weight of 2 mols of free NCO (two mols of –NCO still active after the above reaction) = 84. The percent free NCO = 19.81.

(3) Gram molecular weight of three mols of TDI reacted with two mols of propylene glycol = 674. The percent free NCO = 12.46.

Following this process to a logical end might produce a gigantic molecule of a rather high molecular weight, high viscosity, and low percentage free NCO which will require very little crosslinking to produce a film. The viscosity will be a problem, but these high molecular weight, high viscosity molecules are generally carried in solvents. This is the basis for moisture cured systems.

The above listed propylene glycols can be substituted with other polyols of different configurations to achieve different goals.

Thermoplastic Pellets or Millable Gums

Using the same logic as explained for prepolymer products with terminal isocyanates, adducts with terminal hydroxyls can be produced. Generally, thermoplastic polyurethane pellets or "millable gums" are produced using this line of logic. Thermoplastic materials are usually based on MDI since the biuret linkage in the TDI prepolymer is more thermally stable than the allophanate linkage of the MDI and does not readily lend itself to the thermoplasticity.

ADVANTAGES OF ADDUCTION

(1) The adducted higher molecular weight isocyanate

component will have viscosities which are more compatible with the polyol components.
(2) The resulting diisocyanate product is more stable with less tendency to crystallize.
(3) The tendency for molecules to initially resist initiation of the reaction (induction) is overcome, and results in a faster reacting system.
(4) Less exotherm producing components can be compounded.
(5) Water impurities are reacted off during the adducting process, and since the prepolymers are vacuumed while they are still hot to remove dissolved gases, less foam-producing component results.

RANGE AND TYPES OF POLYURETHANE PRODUCTS

The very versatile polyurethane precursor products are available as liquids but they can be designed to produce myriads of finished products such as foams of a very light density to a heavy microcellular form in both flexible or rigid variety. A no foam product—a solid elastomer—of varying hardness and toughness from a very soft rubberlike product to a hard tough rocklike product can be obtained (Table 7-1). No other plastic has such latitude and versatility.

Foams:	Lower density being cellular foam and the latter high density being microcellular or no foam variety.
Weight range:	1.3 to 70+ lb/ft^3.
Types:	Very rigid to very flexible.
Hardness:	Very low; Shore A to 90-95 Shore D.

Generally polyurethanes are:
- Resistant to oils and greases.
- Good in ozone resistance.
- Compoundable to be tough and very scuff and wear resistant.
- Elastomeric or rubberlike to hard and rocklike.
- Flexible at low temperatures.
- High in impact strength.
- High in hysteresis (high conversion rate of mechanical energy to heat).
- Good in heat insulating properties, and poor in heat conductivity.

Table 7-1: A Sample List of Available Polyurethane Elastomers Indicating Various Hardnesses

CALTHANES (Polyurethanes)

PROPERTIES

USES*

PRODUCTS		HARDNESS (Shore)	COLOR	TENSILE psi	ELONGATION %	
1.	NF1300-60A	60A	clear	480	240	1
2.	NF310W	60A	clear	5,300	500	1,2
3.	NF1200	73A	clear	4,900	370	2
4.	NF1300	83A	white	2,000	300	2
5.	NF2300	88A	clear	2,620	140	2,3
6.	1900	50A	tan	550	300	1
7.	1700	70A	tan	990	180	2
8.	1400	80D	brown	4,500	80	4
9.	ND2300	65A	transparent	800	180	5,1
10.	ND1100	93A	transparent	2,300	130	5,2,3
11.	ND 3200	80D	transparent	6,300	10	5,4

*USES

1. Flexible mold materials which will duplicate the finest of details.
2. Suitable for rugged application: Metal forming, parts, rollers, vibration pads, wheels, wear pads, and diaphragms.
3. Suitable for very tough applications: Wear deck plates, snakes, impellers, drop hammer dies, and couplings.
4. Suitable where resistance to severe abrading, impact, and load: Foundry equipment, metal forming rolls, drop hammer faces, and cable guides.
5. ND systems are tough, chemical resistant polyurethane elastomers with excellent non-discoloring characteristics.

CALTHANES are two component liquid systems, which when mixed produce a variety of urethane elastomers with a wide range of properties.

Courtesy of Cal Polymers, Long Beach, California.

The properties of some polyurethanes are listed in Tables 7-2, 7-3 and 7-4.

POLYURETHANE USES (Table 7-5)

Since polyurethane is elastomeric, rubberlike, and has an unmatched resistance to scuff wear, the natural tendency is to think of it as being used for automobile tires. Experience has shown that for high speed tires, polyurethane is unsuitable, for it has a very high hysteresis factor. Mechanical energy is converted to heat at high conversion rates and being a poor heat conductor, the heat generated by high speed flexing is trapped within the polymer and causes the polymer to reach its melt temperatures. For low speed tires such as those found on fork lifts and on casters, the heat generated is insufficient to produce melt temperatures so that polyurethane elastomers are ideal products for this use. The MDI based elastomer is preferred

Table 7-2: Comparison of CYANAPRENE Elastomeric Material with Conventional Materials

Environment	CYANAPRENE Polyester Urethane	Polyether Urethane	Natural Rubber	Neoprene Rubber
Heat	G	F	F	G
Cold	G	G	E	G
Weather	E	E	P	G
Ozone Resistance	E	E	P	F
ASTM No. 1 Oil	E	F	P	G
ASTM No. 3 Oil	E	P	P	G
Aliphatic Solvents e.g. Heptane	E	F	P	G
Chlorinated Solvents e.g. Trichloroethane	F-G	P	G	P
Aromatic Solvents e.g. Toluene	F	P	P	F
Dilute Acids e.g. 5% HNO$_3$	P-F	F	G	G
Dilute Alkalies e.g. 5% NaOH	P-F	F	G	G

E — excellent
G — good
F — fair
P — poor

Courtesy of American Cyanamid Co., Wayne, New Jersey.

over the TDI based elastomer for two reasons: (1) the MDI elastomer has faster recovery or resilience; (2) the quicker recovery is also accompanied by less heat generation.

Flexible and rigid urethane foams have become common items in our society today as cushions and insulations. Stuffing of tires with flexible urethane foams is a general practice in the fork lift tires. Foam stuffed pneumatic tires are insensitive to punctures by nails and cuts (Figure 7-2).

Table 7-3: Electrical Properties of CYANAPRENE Urethane Elastomers (CYANASET M Cures)

TESTS AND CONDITIONS	CYANAPRENE A-8	CYANAPRENE A-9	CYANAPRENE D-6	CYANAPRENE D-7	CYANAPRENE D-7 plus 100 phr Aroclor 1254
Resistivity					
Volume Resistivity 23°C 50%RH Ohm-Cm Impressed Voltage—500	1.5×10^{13}	8.3×10^{10}	1.1×10^{12}	6.5×10^{12}	4.4×10^{12}
Surface Resistivity					
23°C 50%RH Ohms Impressed Voltage—500	1.8×10^{12}	4.0×10^{12}	4.7×10^{12}	1.4×10^{15}	9.5×10^{14}
Dielectric Constant					
23°C 50%RH Frequency 60 cps	9.53	8.66	7.29	5.60	5.73
10³ cps	8.51	7.82	6.49	5.12	5.25
10⁶ cps	6.87	6.0	5.23	4.53	4.18
Dissipation Factor					
23°C 50% RH Frequency 60 cps	0.230	0.089	0.071	0.055	0.046
10³ cps	0.054	0.060	0.065	0.045	0.048
10⁶ cps	0.082	0.087	0.071	0.044	0.102
Dielectric Strength					
Electrodes 1 inch apart 28°C oil temperature 500 vps rate of rise Volts/mil	310	330	250	285	230

Courtesy of American Cyanamid Co., Wayne, New Jersey.

Table 7-4: Properties of a Polyurethane Elastomer

TYPICAL PROPERTIES: UI-5780

EFFECTS OF IMMERSION IN $100^\circ C$ WATER:

	ORIGINAL	3 DAYS	6 DAYS	9 DAYS
HARDNESS, SHORE A	87	83	81	80
TENSILE STRENGTH, PSI	1280	1500	1440	1340
100% MODULUS, PSI	838	791	747	686
ULTIMATE ELONGATION, %	246	347	371	394
TEAR STRENGTH, PLI	253	231	262	226

EFFECTS OF AGING IN FORCED DRAFT OVEN @ $100^\circ C$:

	ORIGINAL	3 DAYS	6 DAYS	9 DAYS
HARDNESS, SHORE A	87	86	86	87
TENSILE STRENGTH, PSI	1440	1420	1430	1410
100% MODULUS, PSI	830	890	830	890
ULTIMATE ELONGATION, %	300	290	290	270
TEAR STRENGTH, PLI	250	230	260	280

EFFECT OF HUMIDITY AND TEMPERATURE ON ELECTRICAL PROPERTIES:

	ORIGINAL	*30 DAYS @ 100% R.H. & $40^\circ C$
DIELECTRIC STRENGTH, VOLTS/MIL	450	400
DIELECTRIC CONSTANT, 100 KH$_z$	2.6	2.9
VOLUME RESISTIVITY, OHM-CM	1×10^{16}	4×10^{15}
SURFACE RESISTIVITY, OHM-CM	5×10	4×10

*AFTER RETURNING TO AMBIENT CONDITIONS FOR 24 HOURS, ALL SPECIMENS EXHIBITED A 95% RECOVERY OF ORIGINAL VALUES.

EFFECT OF TEMPERATURE OF ELECTRICAL PROPERTIES:

VOLUME RESISTIVITY, OHM/CM	ASTM D257	SURFACE RESISTIVITY, OHMS	ASTM D257
$-80^\circ C$	3.70×10^{13}	$0^\circ C$	1.12×10^{16}
$0^\circ C$	3.40×10^{16}	$25^\circ C$	6.02×10^{15}
$25^\circ C$	3.02×10^{16}	$100^\circ C$	9.67×10^{13}
$100^\circ C$	2.34×10^{14}	$125^\circ C$	7.09×10^{13}
$125^\circ C$	3.32×10^{13}		

DIELECTRIC CONSTANT	ASTM D150
$-80^\circ C$	3.72
$0^\circ C$	3.31
$25^\circ C$	3.21
$100^\circ C$	3.09
$125^\circ C$	2.92

DISSIPATION FACTOR	
$0^\circ C$	0.0271
$25^\circ C$	0.0137
$100^\circ C$	0.0030
$125^\circ C$	0.0025

EFFECT OF FREQUENCY

DIELECTRIC CONSTANT, $25^\circ C$	ASTM D150
60 HZ	3.19
10^3 HZ	3.03
10^6 HZ	2.97

Courtesy of Urethane Industries, Placentia, California.

Table 7-5: Suggested Applications of Polyurethane Elastomers

— A —

Abrasion-resistant parts and pads
Adhesives for polycarbonate, ABS, Fiberglass/resin composites, and other plastics
Adhesives for dissimilar materials
Art objects and decorative plaque molds
Athletic track surfaces
Automobile ball joint lines and seals
Automobile bumper parts and pads

— B —

Binder for aggregates
Bonding agents for water desalination filters
Bumpers, guide blocks and pads used in launching and landing devices
Bushings for moving parts and linkages

— C —

Cast industrial parts
Casting molds
Caulking materials
Chafing pads and bumpers between metal parts where constant vibration causes wear
Check fixtures
Chemical resistant parts
Chemical resistant coating
Computer parts
Concrete molds
Concrete waterproofing
Conveyor belts and rollers
Core box liners
Couplings
Cushioning element in a composite railroad wheel
Cushioning pads used in the mounting of tractor treads

— D —

Design prototypes
Diaphragms
Die blocks used for metal forming applications in automotive, aircraft and truck body industries
Drop hammer die faces
Drum lining for use in deburring and polishing metal stampings & castings where the parts are vibrated with an abrasive

— E —

Encapsulation of electronic parts
Encapsulation of electrical components

— F —

Feed spirals
Filling hoses in castings
Flexible parts
Foundry and slinger lugs
Foundry forming and stamping pads
Foundry patterns and core boxes

Foundry squeeze boards used to pack sand under high pressure around a pattern
Fuel line parts and fittings
Furniture Molds

— G —

Gears
Gaskets
Gym floor coatings

— H —

Hammer forms
Hatch edge pieces to act as seals when hatch is closed
Helicopter blade covering sheet material to protect against abrasion
Helipad waterproofing and surfaces
Hydraulic seals and rod scrapers
Hydroform pads

— I —

Impact blankets and pads
Impellers
Industrial wheels & parts
Insulation materials

— J —

Jet engine vane assembly parts

— L —

Lining chutes, hoppers, tanks and tumbling barrels
Liners for sheaves handling wire rope cables
Low torque spring shackle
Lugs on sand slingers in foundries

— M —

Maintenance and repairs of equipment
Mechanical equipment room and roof waterproofing
Medical parts
Metal foaming pads
Military training aids
Mold material where flexibility is desired in order to remove complicated shaped parts such as some automotive crash pads
Molds for urethane foam, polyester and epoxy resins and water emulsion polyester

— N —

Non-skid surfaces
Noise abatement

— O —

Oceanographic parts

— P —

Pads for fifth wheel type mounting on railroad car trucks
Pedestrian traffic waterproofing

Polyester and epoxy resin molds
Potting compounds for electrical components
Printing rolls
Production molds
Production parts
Protecting sheathing on hydraulic tubing
Protype parts
Pulsation dampeners for hydraulic systems
Pump impellers and parts
Pump liners

— R —

Railroad track end posts
Ramp tie pieces where ramps lie against the ground
Reenterable encapsulants
Replacements for rubber parts
Rollers
Roller facings
Roof membrane waterproofing

— S —

Sealants for concrete and wood
Sheeting used to protect areas subject to severe abrasion
Shock absorbing elements
Skateboards
Skateboard wheels
Skin molds
Snakes for stretch forming
Snow plow blade edges
Snow vehicle sprockets
Sound dampening pads
Specialty coatings
Stretch dies
Swim fins and goggles

— T —

Telecommunication splice encapsulant
Tennis court surfaces
Tires and casters
Torsion bushings for suspension systems
Traction drive belts

— U —

Universal joint seals

— V —

Valve parts in air and hydraulic systems
"V" belts and cog belts
Vehicular traffic waterproofing
Vibration mountings and shock pads
Void fillers

— W —

Washers and gaskets
Water-emulsified polyester molds
Waterproofing structures
Wear deck plates
Wheels
Windshield wiper bushings

Courtesy of Urethane Industries, Placentia, California.

Figure 7-2: Fork lift with polyurethane foam filled tires. (Photo courtesy of DFC Company, Los Angeles, California.)

The harder and tougher polyurethanes have been used in tooling where they have substituted the female portion of the matched dies (Figure 7-3). Instead of a cavity, the female portion is simply a block of polyurethane with a harder polyurethane layer laminated to the top portion which comes into contact with metals. This type of material has been successfully used to form brittle metals which would break in the normal matched dies. The hard layer presses against every minute portion of the forming metal and prevents the metal from splitting, cracking, and breaking.

Polyurethanes have also been successfully used as bumpers.

The stiffer urethanes have found use as drive belts (Figure 7-4). The softer urethane elastomers are currently being used as printing rollers (Figure 7-5). A new significant use of polyurethane elastomers, today, is as a cast flexible mold for concrete slabs used in building constructions (Figures 7-6 and 7-7).

Polyurethanes have become popular for use in reaction injection molding (RIM). In the RIM process, the mixed urethane product is injected into a strong metal mold under pressure, filling the mold almost full. The product is allowed to expand in the mold and in doing so, the portion that comes in contact with the mold is essentially foam free. A smooth finished surface results because the bubbles are broken and eliminated as the materials scrape along the mold contact surface. The center may or may not contain foam depending upon the formulation. Ski boots are one of the end products of the RIM process.

200 Handbook of Thermoset Plastics

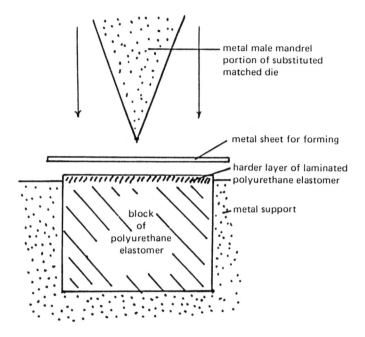

Figure 7-3: Tooling polyurethanes—metal forming.

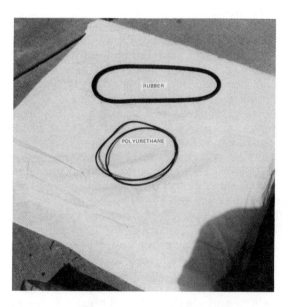

Figure 7-4: Polyurethane elastomer drive belts generally have smaller cross section than rubber drive belts. (Photo courtesy of DFC Company, Los Angeles, California.)

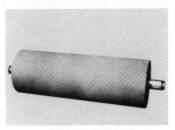

Figure 7-5: Polyurethane roll covers. (Photo courtesy of American Cyanamid Co., Wayne, New Jersey.)

Figure 7-6: Cast flexible polyurethane mold for concrete. A current large use of polyurethane elastomer. (Photo courtesy of Conap, Inc., Olean, NY.)

Figure 7-7: Concrete slab with surface features transferred from the polyurethane mold. (Photo courtesy of Conap, Inc., Olean, NY.)

Adhesives are being compounded with polyurethanes (Figure 7-8). They are used for adherence of the following surfaces: neoprene to neoprene, neoprene to concrete, plastics to glass. Glass is normally laminated with polyvinyl butyral (PVB), but in laminating polycarbonates to glass with PVB the polycarbonate develops stress cracks and urethanes do not. (The former is the basis for bulletproof glass.) Polyurethane sealants have become an integral part of automobile production since they can be compounded to meet the following production requirements even though they are slightly more costly than vinyl plastisol sealants:[2]

(1) Sag resistant application in production environment.
(2) Flow under shear conditions to adequately fill the window cavity.
(3) Adjustable gel time from 2 minutes to 4 hours for compatibility with production line timings.
(4) Good early tack and green strengths.
(5) Adhere to the substrate before cure.
(6) Resist environmental stresses and maintain sealant integrity upon aging.

NEOPRENE LUBRICANT ADHESIVE #106

DESCRIPTION

D.F.C. NEOPRENE ADHESIVE NO. 106 IS A ONE COMPONENT MOISTURE CURING POLY-URETHANE ADHESIVE DESIGNED TO LUBRICATE THE WALLS OF THE JOINT FOR EASY APPLICATION OF THE NEOPRENE SEAL. UPON CURING, NEOPRENE ADHESIVE #106 SETS UP TO A TOUGH DURABLE ADHESIVE WHICH HOLDS THE SEAL IN PLACE AND PREVENTS WATER AND DIRT FROM GETTING BETWEEN THE SEAL AND THE CONCRETE.

SURFACE PREPARATION

REMOVE ALL DEBRIS FROM SIDES OF JOINT USING MURIATIC ACID ETCH OR SANDBLASTING. IF MURIATIC ACID IS USED, FLUSH JOINT WITH HIGH PRESSURE HOSE AND PERMIT TO DRY THOROUGHLY BEFORE APPLYING NEOPRENE ADHESIVE #106.

APPLICATION

APPLY NEOPRENE ADHESIVE #106 GENEROUSLY TO BOTH SIDES OF JOINT IN ADDITION TO THE SIDES OF THE NEOPRENE SEAL.

SPECIFICATION

THIS PRODUCT MEETS OR EXCEEDS THE CALIFORNIA STATE SPECIFICATION No. 8040-61J-04 FOR LUBRICANT-ADHESIVE.

CLEANLINESS AND SAFETY

AVOID CONTACT WITH SKIN AND APPLY UNDER GOOD VENTILATION. EQUIPMENT MAY BE CLEANED WITH HYDROCARBON BASED SOLVENT BEFORE ADHESIVE HARDENS.

TYPICAL PHYSICAL PROPERTIES

WEIGHT PER GALLON	9.2	VISCOSITY	16,000-450,000 CPS
SOLIDS CONTENTS BY WT.	65% MIN.	ELONGATION	250%
LUBRICATING LIFE @ 77°F. (MINIMUM)	1 HOUR	FILM STRENGTH	4,000 PSI
		SHEAR RATIO, MIN	2.5
		COLOR	GREY

TECHNICAL DATA

IF ADDITIONAL TECHNICAL INFORMATION OR MARKETING INFORMATION IS DESIRED, PLEASE CALL OR WRITE:

D.F.C. COMPANY
13340 So. CENTRAL AVENUE
LOS ANGELES, CALIFORNIA 90059

WARRENTY

MANUFACTURER'S ONLY OBLIGATION SHALL BE TO REPLACE SUCH QUANTITY OF THE PRODUCT PROVED TO BE DEFECTIVE. USER SHALL DETERMINE THE SUITABILITY OF THE PRODUCT FOR HIS INTENDED USE AND ASSUME ALL RISK AND LIABILITY IN CONNECTION THEREWITH.

* * * * * *

13340 SOUTH CENTRAL AVENUE   LOS ANGELES, CALIFORNIA 90059

(213) CALL DFC
(213) 636-1214

FROM NORTHERN CALIFORNIA (800) 262-4602
FROM OUTSIDE CALIFORNIA (800) 421-5313

Figure 7-8: A sample data sheet of a polyurethane adhesive. (Courtesy of DFC Company, Los Angeles, California.)

(7) Compatible with paint applied to its surface without plasticizer bleeding to degrade the paint.

(8) The sealants contribute to the structural stiffness of the automobile since they actually adhere the windshield to the body as an integral unit.

Because of their lower cost and their natural elastomeric properties in comparison to other 100% solids polymers, the two component polyurethanes have begun to be used as general sealants. When compounded with coal tar, the cost is even lower, and they are being used in jet-fuel-resistant sealants in airport runway cracks and joints.

Polyurethane wood binders and patches are used for filling cutout areas of the surfaces of plywood to remove knots and imperfections. The polyurethane has:[3]

(1) Flexibility in formulations.

(2) Excellent structural properties.

(3) High adhesive and cohesive strengths.

(4) Versatility of cure temperatures.

(5) The ability to bond to surfaces with high moisture content.

(6) No gas emissions.

POLYURETHANE COATINGS

Basically there are two types of polyurethane coatings that are available but each has a specific area of use: (1) Single component moisture cured polyurethane coatings, for thin coating applications similar to regular paint and lacquer applications, and (2) Reacting two component polyurethane coatings which are generally thicker elastomer membrane types for roof and parking decks or for sealant applications.

Single Component Moisture Cured Polyurethane Coatings

Single component liquid moisture cured polyurethane coating systems have become very popular. The coatings can be clear or opaque. The moisture cured variety requires that when a thick coating is to be applied, the coatings must be applied in multiple thin layers with each subsequent coat being applied within hours of putting the previous coat down. Thus the delay is only for evaporation of solvents and allowing moisture to contact the coating to begin the polymerization process, but before the material has become fully polymerized into a hard film. Thick moisture cured coats cannot be applied at one time, because they can turn into a foam mass. The heavy layers will entrap and be unable to dissipate or eliminate the generated carbon dioxide gas from the surface of the film. The original TDI based single component moisture cured coatings showed excellent scuff and wear resistance, chemical resistance, and durability in exposure to outdoor environment. The clear and colored opaque TDI based moisture cured varieties were widely used in seamless floor applications. However, coatings made from TDI have a tendency to yellow with age or upon exposure to sunlight. To overcome this yellowing, industry has developed a nonyellowing coating based on hydrogenated isocyanates, mainly the aliphatic version of the MDI (Figure 7-9) and/or isophorone diisocyanate (IPDI). However, in overcoming the yellowing of the original TDI based coating, a degree of scuff resistance, and low cost were sacrificed. Since the nonyellowing moisture cured urethane coatings will show superior performance in the area of scuff resistance and outdoor durability in comparison to other coatings made from other polymers, it is still the polymer of choice.

Figure 7-9: Samples of moisture cured polyurethane coatings. Left, TDI based product; right, nonyellowing saturated diisocyanate product. (Photo courtesy of DFC Company, Los Angeles, California.)

Moisture cured coatings are so hard, tough and solvent resistant that other coating materials and solvents can not penetrate into nor adhere to them. These are desirable properties for use in seamless floor applications if properly applied. To the dismay of many manufacturers, however, this seemingly-simple-to-apply coating previously sold over the counter to the general public, has had a very high

206 Handbook of Thermoset Plastics

failure rate. Failures indicated that this type of application is not for novices and requires the services of fully knowledgeable professionals. Many suppliers have left the field and those remaining in the seamless floor field will insist on application through known experienced licensed applicators only (Figure 7-10).

Figure 7-10: A sample of polyurethane seamless flooring. (Photo courtesy of Advanced Coatings & Chemicals, Temple City, California.)

Polyurethanes, in general, whether they be of the moisture cured variety or the two component variety, have difficulty in adhering to various substrates and require cleaning to a point where the substrates must be almost surgically clean and completely dry. When adhesion failures occur, the film is so strong that, unlike other coatings that crumble into bits and pieces, the polyurethane films will delaminate from the surface in strips and sheets.

The use of adhesion promoters applied to the substrate after proper surface preparation and prior to the application of the polyurethane coatings have significantly reduced coating failures.

Being so resistant to penetration by solvents and other coatings, the removal of the moisture cured coating was thought to be difficult. This problem lately has been overcome by using methylene chloride strippers which contain phenols.

Polyurethane coatings presently are used as premium grade lacquers in airplane, automotive, and other high grade coating applications (Figures 7-11, 7-12, 7-13).

ADVANCED
COATINGS AND CHEMICALS
4343 TEMPLE CITY BLVD.
TEMPLE CITY, CALIFORNIA 91780
TELEPHONE (213) 579-6270

TECHNICAL DATA SHEET

NO. 121

LIGHT STABLE URAFILM
THE WEATHERABLE COATING
MARINE • AIRCRAFT • TRANSPORTATION

DESCRIPTION
URAFILM, a light stable, two component, aliphatic polyurethane has been developed to eliminate costly maintenance problems especially in the marine, aircraft and transportation field, but can also find use for home and industry. URAFILM is the ultimate in air dry coatings in weatherability, abrasion resistance, chemical resistance, corrosion resistance, flexibility (retention of flexibility), fresh and salt water resistance and fuel resistance.
URAFILM exceeds MIL-C-81773, MIL-C-83019, and MIL-C-83286.

CHARACTERISTICS
It's tough, (at least 3 times tougher than epoxy in abrasion resistance).
It's color fast, (dry film of 3 mils has been exposed in Florida for 24 to 36 months with nominal change in color and gloss).
It cleans easily.
It can be washed, waxed or repainted.
It is chip and mar resistant.
It resists steam cleaning and chemicals.
Dust and dirt do not cling to surface.
Brush and spray versions available.
URAFILM can be custom formulated for a heat cure finish or to meet specific resistance or performance properties.

RESISTANCE & PHYSICAL PROPERTIES
Tensile p.s.i.: 4000 + lbs.
Elongation (free film): 60% + @ 5 mils
Impact: 100 in. lbs.
Distilled Water Immersion: Outstanding
Fuel & Lubricants (Including Skydrol): Outstanding
Weather Resistance: Superb
Salt Spray Resistance (1500 hours): Outstanding
Humidity Resistance (1500 hours): Outstanding

_{Advanced Coatings and Chemicals Corporation warrants that its goods are free from defect in materials and workmanship. Liability of Advanced Coatings and Chemicals Corporation under all warranties, expressed or implied, shall be limited to replacement of defective goods.}

Figure 7-11: A sample data sheet on premium high performance polyurethane coating. (Courtesy of Advanced Coatings & Chemicals, Temple City, California.)

Two Component Polyurethane Coatings

When 100% solids, two component polyurethanes are used as coatings, the fine line between the cast elastomer and two component polyurethane coatings becomes blurred. Generally the 100% solids, two component polyurethane coatings are opaque and elastomeric. In addition they must be compounded to paint standards and must possess the ability to minimize any bubble formation in the thick layers, "wet" and "adhere" to substrates, level out low spots (leveling), and resist degradation from ultraviolet exposure. Interestingly, a large use for the process has been in the application of athletic running tracks in competition with products such as synthetic "turfs" (Figure 7-14).

208 Handbook of Thermoset Plastics

Figure 7-12: An example of applied polyurethane coatings. (Photo courtesy of Advanced Coatings & Chemicals, Temple City, California.)

Figure 7-13: Another example of applied polyurethane coating. (Photo courtesy of Advanced Coating & Chemicals, Temple City, California.)

Figure 7-14: Polyurethane elastomer athletic running track. (Photo courtesy of Sports Technology Group, Inc., Placentia, California.)

Two component polyurethanes have been designed and developed to coat and resurface concrete bridges, but upon their use for this purpose, the authorities discovered that the polyurethane deck coatings applied to old concrete bridges did not protect and preserve the bridge as intended. The coatings actually accelerated the degradation process by trapping moisture and other bridge-destroying water-soluble chemicals that have penetrated into the concrete. Normally moisture would evaporate from the top surface of the bridge concrete. Once the coating has been applied, the moisture and chemicals must pass through the bridge concrete and out the bottom. These chemicals in passing through the bridge concrete helped destroy the reinforcing steel and caused premature bridge failures. After this discovery, bridge authorities discontinued recoating old bridges with urethanes.

Currently, the surfaces of bridges are repaired so that moisture can evaporate from the top surface of the concrete.

COMPONENTS FOR POLYURETHANES

The beginning of commercial polyurethanes in the United States was in their use during World War II by Lockheed to fill the cavities in airplane propellers with rigid polyester urethane foams. As with most beginnings, polyurethane materials available in the United

States at that time were crude in terms of today's standards, and today's catalysts were then unknown. The materials available were only suitable for foam production and very undesirable for producing solid elastomers. Early polyurethane products were generally polyester urethanes, mainly based on dibasic acids and polyhydroxyl compounds. The most popular choice of dibasic acids was the straight chain six carbon molecule, adipic acid. Adipic acid was plentiful due to the requirements of the ongoing nylon industry. Polyesters are generally prepared in a reactor at temperatures of 475°F to 500°F for 8 to 24 hours. This preprocessing added cost to polyester polyurethanes on top of the normal cost of producing the final polyurethane product.

Listed below are currently used "common" polyurethane ingredients. For orientation purposes, only the more rudimentary and commonly used ingredients are listed. See the appendix (starting on page 242) for structures, data and other pertinent information.

Diisocyanates

(1) Toluene diisocyanate (TDI) and its hydrogenated (saturated) counterpart.

(2) Methylene diphenyl diisocyanate (MDI) and its hydrogenated (saturated) counterpart.

(3) Isophorone diisocyanate (IPDI).

(4) Polymeric isocyanates.

TDI: The most prevalent TDI molecules consist of 2,4- and 2,6- isomers at 80% and 20% respectively. Other ratioed TDIs (35% to 65%) are also available.

2,4-TDI   2,6-TDI

The ortho $-NCO$ group at room temperature is eight times slower in activity than the $-NCO$ group at the para position. [At elevated temperatures of 212°F (100°C) and above, the rates of the two $-NCO$ groups become approximately equal but at these high temperatures, the TDI molecule may crosslink the polyol prematurely.] After one of the NCOs has reacted, the second one has only one-third of its initial activity. Compounders can apply this room temperature difference in the speed of reactivity to their advantage. If a longer pot life is desired, the reactivity can be retarded to a degree by carefully adjusting the compound to where only the 2,6- isocyanates remain unreacted. (Pure TDI that consists only of the 2,6-isomers or even the 2,4- isomers is available at higher cost.)

When TDI is used to produce an adduct, a nontoxic solution is defined as that which contains less than 1.0% free toluene diisocyanate monomers.

Isocyanates reacted with water form a product (carbamic acid) which decomposes immediately to carbon dioxide and amine. (See equations on page 186.)

Isocyanates reacted with carboxylic acid decomposes at slightly elevated temperatures to an amide and carbon dioxide. (See equations on page 186.)

MDI: The MDI molecule and its reaction rate do not allow compounders the latitude of adjusting reaction rates depending upon the position of the isocyanates.

$$OCN-\!\!\!\!-CH_2-\!\!\!\!-NCO$$

4,4'-Methylene Diphenyl Diisocyanate (MDI)

Currently liquid polyurethane products compounded with MDI are the desirable prepolymers by many users and industrial hygienists, since MDI has not been classified as being seriously dangerous and TDI has been classified as a poison.

Saturated Diisocyanates: The hydrogenated or saturated varieties of diisocyanates (aliphatic diisocyanates) are very slow reacting products and generally require the aid of a catalyst for them to react. They impart nonyellowing characteristics to products made from them. Products made with aliphatic diisocyanates can be used at higher temperatures than products made similarly with the unsaturated variety.

Polymeric Isocyanates: Generally polymeric isocyanates are used in rigid polyurethane foam formulations.

Isocyanate Index (NCO/OH): Experience has shown that in order to make a reproducibly designed polyurethane product, the compounder should use a slight excess of the isocyanate component so that sufficient amounts of reacting isocyanates are present and available. The remaining excess isocyanate will be conveniently reacted fully by moisture, or reaction products of ureas, or urethanes, and form biurets or allophanates.

The Isocyanate Index is the equivalents ratio of the reacting isocyanate compound to the polyol. An Isocyanate Index of 1.0 means isocyanate and hydroxyl reactants at stoichiometry. An Isocyanate Index of 1.05 means half a percent excess of equivalents of the isocyanate component. The popular NCO/OH index is 1.05.

Experiments should be run to determine the best Isocyanate Index to produce the most satisfactory end product. Tests are run at Isocyanate Indexes of 0.95, 1.00, 1.05, and 1.10. One should use the spread stated since any index values closer together will run the risk of measuring the same point due to experimental errors.

Catalysts

Tin catalysts	Stannous octoates, dibutyl tin dilaurates.
Morpholines	N-methyl and N-ethyl morpholines.
Tertiary amines	Triethylenediamine (DABCO), QUADROL, 2,3,6,-tri(dimethylaminomethyl)phenol (DMP 30).
Metal catalysts	Lead octoates, lead naphthanates, mercury octoates and oleates, phenyl mercuric oleates.
Proprietary catalysts	Supplied by various companies (listed on page 263).

Catalysts play a very important role in urethane reactions and the evolution of the polyurethane industry can be correlated with the development of the catalysts. Early catalysts were tin and the morpholines. Mercury is a more recent catalyst. TDI reacts with amines, alcohols, water, mercaptans, phenols, carboxylic acids, amines, ureas and urethanes. These reactions occur at different rates and these rates can be influenced by tertiary amines and metallic catalysts.

Catalysts for urethane reactions can be specific for the type of reaction desired. The isocyanate-hydroxyl reaction will proceed fairly readily at room temperature but is slow enough to allow a few hours for application. Tin, lead, morpholines, triethylenediamine types catalyze this reaction and the isocyanate and water reactions.

The mercury catalysts preferentially catalyze the hydroxyl-isocyanate reactions and do not catalyze the water-isocyanate reaction. Change in temperature has a more pronounced effect upon the reaction of H_2O-TDI than polyol-TDI and thus influences the rate of carbon dioxide evolution much more than polymerization.

The most satisfactory bases for catalysts are the tertiary amines: i.e., DABCO, N-methylmorpholine, diethylethanolamine, etc. Catalysts are used at levels of 1.0 part or less per 100 parts of the nonvolatile portion of the polyols.

Special Situation Catalysts: The following are examples of unusual products becoming catalysts:

(1) 4,4'-methylenebis(o-chloroaniline) (MOCA) in special situations can be catalyzed by adipic acid.

(2) Catalyst for castor oil MDI is ethyldiethanolamine or N,N,N',N'-tetrakis(2-hydroxypropyl)ethylenediamine.

Aromatic Amines

Aromatic amines are popular curing agents for polyisocyanate products. As a group, they impart maximum physical properties such as hardness, toughness, tear resistance, and impart higher heat deflection temperatures to any prepolymers. Generally the aromatic amines are solid or crystalline at room temperature and require heating and melting prior to use. Some of the more popular aromatic amines are:

(1) Methylenedianiline (MDA).
(2) MOCA, 4,4'-methylenebis(o-chloroaniline) or Bis-Amine A (Japanese MOCA).
(3) Cyanacure, 1,2-bis(2'-aminophenylthio)ethane (American Cyanamid Co.).
(4) Poly-Cure 1,000, methylenebis(methyl anthranilate) (PTM & W).
(5) Polarcure 740M, trimethyleneglycol di-p-aminobenzoate (Polaroid Corp.).

The fully active diamine (two nitrogens with four active hydrogens) is considered to have a functionality of two in normal urethane reactions.

From the chemical standpoint, all aromatic amines are suspected of being hazardous and can be expected to be on a future suspected toxicity list.

Structurally speaking, methylenedianiline (MDA) would be the ideal product to impart higher aromaticity to the polyurethane polymer and to raise the toughness and the practical service temperature to 250°F (121°C). However, MDA as a practical curative for isocyanate prepolymers has not been popular since the reaction of isocyanates with the amines on the MDA aromatic ring is uncontrollably fast and does not allow for mixing or handling time. In an attempt to slow down this rapid reaction, large chlorine molecules were introduced onto the MDA's aromatic rings adjacent to the amine radicals. The chlorine sterically hinders the amine radical because of the large size of the atom and causes the dislocation of the electron activity centers.

$$H_2N-\underset{Cl}{\underset{|}{C_6H_3}}-CH_2-\underset{Cl}{\underset{|}{C_6H_3}}-NH_2$$

MOCA®

The result is a brief but sufficient slowing of the amine reaction to allow the MDA molecule to become a useful aromatic curative. The product is called MOCA (4,4'-Methylene bis Ortho Chloro Aniline). MOCA has produced urethane elastomers with remarkable properties. Unfortunately, the chlorine in the MOCA has caused the product to be classified as being a suspect carcinogen. With the idea that the elimination or substitution of the chlorine will remove this aromatic amine from the toxic chemical list, other producers have substituted different chemical radicals or removed the chlorine and modified the configurations of the aromatic amine to produce aromatic products to duplicate the properties of MOCA. Other radicals were introduced next to the amines on the MDA radical or chemical radicals were inserted in between the two aromatic rings as in Polarcure 740. As far as this writer is concerned, the products made with these alternate structures still do not match products made with MOCA.

However, despite the disadvantage of the rapid reaction of MDA, chemists have used this disadvantage as an advantage. A calculated *small* amount of MDA is incorporated into the liquid urethane curative component. The components of castable polyurethane components containing MDA can be very fluid liquids but when the two liquid components are mixed, the MDA reacts instantly and causes the mixed liquids to become a useful workable thixotropic paste with enough body to be used on vertical surfaces without sagging. As for the toxicity of MDA, animal studies have produced only inconclusive results regarding possible carcinogenicity.[4]

Normally the aromatic amine MOCA is solid at ambient temperatures and requires heating to 200°F to 205°F (93°C to 96°C) to be melted into a usable liquid. The disadvantage of the hot aromatic amine is that it also requires that the isocyanate portion (the prepolymer) be heated to a similar temperature so that when the melted aromatic amine is added, it will remain liquid and not turn into crystals upon introduction into a cold isocyanate system. Unfortunately, the high temperature causes the reaction to take place rapidly to a point where working time becomes very limited.

A patent[5] points out a means to produce a useful room temperature liquid aromatic amine with N-methylpyrrolidone. Although elastomers with remarkable properties can be produced using these amines, the final diluted product is produced at a sacrifice of about 50% of the potential physical properties that could be produced if used hot and directly without dilution. Some of the properties sacrificed are: tensile strength, compression set, tear resistance, and lower heat deflection temperature.

Polyether polyol prepolymers cured with aromatic amines generally can be used at service temperatures of around 250°F (121°C), whereas, the same prepolymer cured with polyether polyols has a maximum service temperature close to 180°F (82°C).

If less than the theoretical amount of aromatic amine curing agent is used, the excess isocyanate will react with a urea group hydrogen to form a biuret. Biuret crosslinking increases modulus and resistance to compression set and adversely affects resistance to abrasion and tearing. As the curing agent approaches stoichiometric proportions, the structure is more nearly linear with an attendant high level of hydrogen bonding. These reactions affect physical properties as listed below[6] and shown in Table 7-6.

Highest tensile strength and lowest compression set is achieved with 90% to 95% of the theoretical amount of amine curing agent. Increasing the curing agent level decreases tensile strength and increases elongation. Compression set increases significantly at higher concentrations.

Good abrasion resistance is obtained over the range of 90% to 105% of theoretical. However, increasing concentration to 110% results in a significant decrease in this important property.

Optimum flex resistance is obtained at 100% to 105%.

Tear strength is directly proportional to curing agent concentration.

Table 7-6: Effect of Curative Concentration on Elastomer Properties

Curative: 4,4'-Methylenebis(o-chloroaniline)
Prepolymer: QO POLYMEG 1000—80/20 TDI
TDI/QO POLYMEG Ratio: 2:1

Curative, % of Theory	90	95	100	105	110
Hardness (Shore A)	95	96	95	93	95
Tensile, psi	3460	5300	4325	4140	2700
Modulus 100%, psi	1640	1750	1630	1570	1520
300%, psi	2870	3000	2570	2430	1900
Elongation, %	330	420	460	525	525
Tear Resistance, Split, pli	130	225	250	450	450
Die C, pli	365	440	425	525	525
Resilience, Bashore, %	48	47	45	46	44
Ross Flex (cycles to failure)	—	5900	20000	25000	—
Low Temperature, T_f (°C.)	−57	−55	−55	−54	−54

The reactivity of prepolymers varies with the percent NCO, amount of free TDI and TDI type (80/20 vs. 100%). Elastomers made from 80/20 TDI prepolymers and 4,4'-methylenebis(o-chloroaniline) have shorter pot lives than their counterpart TDI 100% prepolymers and, therefore, mixing at lower temperatures is advised.

Courtesy of QO Chemicals, Inc., Oak Brook, Illinois.

Some aromatic curing agents are:

XA, Bis beta-hydroxyethyl ether of hydroquinone (Eastman Chemical).

HER, Isomer of XA (Eastman Chemical).

HQEE, Hydroquinone di-(β-hydroxyethyl) (Eastman Chemical).

Conditioners

Driers or Moisture Absorbers: Examples of these are: Molecular sieves (multisieve sizes available), Drierite® (calcium sulfate, multisieve sizes available) and CaO, MgO.

Elimination of the water-isocyanate reaction which produces carbon dioxide gas can be accomplished by the addition of the above driers to the components. The above driers have a unique property in that once the product absorbs the moisture, the absorbed water is no longer available for reaction with the isocyanates. Drierite® uses the water as a molecule of crystallization-hydration. Molecular sieves absorb the moisture preferentially into their structure. However, the amount of moisture absorbed is limited to about 10% of their original weight. Temperatures close to 600°F (315°C) are required to eliminate the entrapped moisture. Many compounders use the molecular sieve or Drierite® system where heating to eliminate the moisture cannot be accomplished without destroying the system, or where facilities are not available to accomplish the heat drying process. Normally, the time consuming dehydration process to remove water from the wet product is accomplished at temperatures of 280°F (237°C) and 30 inches of vacuum for 8 to 16 hours.

Testing for the amount of moisture eliminated by the driers, in the absence of equipment, is done empirically by mixing the polyol or curative compound with the prepolymer or isocyanate-containing compound for foam development. Periodic tests are run after each addition of the drier compound until the desired level of nonfoaming is obtained.

The more sophisticated and recommended approach to test for water content is by the Karl Fisher method for determining the amount of moisture in the products. The desirable and practical target for the permissible amount of moisture in components to produce solid elastomers is 0.01% or less by weight, and preferably, 0.005% or less for a no foam solid elastomer variety.

Molecular sieves have another very interesting property, with many uses and ramifications. They can be preloaded with other liquid chemicals. The loaded molecular sieves upon meeting moisture will preferentially absorb the moisture and release the preloaded chemical.

An expensive but effective means of removing moisture from components is to pass these liquids through a column (of sufficient length) filled with 8 mesh molecular sieves. Products are simply poured into the column at the top and the materials emerging at the bottom should have sufficiently low water content to meet the requirements of no foam elastomers.

Surfactants: Silicones are the surfactants that are normally used in the one shot flexible and rigid foam formulations to facilitate the emulsification of water, polyols, amines, catalysts, fluorocarbons, and TDI. As a rule, they are not used in the production of elastomers.

Silicones are used because organic emulsifiers have been unable to sufficiently and efficiently lower the surface tension of the polyurethane foam ingredient mixes during production or during the foaming process.

Silicone surfactant level of use is approximately 1 part by weight (pbw)/100 of polyol.

Adhesion Promoters: The adhesion promoters used to facilitate adhesion of urethane polymers to inorganic substrates are the amino functional silanes. If the substrate surface is not almost surgically clean, using the adhesion promoter will be useless.

Mold Release Agents: The high molecular weight silicone waxes perform well as mold releases for polyurethanes, especially those that are put up in aerosol containers. Silicone greases are also effective release agents, but leave greasy undesirable residues both on the mold surfaces and on the molded articles.

Fillers: Generally, fillers are added to lower the cost per unit volume of the final product. The fillers naturally contain a high percentage of moisture. This residual water must be removed by baking at high temperatures prior to addition to the polyurethane systems or must be cooked out after addition to the polyols. Powdered fillers are incorporated into the urethane systems but present a difficulty in maintaining their suspension in isocyanate systems since the current technological method of suspending particles in the liquid medium requires the particles to be coated with active chemical compounds to prevent settling as a dry caking on the bottom of the container during prolonged storage. The suspending chemical is neutralized by the presence of the isocyanates and particles still tend to settle on the bottom of the container upon storage. The following materials are used as fillers:

(A) Barytes (barium sulfate)
(B) Calcium Carbonates
(C) Talcs
(D) Clays

Barytes have been added to urethane foams to increase foam densities so that they will "feel" similar to the weights of foam rubber or cotton batting at rates below 160 parts by weight (pbw) of the polyol. Barytes are also tremendous heat sinks.[13]

Loading of barium sulfate into RIM part in the range of 40% has been used as sound deadening car floor boards.[9]

The addition of 0.5 to 30 wt % of CaO to one-shot polyurethane prepolymers forms products with little blistering and better dimensional control in caulking and sealing applications. Mixtures of CaO and MgO can be used but particle sizes of 350 mesh or finer are preferred.

Extenders and Diluents: Examples of these are:

(A) Ester of triethylene glycol, dicaprate and dicaprylate.
(B) Coal tar.
(C) Dipropylene Glycol Dibenzoate, (Benzoflex 9-88 SG).
(D) Di-(2-methoxyethyl phthalate).

The addition of plasticizers slows reaction and gelation time, decreases the viscosities of the incorporated products and improves the mixture handling characteristics. Usually, it lowers the cost of the compound.

The additions of plasticizers affect the elastomers as follows: (1) Reduces hardness, tensile and tear strength, and modulus values. (2) Increases the compression set and elongation, flexibility of the elastomer at lower temperatures.

Prepolymers made from high molecular weight polyether polyols with the addition of coal tar is presently used widely in sealants. Adding coal tar to urethanes facilitates adhesion to many surfaces. Even here the use of silane adhesion promoters, or polyamide-epoxy primers will improve adhesion.

Amines: Examples of these are:

(A) TMBDA (tetramethylbutanediamine).
(B) Dimethylethanolamine.
(C) Triethanolamine.

The common aliphatic primary amines by themselves generally do not play an active part in producing solid casting elastomeric polyurethanes. The amine and isocyanate reaction is so rapid at room temperature that the complete reaction will take place within seconds. The rapidity of the reaction prevents its use in production casting methods.

The aliphatic primary amines are used in one step foam production.

Innovative amine formulations have been introduced in RIM processes where the fast reaction and set time is an asset.

Secondary amines require heat or catalysts to react with isocyanates.

The tertiary amines are useful as catalysts. (See the section on Catalysts.)

Polyols

Polyhydroxy compounds are the large family of chemicals that convert the isocyanates into polyurethanes. At this point, the discussion on these compounds will be limited to the simple and the most common polyols, polyether backboned polypropylene polyols and polyester polyols.

Simple Glycols:

(A) 1,4-butanediol.

(B) 1,6-hexanediol.

These diols may be used alone as chain extenders for MDI based prepolymers, but are considered too weak to be used alone as chain extenders in TDI based prepolymers. Normally these diols are used in the extender components with other diols, triols, and ingredients.

Simple Triols:

(A) Glycerine.

(B) 1,2,6-hexanetriol.

(C) Trimethylolpropane.

Other Polyols:

Pentaerythritol.

Polyols with Polyether (Polypropylene) Backbones or Polyoxypropylene Derivatives of Trimethylolpropane, etc.: These are the most common types of polyols used by the polyurethane industry and many different types of spatial configurations are available. Only the very basic linear ones will be discussed in this section. The more complicated varieties will be discussed in a later section.

Polyether polypropylene polyol (diol) molecules have the following configuration:

$$H-\underset{\underset{H}{|}}{\overset{\overset{H}{|}}{C}}-\underset{\underset{H}{|}}{\overset{\overset{OH}{|}}{C}}-\underset{\underset{H}{|}}{\overset{\overset{H}{|}}{C}}-O-[\underset{\underset{H}{|}}{\overset{\overset{H}{|}}{C}}-\underset{\underset{H}{|}}{\overset{\overset{H}{|}}{C}}-\underset{}{\overset{\overset{H}{|}}{C}}-O]_x-\underset{\underset{H}{|}}{\overset{\overset{H}{|}}{C}}-\underset{\underset{H}{|}}{\overset{\overset{OH}{|}}{C}}-\underset{\underset{H}{|}}{\overset{\overset{H}{|}}{C}}-H$$

$x = 6 - 20$

Polypropylene polyether backbone polyols are made by the reaction of propylene oxide with initiator compounds to give polyols of desired molecular weight and configurations. The polyols produced by this method normally have hydroxyls attached to the second carbon from the end of the chain. Since radicals attached to the primary carbon or the end carbon have the fastest reactivity, manufacturers have capped some of the polypropylene polyether polyol molecules with ethylene oxide and have provided polyols with hydroxyl radicals attached to the primary carbon.

The spatial molecular configuration of the polyether (polypropylene) backbone is similar to a coiled spring and contributes to the bounce, flexibility, and elongation of the final urethane products. The longer the space between the hydroxyls, the more elongation and flexibility. Generally, triols have a tendency to lessen elongation, contribute to the formation of an elastomeric solid, and contribute to lack of tear strength.

The polypropylene polyether diols with a molecular weight range in the vicinity of 2,000 are the most commonly used products. Popular polyether triols are those with molecular weights of 3,000, 4,000, and 6,000.

Polyether polyols with *Polyethylene* backbones are generally avoided for durable polyurethane products by the polyurethane industry since they are subject to degradation by moisture and are sensitive to moisture.

During the development of polyurethane compounds one encounters a phenomenon in which repeated experiments with different batches of the polyether polypropylene polyols do not give the same results. Careful analysis shows that although the average molecular weights are the same there is a molecular weight distribution change and it is this change that may be the reason for the unexpected results.

Polyester Based Polyols: While the polyester configuration contributes sluggishness of resilience recovery to the final product, it also can contribute toughness and tear resistance.

Reversion of polyurethanes—Choice of polyesters for use as an extender for isocyanates must be made carefully since polyesters can contribute to hydrolytic *in*stability of the final elastomers. Crashes of F-111 fighter bombers in Viet Nam have been attributed to the reversion of the polyester polyurethane sealants in humid environment. Tests have been developed to test for reversion of polyurethane compounds in an attempt to minimize this problem.

INDUSTRIAL MATHEMATICS FOR POLYURETHANES

The polyurethane industry has come up with hyroxyl numbers, isocyanate equivalents, etc., which seemingly are mysterious means to calculate the way to compound polyurethane products. Listed below are some of these mathematical terms. Derivations and explanations are found in the appendix. Polyurethane industrial math, as explained in the appendix, is actually a shortcut method based on familiar fundamental chemical calculations. The other terminologies are explained here to clear up questions that may have arisen in reading the chapter to this point. The math is placed here to prepare the reader for later sections of this chapter.

(1) *Hydroxyl Number, (OH #)*

Definition: Milligrams of KOH equivalent to hydroxyl content of one gram of polyol.

Mathematical expression is:

$$OH \# = \frac{(\text{functionality of Polyol}) \times 56.1 \times 100}{(\text{MW of polyol})}$$

(2) $0.155 \times OH\# =$

    Definition: Amount (grams) of pure TDI required to react stoichiometrically with 100 grams of polyol with specified OH #.

    or

    $0.155 \times OH\# \times \text{Isocyanate Index} =$

    Definition: Amount (grams) of pure TDI required to react with 100 grams of polyol with specified OH # at specified Isocyanate Index (NCO/OH).

(3) $\dfrac{0.75 \times (OH\#) \times (\text{Isocyanate Index})}{(\%\ Free\ NCO)} =$

    Definition: Amount (grams) of Isocyanate product with certain percentages free —NCO required to react with 100 grams of polyols with specified OH # and used at specified Isocyanate Index (NCO/OH).

(4) $0.22 \times OH\# \times (\text{Isocyanate Index}) =$

    Definition: Amount (grams) of pure MDI required to react with 100 grams of polyol with specified OH # at specified Isocyanate Index (NCO/OH).

(5) $33 \times (\%OH) = OH\#$

    Definition: Calculating the hydroxyl number (OH #) when percent hydroxy by weight (%OH) is known.

(6) $(MW\ R'OH) = \dfrac{(func.\ R'OH) \times 56100}{(OH\#)}$

    Definition: Calculation of molecular weight (MW) of the polyol when the hydroxyl number and functionality is known.

    or

    $(OH\#) = \dfrac{(func.\ R'OH) \times 56100}{(MW\ R'OH)}$

    Definition: Definition of OH # as in #1. Calculation of OH # of the polyol when the molecular weight and functionality is known.

(7) $\%\ Free\ NCO = \dfrac{(func.\ of\ the\ RNCO) \times (MW\ -NCO) \times 100}{(MW\ of\ RNCO)}$

Definition: Calculation of the percent isocyanate by weight in an isocyanate containing molecule.

TERMINOLOGY

1. Functionality: (func.)	The number of chemically active atoms or groups per molecule for the designated reaction.
2. Molecular Weight: (MW)	The sum of the weights of all the atoms in a molecule expressed in grams and based on oxygen = 16.0.
3. Isocyanate Equivalent: (NCO eq.)	The gram molecular weight of a unit containing one isocyanate group.
4. Isocyanate Index: (NCO/OH)	The ratio of isocyanate to hydroxyl equivalents. At stoichiometry the number is 1. At 5% excess of the isocyanate equivalent, the number is 1.05.
5. Prepolymer: (Adduct)	Commonly means the reaction product of excess isocyanate with polyols where the terminal radicals are all isocyanates. *Or* the reaction product of excess polyol with isocyanates where the terminal radicals are all hydroxyls. A prepolymer product is usually of low molecular weight in comparison to the final polymer.
6. Glycols (Diols):	Compound that contains two hydroxyl groups.
7. Polyols:	Compound that contains two or more hydroxyl groups.
8. Triols:	Compound that contains three hydroxyl groups.
9. Exotherm:	The natural heat generated by the chemical reaction.
10. Endotherm:	A reaction that causes heat to be lost or absorbed. Opposite of exotherm.
11. Elastomer:	A coined terminology for plastics or synthetics which are rubberlike and resilient.
	The difference between rubber and elastomer: The rubber industry does not like it when synthetics (plastics)

are called rubber. Furthermore, it is claimed that rubber has characteristics (like "snap back") which do not exist in plastic materials. The coined word "elastomer" has sufficiently alleviated disagreements.

12. Monomer: A simple reactive substance, i.e., TDI., a glycol.

13. Polymer: A compound prepared by the linking together of one or more simple monomers.

14. TDI: Toluene diisocyanate.

15. Thermosetting: A chemical crosslinking of a polymeric resin or plastic which cannot be reversed by heating or cooling.

16. Thermoplastic: A resin or plastic which can be repeatedly softened or melted and which will harden to the new shape when cooled.

17. Acid Number: Represents the number of milligrams of KOH equivalent to the acidity of 1 gram of the polyol. It is determined by titration with a standard alkali.

18. Hydroxyl Number: Represents the number of milligrams of KOH equivalent to the hydroxyl content of 1 gram of the polyol. It is determined by phthalation and acetylation methods.

19. Reactive Number or Equivalent Number: See functionality above.

20. Equivalent Weight: Calculated from the reactive number as follows:

$$\text{Eq. Wt.} = \frac{\text{MW in grams}}{\text{Reactive number}}$$

21. Chain Stoppers: Generally a molecule containing a single reactive radical which will stop polymerization. Simple alcohol is a chain stopper for polyurethane reactions.

GUIDELINES AND THEORIES IN COMPOUNDING POLYURETHANE ELASTOMERS

The following are the ideas that may be considered when monomers are being assembled to formulate an elastomer:

(1) Generally polyether polyols based on *polyethylene* backbones are to be avoided for durable thermoset polyurethanes since they yield products of increased water sensitivity, even though they produce polymers with better oil resistance than other polyols.

(2) Polyols, other than simple polyols based on polypropylene polyether polyols, are available for consideration for the hydroxyl portion of the elastomer formulations. They can be propylene oxide extended polyols based on trimethyol propane, pentaerythrytol, and sucrose bases. Tetramethylene oxide backboned polyols are also available.

(3) Order of descending activity with –NCO; primary amines, primary alcoholic hydroxyl groups, water, urea, secondary and tertiary alcoholic hydroxy groups, urethane hydrogen, carboxylic acid groups, amides.

(4) Smaller molecules, in comparison with larger counterparts, react extremely fast, produce products with the least elongation, produce harder and tougher materials by providing tighter crosslink densities. Increasing the size of the molecules will produce a softer and less tough product with greater elongation. Linear high molecular weight diols contribute to softness, elongation, and flexibility.

(5) *Hardness* of the product can be defined in terms of how closely the chains are crosslinked. In general, harder and tougher products are obtained as the number of urethane groups are increased. The more allophanate crosslinking the lower the elongation and tensile strength while hardness and load bearing will increase. In general product properties will be similar as long as the molecular weight of the diols are the same.

(6) Physical properties of a urethane structure are affected by the amount of curing agent present which controls the extent of crosslinking or chain branching.

(7) Aromatic amines are popular curing agents and develop the maximum toughness properties possible of a simple prepolymer.

(8) *Blocked polymers*—Isocyanates reacted with phenols form "blocked" isocyanates. Upon heating to certain critical temperatures, these blocked diisocyanates unblock and regenerate the diisocyanate and an endothermic reaction occurs. The regenerated diisocyanates are now free to react normally with compounds in the recipe as evidenced by developing an exotherm. This is a useful method for making "blocked" isocyanates recipes for baked coatings. MDI can be blocked with methyl ethyl ketoxime, ϵ-caprolactam, and benzotriazole.[14]

(9) The general practice in the polyurethane industry is to use the two component prepolymer system to eliminate the problems of compatibilities, viscosities, generated exotherms, etc.

(10) TDI for use in the prepolymer system—Of the two types of 80/20 TDI available, the high acid TDI is required for reaction stability, prevention of runaway reactions, and for shelf stability.

Increasing acidity by the addition of benzoyl chloride to retard the TDI reactions is common. However, increasing the residual acids may lead to product degradation.

(11) Selective catalysts can be effectively used to favor the linear polymerization reaction over carbon dioxide evolution. The catalyst must not be added to the prepolymer. The addition of the catalyst will have a detrimental effect on the shelf stability of the prepolymer by forming diazides and triazides, or causing premature polymerization of the liquid prepolymer in storage, and resulting in gelation of the prepolymer.

(12) A useful and popular catalyst used by the polyurethane industry is *triethylenediamine*, a tertiary amine catalyst, better known by its commercial name, DABCO.

$$\begin{array}{c} \diagup CH_2-CH_2 \diagdown \\ N-CH_2-CH_2-N \\ \diagdown CH_2-CH_2 \diagup \end{array}$$

The pure form is crystalline and the schematic and formulation is given above. DABCO is also available as LV 33. It is 1/3 DABCO and 2/3 dipropylene glycol and is a liquid.

(13) Prepolymers made with diols of molecular weight of 1,000 or lower will produce elastomers when cured with other diols. Prepolymers made with diols of higher molecular weight, beginning in the critical 2,000 to 3,000 molecular weight range, will not solidify when cured with other diols. Experimentation shows that these higher diol products require incorporation of triols to become a solid. In order to produce an elastomer with polyols in this critical molecular weight range the minimum combined hydroxyl functionality of 2.2 is required to form a solid.

(14) In theory, the production of prepolymers seems to be very simple and straightforward. However, developers of urethane components sometimes have difficulty in following through with their experiments due to unexpected pitfalls. Theory designates that in order to produce a hard and tough system a small molecular weight polyol is the basis for the product. In proceeding to adduct it with isocyanates the experiment produces one of many pitfalls in compounding. The adduct has an unexpectedly high unacceptable viscosity which is unsuitable as a practical component. The formulator would have to resort to using additives, or using excess isocyanates, or raising mixing temperatures to lower viscosities to acceptable ranges. The developer must redesign the system by putting the small molecular weight polyol in the curing agent component and adducting the larger molecular weight based polyol that may have been in the curing agent. Or, he can use another polyol of different backbone but of similar molecular weight.

Fortunately, in the case of 500 to 700 molecular weight polyols there is a classic substitution example. The prepolymer of polypropylene polyether polyols in the molecular weight range of 600 to 700 (hydroxy number of around 190) produces a product whose viscosity is unacceptably high as a component, but if one substitutes a polyol with a tetramethylene oxide backbone of similar molecular weight, a prepolymer with an acceptable viscosity can be produced. Furthermore, the tetramethylene oxide backbone product will produce a product with better physical properties than those provided by the polypropylene polyether backbone products.

(15) Elastomers made with MDI-polyether polypropylene polyol exhibit exceptional resilience and are accompanied by lower heat generation for parts under dynamic strain. Elastomers based on MDI-polyether have better resilience than ones based on TDI-polyether. The elastomers compounded with TDI-polyester show slow and sluggish recovery.

(16) The prepolymer system is very versatile. Using a single prepolymer, and by curing it with different "hardener systems," one can obtain elastomers with different hardnesses and different physical properties. For example, the simplest prepolymer for general purpose elastomeric urethane consists of 2,000 molecular weight polypropylene polyether polyols with a hydroxyl number of about 56 reacted with the appropriate amount of TDI. The normal hardness of the above prepolymer cured with aromatic amines can be as high as Shore A of 75 to 80 with elongation of about 400%. If the prepolymer is cured with polypropylene polyether polyol of 4,000 molecular weight, the hardness could be as low as Shore A of 35 with elongation of about 800% or more.

(17) Heat curing or tempering of cast thermoset polyurethanes. Cast thermoset urethane elastomer properties can be influenced by the procedures used to cure or post cure the cast products.

The three optional steps normally taken to cure products:

(1) Begin 140°F oven curing at the later stages of polymerization or immediately after the exotherms subside.

(2) Allow the exotherm to subside and cool at room temperatures up to two or three hours and then post cure in a 140°F oven several hours or overnight.

(3) Allow the cast elastomer to age at room temperature overnight and then post cure in a 140°F oven for several hours or overnight.

Even though the three resulting elastomers were cast from one mix, generally, the final products are *not* identical in properties. Step 1 produces some tempering and increase in physicals. Step 2 produces changes not as significant as in Step 1. Step 3 produces an elastomer with the softest and greatest elongation.

Heat tempering of polyurethane elastomers similar to that used in tempering of steels (heating the metal red hot and quickly cooling in cold oil or water quenching.) has met with some success in producing elastomers with improved toughness and hardness.

(18) The room temperature curing elastomer recipes generally must be vacuumed before or after casting to be bubble free, and the recipes must include antifoams to facilitate the vacuuming.

(19) Amines as a group have been ignored by the mixing and casting minded compounders since the reaction rate does not allow for sufficient working time, but this type of product could be useful in the RIM processes where a quick reaction time may be essential.

(20) Generally, the MDI prepolymer systems require heating for processing. MDI must be heated to 122°F (50°C) for a period of 4 to 8 hours to allow for complete melting and for dimer crystals to precipitate and be separated from the mix prior to the use of melted MDI in the making of the prepolymer.

(21) Higher densities for MDI foams is assumed. Higher investment costs are required for processing MDI materials.

(22) Recent MDI technology has improved the working properties to a point where properties such as demolding time, cream time, cure period, etc. are more favorable than those obtained from TDI based products. For example, the cure time for MDI elastomers is 2 hours vs 24 hours now cited for TDI elastomers.[8]

(23) Solvents for urethanes—Use only solvents which have been dried either by commercial solvent houses which are familiar with the requirements of the polyurethane industry, or which have been passed through molecular sieve drying beds. Most solvent suppliers are not acquainted with the urethane requirements for extremely dried solvents and when asked for solvents with water contents in the vicinity of 0.01% complain that the customer is asking for reagent grade quality solvents in bulk quantities, and that they do not carry them in bulk and will attempt to substitute.

Again, the solvents used in the polyurethane formulations *must* be protected from exposure to atmospheric air to prevent them from absorbing moisture in this manner.

(24) Simple alcohol is a chain stopper.

(25) A good clean up solvent for liquid urethanes is 50% alcohol and 50% acetone. The alcohol will prevent the urethane from polymerizing and the acetone is a good solvent for liquid urethanes and their ingredients.

(26) Methylene chloride is the usual solvent for cleaning out reactors. Reactor cleaning grade methylene chloride is available from solvent reclaimers at somewhat lower prices. The reactor is loaded with the cleaning grade methylene chloride for overnight, or over weekend residency. The agitator is turned on in the morning to stir the expanded and gelled mass off the sides of the reactor. The spent solvent, along with the gelled mass, is pumped back into salvaging drums and returned to the reclaimer for reprocessing. The salvage people do not like to see alcohol in the methylene chloride, but the addition of a small amount of isopropyl alcohol will effectively

neutralize any isocyanate remaining in the reactor and facilitate the reactor rejuvenation.

COMPOUNDING OF THERMOSET POLYURETHANE ELASTOMERS

The terminology *"parts per hundred of resin"* or *"phr"* is commonly found in the literature. Calculations and theoretical thinking is based on 100 parts of the main polymer. This is a normal practice employed by polymer people and it shall also be used here. For example, the industrial math described in this chapter is based on phr or 100 grams of polyol.

Prepolymers

Prepolymer, Example I

Simple adduct of 2,000 MW diol with TDI

Adducting 2,000 molecular weight diol with TDI produces a common general purpose base prepolymer with many practical uses.

Calculations:

$$\text{OH \#} = \frac{(\text{func.}) \times 56{,}100}{\text{MW of diol}} = \frac{(2) \times 56{,}100}{2{,}000} = 56.1 = \text{OH \# of diol with MW of 2,000.}$$

0.155 x OH # x (NCO/OH) =	grams of TDI required to react with 100 grams of polyol at a certain Isocyanate Index.
0.155 x 56.1 x 1	= 8.7 grams of TDI is required to react 100 grams of 2,000 MW diol at stoichiometry.
2 x 0.155 x 56.1 x 1	= 17.39 grams of TDI required to adduct 100 grams of 2,000 MW diol into a simple prepolymer with terminal isocyanates as reactive radicals (8.7 grams in excess).

Calculation of theoretical free –NCO of above prepolymer:

Total grams of polyol = 100
Total grams of TDI used = 17.39
Weight (grams) of unreacted –NCO
 (0.4822 x 8.7) = 4.19

$$\% \text{ free } -\text{NCO} = \frac{\text{weight free } -\text{NCO} \times 100}{\text{Total weight}} = \frac{4.19 \times 100}{(100 + 17.39)} = 3.57\%$$

The above prepolymer can be prepared by the method described in the appendix. The ingredients listed above will generate their own exotherms when the reactions take place, and may require the addition of water to a certain level in the cooling jacket to act as a heat sink to maintain temperatures below 185°F (85°C).

The component II which can be used with the above prepolymer to produce elastomers of varying properties is described later in this section.

Prepolymer, Example II

Simple combination adduct of 2,000 MW diol and 3,000 MW triol with TDI

A prepolymer can be prepared by using a mixture of 1 part 2,000 molecular weight polypropylene polyether diol and 1 part 3,000 molecular weight polypropylene polyether triol reacted with TDI. A prepolymer, by combining diols and triols, allows greater latitude in the selection of curing agents and facilitates compounding of the final product.

Calculations:

OH # of the 2,000 MW diol (from above example) = 56.1.

OH # of 3,000 MW triol:

$$\text{OH \#} = \frac{(\text{func.}) \times 56{,}100}{\text{MW of triol}} = \frac{3 \times 56{,}100}{3{,}000} = 56.1$$

0.155 x OH #	= grams of TDI required to react with 100 grams of polyol with certain OH #.
0.155 x 56.1 x 1	= 8.7 grams of TDI required to react with 100 grams of diol with MW of 2,000.
0.155 x 56.1 x 1	= 8.7 grams of TDI required to react with 100 grams of triol with MW of 3,000.
(2 x 0.155 x 56.1) x 2	= 34.8 grams TDI to adduct 100 grams of 2,000 and 100 grams of 3,000 molecular weight polyols into a simple prepolymer with reactive terminal isocyanates.

Calculation of theoretical free NCO of above prepolymer:

Total grams of polyols = 200
Total grams of TDI used = 34.8
Weight (grams) of unreacted
 –NCO. (0.4822 x 17.4) = 8.39

$$\% \text{ free -NCO} = \frac{\text{weight free -NCO} \times 100}{\text{total weight}} = \frac{8.39 \times 100}{100 + 100 + 34.8} = 3.57$$

The above prepolymer can be prepared by the method described in the appendix. The reaction of ingredients listed above will generate its own exotherms and may require the addition of water to a certain level in the cooling jacket to act as a heat sink to maintain temperatures below 185°F (85°C). Component II which can be used with the above prepolymer is also discussed later in this section.

Prepolymer, Example III
Simple adduct of 6,000 MW triol with TDI

6,000 MW polypropylene polyether triols are currently a very common base for many polyurethane sealants and caulkings where the emphasis is elasticity, quick gelling, minimum shrinkage, low Shore A durometer hardness, and where great strengths are NOT required.

Calculations:

OH # of the 6,000 MW triol.

$$\text{OH} \# = \frac{(\text{func.}) \times 56{,}100}{\text{MW of polyol}} = \frac{3 \times 56{,}100}{6{,}000} = 28.05$$

0.155 x OH # x NCO/OH = grams of TDI to react with 100 grams of polyol of certain OH # at certain Isocyanate Index.

0.155 x 28.05 x 1 = 4.35 grams of TDI required to react stoichiometrically with 100 grams of triol with MW of 6,000.

2 (0.155 x 28.05 x 1) = 8.7 grams of TDI required to adduct 100 grams of 6,000 MW triol into simple prepolymer with reactive terminal isocyanate (4.35 grams in excess).

Calculation of theoretical free NCO of above prepolymer:

Total grams of polyol = 100
Total grams of TDI used = 8.7
Weight (grams) of unreacted
  –NCO. (0.4822 x 4.35) = 2.1

$$\text{\% free -NCO} = \frac{\text{weight free -NCO} \times 100}{\text{total weight}} = \frac{2.1 \times 100}{108.7} = 1.93$$

The above prepolymer can be prepared by the method described in the appendix. The ingredients listed above will require the assistance of hot water in the cooling jacket to initiate reaction. Remove the hot water at a critical period to prevent the reaction temperatures from rising above 185°F (85°C).

Prepolymer, Example IV
1,000 MW diol reacted with MDI
A simple basic MDI general purpose prepolymer

Calculations:

Hydroxyl number of 1,000 MW diol.

$$\text{OH \#} = \frac{(\text{func.}) \times 56{,}100}{(\text{MW diol})} = \frac{2 \times 56{,}100}{1{,}000} = 112.2$$

0.22 x OH # x NCO/OH	= grams MDI required to react with 100 grams 1,000 MW diol.
0.22 x 112.2 x 1	= 24.68 grams required at stoichiometry.
2 (0.22 x 112.2 x 1)	= 45.36 grams required to cap diol with MDI.

Calculation of theoretical free NCO of above prepolymer:

Total grams of polyol = 100
Total grams of MDI used = 45.36
Weight (grams) of unreacted
  –NCO. (0.335 x 24.68) = 8.26

$$\text{\% free NCO} = \frac{(0.335 \times 24.68) \times 100}{\text{MW}} = \frac{0.335 \times 24.68 \times 100}{100 + 45.36} = 5.7\%$$

The above prepolymer can be prepared by methods described in the appendix. The materials will require melting by heating prior to addition into the reactor. The heating jacket will also require hot water to initiate the MDI-polyol reaction. The heat source must be removed at a critical time, and may require a cooling bath to prevent the exothermic heat from rising above 185°F (85°C).

Prepolymers and Curing Compounds

Aromatic Amine (MOCA): MOCA as the curing agent will produce an elastomer with the maximum toughness possible for each of the prepolymers above.

Calculation of required amount of MOCA:

$$CL = \frac{(func.) \times 56{,}100}{(MW \text{ of MOCA})} = \text{Curative Level}$$

$$= \frac{(2) \times 56{,}100}{267} = 420$$

$$X = \frac{7.5 \times CL \times (NCO/NH_2)}{\% (NCO)} = \text{grams of prepol at certain \% NCO to react with 100 grams of pure MOCA at certain Isocyanate Index } (NCO/NH_2).$$

$$= \frac{7.5 \times 420 \times 1}{3.57} = 882 \text{ grams of prepolymer with 3.57 \% free NCO to react with 100 grams of } \textit{pure} \text{ MOCA at stoichiometry.}$$

Since MOCA has 95% purity, 882 grams of prepolymer with 3.57% free NCO are required to react with 100 grams of *pure* MOCA. The use of commercial grade MOCA at 95% fortunately calculates out to an Isocyanate Index of 1.05.

Ratio of Use:
 100 grams of 3.57% Example I and II Prepolymers.
 11.3 grams of MOCA.

To produce the urethane elastomer using above materials requires both components to be heated to 205°F (96°C) prior to mixing. The theoretical usable potlife of the hot liquids will be from 5 minutes to 8 minutes at 205°F (96°C). The complete elastomer formulation requires the addition of antifoam agents and vacuuming to produce a bubble free elastomer. Normally, a catalyst is not required for the reaction of the above hot prepolymers with hot MOCA.

The theoretical hardness of the product will be around Shore A of 60 to 80 with elongation of 200 to 400%. Useful service temperature will be under 250°F (121°C) with good low temperature flexibity. The above elastomer will have good overall physical properties, with especially good scuff resistance.

The physical properties of the final elastomer can be changed within a limited range by varying the amount of MOCA in the ratio. Increasing the urethane groups, *(using less MOCA)*, properties of the elastomer are influenced as follows (see Table 7-7).

Higher Shore A hardness
Higher tensile strength
Higher heat deflection temperatures
Higher load bearing properties

By increasing the MOCA content, the resulting properties of the elastomer generally are:

Lower Shore A hardness
Increase in compression set
Less tensile strength
Increased elongation

Using similar calculation methods one can obtain the ratio of use of other aromatic amine curatives. Note: Even though there are four active hydrogens on the MOCA radical, the functionality is considered *two* for MOCA and other aromatic amines with two fully active primary amine groups.

Ratio to produce elastomer with MOCA and 1.93% free -NCO prepolymer:

$$X = \frac{7.5 \times CL \times (NCO/NH_2)}{(\% NCO)}$$ = Amount of % NCO required to react with 100 grams of pure MOCA at certain isocyanate index (NCO/NH_2).

$$= \frac{7.5 \times 420 \times 1}{1.93}$$ = 1,632 grams of 1.93% prepol to react with 100 grams of pure MOCA.

Result of calculation:

100 grams of 1.93% Prepolymer (Component I)
6.2 grams of MOCA (Component II)

The materials will require melting by heating prior to addition

Table 7-7: A Typical Properties Table of a Suppliers Polyurethane Elastomers

PROPERTY	ASTM TEST METHOD	UI-5800	UI-5830	UI-5860	UI-5880	UI-5900	UI-5910
SHORE HARDNESS (SHORE A)	D 2240	80 ± 2	83 ± 2	86 ± 2	88 ± 2	90 ± 2	91 ± 2
MIX RATIO BY WEIGHT	UI R & D	100A/62B	100A/80B	100A/91B	100A/93B	100A/103B	100A/95B
BY VOLUME	UI R & D	1A/0.51B	1A/0.62B	1A/0.76B	1A/0.8B	1A/0.886	1A/0.78B
VISCOSITY - PART A (POLYOL) - CPS	D 2393	530	530	640	640	530	530
PART B - CPS	D2393	400	400	400	400	400	400
WEIGHT PER GALLON - PART A - LBS	UI R & D	8.2	8.6	8.4	8.6	8.6	8.2
PART B - LBS	UI R & D	10.0	10.0	10.0	10.0	10.0	10.0
MIX TIME - SECONDS	UI R & D	12-15	12-15	12-15	12-15	12-15	12-15
GEL TIME (@ 100-105°F) - SECONDS	D 2471-71	30-40	30-40	30-40	30-40	30-40	30-40
DEMOLD TIME (140-160°F) - SECONDS	UI R & D	90-120	90-120	90-120	90-120	90-120	90-120
CURE TIME - 95% - HOURS	UI R & D	24	24	24	24	24	24
- COMPLETE - DAYS	UI R & D	2-3	2-3	2-3	2-3	2-3	2-3
TENSILE STRENGTH - PSI	D 412-80	4470	2250	2400	1500	2700	5680
ELONGATION, ULTIMATE - %	D 412-80	540	410	390	400	380	420
MODULUS (100%) - PSI	D 412	750	970	1000	1050	1150	1885
COMPRESSION SET - %	D 695	17	13	10	11	15	36
TEAR STRENGTH, DIE C - PLI	D 624-81	350	240	285	290	320	420
TEAR STRENGTH, SPLIT - PLI	D 470	62	36-37	37	38	42	95
ABRASION RESISTANCE (TABOR)-MG LOST 1000 GM/1000 REV	CS-17	0.09	4.5	4.5	5.	0.12	0.06
SPECIFIC GRAVITY (25° C)	UI R & D	1.12 PART A/ 1.2 PART B	1.12 PART A/ 1.2 PART B	1.12	1.12	1.12 PART A/ 1.2 PART B	1.12 PART A/ 1.2 PART B
RESILIENCE (REBOUND) - %	BASHORE	70	59	58	58	54	65
LINEAL SHRINKAGE - %	D 2566	1.5	1.6	1.6	1.6	1.6	1.5
LINEAL COEFFICIENCY OF EXPANSION - IN/IN°F	D 2566	1.03×10^{-4}	1.03×10^{-4}	1.03×10^{-4}	1.03×10^{-4}	1.03×10^{-4}	1.03×10^{-4}
SHELF LIFE - MO	UI R & D	3	3	3	3	3	3
SOLIDS - %	UI R & D	100	100	100	100	100	100

Courtesy of Urethane Industries, Placentia, California.

into the mixer. The final elastomer will be lower in Shore A durometer reading, and have an elongation of 400% plus.

Polyols: A variety of polyols and or combination of polyols, along with anhydrous liquids and powder extenders, plasticizers, antifoams, catalysts, can be used to produce a room temperature curing elastomer of varying properties with the above prepolymers. The choice of molecular weights, and the decision to use all diols or all triols or a combination of triols and diols, in the polyol curing agent portion are generally governed by the physical requirements for the elastomer. The choice is unlimited.

Generally the recipe of the very basic ingredients in the polyol curing agents for clear unfilled elastomer systems is represented by the following:

Polyols (polyols)	100
Catalysts	0.4-0.6
Antifoam	0.005

Curing the Example II prepolymer with large molecular weight (4,000) triols.

Calculation of ratio:

$$\text{OH \#} = \frac{(\text{func.}) \times 56{,}100}{\text{MW of polyol}} = \frac{3 \times 56{,}100}{4{,}000} = 42$$

$$X = \frac{7.5 \times \text{OH \#} \times (\text{NCO/OH})}{\% \text{ NCO}} = \text{grams of prepolymer with certain free NCO to react with 100 grams of polyol with certain OH \# at specified Isocyanate Index (NCO/CH).}$$

$$X = \frac{7.5 \times 42 \times 1.05}{3.57} = 92.6 \text{ grams of prepolymer with 3.57\% free NCO required to react with 100 grams of polyol with MW of 4,000 and OH \# of 42 at Isocyanate Index of 1.05.}$$

Ratio of use:

4,000 MW polyol, 100 grams + (wt. of catalyst, antifoam, etc.)
Prepol/3.57%,   92.6 grams

A simple complete room temperature curing Component II will require in addition to the 4,000 MW polyol, a catalyst, either a metal or tertiary amine type, antifoam, and fillers and or other extenders. The product will have a Shore A hardness in the vicinity of 35 to 45 with elongation in the range of 400 to 600%. Useful service temperature is below 180°F (82°C) and it will have a good low temperature flexibility.

Curing of MDI Base Prepolymers with Polyols:

General guidelines—

Generally heat both components to 176°F (80°C) before use.
MDI prepolymers held at 176°F (80°C) for 8 hours show very little change.
After 24 hours at 176°F (80°C), one notices a small change.
At 248°F (120°C), the prepolymer is fairly stable for 2 hours.
Mold temperature of 230°F (110°C).
Post cure at 230°F (110°C).

Example using 1,4-butane diol as the curing agent at Isocyanate Index of 1.05 and Example IV Prepolymer.

$$\text{OH \#} = \frac{(\text{func.}) \times 56{,}100}{\text{MW}}$$

$$\text{OH \#} = \frac{2 \times 56{,}100}{90} = \text{OH \# of 1,4-butanediol (MW = 90)} = 1{,}246$$

$$X = \frac{7.5 \times \text{OH \#} \times (\text{NCO/OH})}{\% \text{ NCO}} = \text{Grams of \% NCO MDI prepol required to react with 100 grams of polyol of certain OH \# at certain Isocyanate Index.}$$

$$= \frac{7.5 \times 1{,}246 \times 1.05}{5.7} = 1{,}721 \text{ grams 5.7\% MDI prepol to react with 100 grams of 1,4-butanediol at Isocyanate Index (NCO/OH) of 1.05.}$$

Mix ratio:

100 grams of 5.7% free NCO MDI Prepolymer
5.7 grams of 1,4-butanediol

MDI vs TDI Based Product Comparison—Literature compares the above 1,4-butanediol cured MDI-polyol with the much heralded TDI-Polytetramethylene oxide polyol product cured with MOCA and concludes that the above type of product is a viable alternative.[6]

GENERAL CONSIDERATION

Now that readers have been introduced to the basics of producing polyurethane elastomers, let us explore how this information can be used as an investigational tool to analyze the components of a polyurethane system.

Before any handling of the raw materials, answers to the following questions can be garnered from the Technical Data Sheets which accompany vendor samples. The information will categorize and more or less identify simple polyurethane systems.

> Is it room temperature mixing, or does it have to be melted prior to mixing?
> What are the end uses of the polyurethane polymer?
> What are the service temperature limitations?
> What is the hardness of the end product?
> What is the resilience of the end product?

Analysis of the liquids will shed further light on the type of systems.

> Are the components both liquids or solids?
> What odors can be detected in the system?
> What is the % free NCO of the prepolymer?
> What is the tear resistance of the elastomer?

If the components require heating, the product can fall into two categories: (1) MDI system, (2) Aromatic amine cured TDI system.

Straight all polyol systems' service use temperature limitation is around 180°F (82°C). The systems containing aromatics will have service use temperature limitations around 250°F (121°C).

Hardness is a function of molecular weights.

Resilience will identify whether the system is MDI/polyether or TDI/polyether, or whether the base polyol is a polyester.

Odor will give clues to the isocyanates used, and to whether solvents or plasticizers have been incorporated into the systems.

The percent of free –NCO can also identify the approximate molecular weights used to produce the elastomer. In the case of TDI, the percent of free –NCO will give clues to the approximate molecular weights, since the current practice is to have less than 1% free

TDI monomer content in order to have the product classified as a nontoxic compound. In the past, the percent of free —NCO was not a means to determine the molecular weight of the polyol since the common practice was to add excessive amounts of free monomer to lower the viscosity of the prepolymer. (MDI is not generally used in this manner, but saturated diisocyanates are used to lower viscosity and to obtain other desirable properties.)

Products made from tetramethylene oxide backbone or polyesters have exceptional tear and wear resistance.

Other Polyurethane Ingredients

The earlier sections of this chapter have dealt with elastomers and coatings generally produced through prepolymer methods using simple polyether polyols and TDI, MDI, and common saturated diisocyanates. There are other materials to be considered when specific properties are stressed.

Castor Oil Derived Polyols: These are an important group of polyols used in presently available polyurethane systems.

Premium grade moisture cured coating systems for indoor and outdoor environment are based on these polyols. They provide for very tough wear resistant surface coatings. One of the successful uses is as aircraft carrier deck coatings. They have become standard materials for floor coatings.

Although castor oil derived polyol prepolymers have not been shown to produce elastomers with exceptionally strong physical properties, they exhibit good dielectric properties and have been extensively used as an electrical insulating medium in telephone cable plugging and potting.

Castor oil is a triglyceride of fatty acids. About 90% of this fatty acid is ricinoleic acid, an 18 carbon fatty acid. There are double bonds in the 9-10 position and hydroxyl groups on the 12th carbon. Other fatty acids are: Dihydroxystearic, palmitic, stearic, oleic, linoleic, linolenic, and eicosanoic.[16]

A prepolymer having a very high percentage of free —NCO designed for electrical use may mean the base is a castor derivative.

Polycaprolactone Polyols: Although technically these are polyesters, their working properties lie between polypropylene polyether polyols and polyester polyols. Their use imparts oil resistance, acid resistance to 50% sulfuric acid,[12] toughness, good hydrolysis resistance, and allows use in rather high temperature environments and has good low temperature flexibility.[7]

Some premium grade moisture cured polyurethane lacquers are based on saturated MDI-caprolactone prepolymers. These coatings can be very tough, hard with a certain amount of resilience, and can have good scuff and wear resistance. One of their successful uses has been as basketball court coatings.

Suitable catalysts for polylactone reactions are: tetrabutyl titanate, Lewis acids and strong inorganic acids.[12]

Graft Polyol: Polyols are copolymerized with styrene and acrylonitrile and are used in flexible foam to enhance maximum load bearing properties.

Reaction Injection Molding (RIM): MDI technology in the United States has been lagging due to:

(1) Customer resistance.

(2) The need for higher investment in equipment.

(3) Lack of all new processing equipment.

(4) Lack of innovative new materials to encourage the MDI urethane lines.

RIM processing has put renewed life into MDI systems because the RIM area of products is seen as the next large area of demand for polyurethane raw materials. The polyurethane industry is seriously undertaking steps to overcome the above problems. (Since TDI has received its derogatory classification, emphasis has been to replace the original TDI based products with MDI based products. These new products showed substantial improvement in split tear strengths.)

The handicap of required long green strength time of previous MDI based products that made handling of the product difficult and time consuming is being overcome in current RIM process with innovative engineering, new products especially designed for RIM, and the appropriate use of heat.

The basic suppliers' laboratories have been busy with their R&D to meet the demands for new products. Almost monthly current literature introduces new offerings for RIM raw materials by the basic suppliers. (Table 7-8)

Recent Advances and New Products

The crop of new materials are modified amine-terminated products. In normal casting, amines were avoided because their end use did not allow sufficient application time. RIM reactions can accommodate this fast reaction time of the amines. Some of the new amine-terminated products are:[8]

(1) Diethyltoluenediamine (American Cyanamid).

(2) Hydroxypropylmelamine (Union Carbide).

(3) Others from Mobay Chemical and Upjohn.

Internal Mold Release (IMR) Formulations: Attention is being focused on IMR formulations in the RIM field. Such formulations are common today and they allow approximately 100 releases from the mold without the use of external mold release agents. The benefits are obvious.[8]

Recent Diisocyanates: Aliphatic diisocyanates—The introduction of new aliphatic diisocyanates have opened up possibilities of pro-

Table 7-8: A RIM Formulation
POLYURETHANE EVALUATION
Rucoflex S-1089 and Rucoflex F-2064 — 1,4-Butanediol Extended

FORMULATIONS:

	1	2	3	4	5	6	7	8	9	10	11	12
Rucoflex S-1089	100.00	100.00	100.00				100.00	100.00	100.00			
Rucoflex F-2064				100.00	100.00	100.00				100.00	100.00	100.00
1,4-Butanediol[1]	10.00	20.00	30.00	10.00	20.00	30.00	10.00	20.00	30.00	10.00	20.00	30.00
Dabco® 33 LV[2]	0.50	0.60	0.80	0.50	0.60	0.80	0.50	0.60	0.80	0.50	0.60	0.80
Isonate® 240 (Ind. 1.05)[3]	74.2	126.8	179.4	84.6	136.8	190						
Isocyanate Blend (Ind. 1.05)[4]							42.8	73.3	103.7	48.9	79.4	109.8

PHYSICAL PROPERTIES	Test Method	1	2	3	4	5	6	7	8	9	10	11	12
Shore A Hardness	D-2240	73-75	81-83	88-90	94-95	92-94	96-97	70-72	78-80	85-87	93-95	92-94	96-97
Shore D Hardness	D-2240	23-25	30-32	40-42	48-50	48-50	60-62	20-22	33-35	36-38	47-50	45-47	60-62
100% Modulus, psi	D-412	730	840	1140	1800	1560	3020	600	770	1170	2260	1650	3530
200% Modulus, psi	D-412	950	1190	1570	2410	2180	3960	990	1310	1670	3440	2350	4990
300% Modulus, psi	D-412	1520	1640	2460	3470	3120	5880	1300	2400	2600	—	—	—
Tensile Strength, psi	D-412	4600	7080	5200	6680	5000	6640	4100	4370	4250	4140	4150	4990
Elongation, % at break	D-412	530	610	460	440	420	340	450	370	370	230	260	200
Die C Tear, pli	D-624	450	410	560	650	670	840	300	330	440	610	600	750
Split Tear, pli	D-1938	130	130	170	240	240	300	30	30	80	160	120	120
Rebound, %	D-2632	41	32	32	37	36	45	25	21	28	30	34	41

[1] In substituting ethylene glycol for 1,4-butanediol, the same number of equivalents was used
[2] Dabco 33 LV-0.25 pph based on total weight
[3] Isonate 240 - quasi prepolymer
[4] Isocyanate Blend 2 pbw Isonate 125M
                     1 pbw Isonate 143L

Courtesy of Ruco Polymer Corporation, Hicksville, New York.

ducing a new generation of optically clear products with their own set of physical properties. Unsaturated polyurethane products became unsatisfactory for medical use when the discovery was made that upon autoclaving at 248°F (120°C), they produced MDA, a suspected mutagen. Products made from aliphatic diisocyanates do not produce MDA upon autoclaving and are touted for medical equipment use.[10]

Along with clarity, the aliphatic diisocyanates impart good weathering properties to elastomers and sealants and avoid the need for painting the surfaces at a later date (Figure 7-15). Some of the recent aliphatic diisocyanates are:

(1) Para- and meta-tetramethylxylene diisocyanate (para and meta TMXDI)[8] (American Cyanamid).

(2) Para-phenylene diisocyanate (Elate 160) (Akzo Chemie America).

(3) Trans-1,4-cyclohexane diisocyanate (Elate 166) (Akzo Chemie America).

Aromatic diisocyanates—

(1) Naphthylene-1,5-diisocyanate (NDI) (Mitsui-Nisso Corp.).

(2) Ortho-tolidine diisocyanate (TDDI) (Mitsui-Nisso Corp.).

Figure 7-15: Calthane ND urethanes—flexible and rigid cast transparencies. (Photo courtesy of Cal Polymers, Long Beach, California.)

Recent Proprietary Catalysts: A variety of proprietary catalysts provides for versatile reaction applications.

(1) X8161, a tertiary amine (Air Products & Chemical Company).

(2) Polycat DBU, SA (Abbott).

(3) Fomrez UL1-1, UL1-28, UL1-29 organo-tin (Witco Chemical Corp.).

(4) Fomrez UL-6, UL-24, UL-32, etc. (Witco Chemical Corp.).

APPENDIX

General Polyurethane Math Development

(Verification that industrial polyurethane math is based on common chemical math principles.)

(I) Hydroxyl Number (OH #)

Definition: Milligrams of potassium hydroxide (KOH) equivalent to hydroxyl content of 1 gram of polyol.

Mathematical expression:

$$\text{OH \#} = \frac{(\text{func. R'OH}) \times 56.1 \times 1{,}000}{[\text{Molecular weight (MW) of R'OH}]}$$

56.1 = Gram molecular weight of potassium hydroxide.

K (39.1) + O (16) + H (1) = 56.1

1,000 = Milligram expression

or simplified,

$$\text{OH \#} = \frac{(\text{func. R'OH}) \times 56{,}100}{(\text{MW R'OH})}$$

or rearranged,

(II) $$\text{MW R'OH} = \frac{(\text{func. R'OH}) \times 56{,}100}{(\text{OH \#})}$$

(III) $$\text{\% OH} = \frac{(\text{func. R'OH}) \times (\text{MW -OH})}{(\text{MW R'OH})} \times 100$$

(IV) $$\text{OH \#} = 33 \times (\text{\% OH})$$

Derivation:

(1) $\text{(MW R'OH)} = \dfrac{\text{(func. R'OH)} \times 56{,}100}{\text{(OH \#)}}$ (from II)

(2) $\text{(\% OH)} = \dfrac{\text{(func. R'OH)} \times \text{(MW -OH)}}{\text{(MW R'OH)}} \times 100$ (from III)

or rearranged,

(2a) $\text{(MW R'OH)} = \dfrac{\text{(func. R'OH)} \times \text{(MW -OH)}}{\text{(\% OH)}} \times 100$

(3) (1) = (2a) = (MW R'OH)
MW of -OH = O(16) + H(1) = 17

$$\dfrac{\text{(func. R'OH)} \times 56{,}100}{\text{(OH \#)}} = \dfrac{\text{(func. R'OH)} \times 17}{\text{(\% OH)}} \times 100$$

Simplified, the above becomes:

$$\text{OH \#} = \dfrac{\text{(\% OH)} \times 56{,}100}{17 \times 100}$$

(IV)   OH # = 33 × (% OH)

(V)   % OH = $\dfrac{\text{(OH \#)}}{33}$ rearranged (IV).

(VI)   % NCO = $\dfrac{\text{(func. RNCO)} \times \text{(MW of NCO)}}{\text{(MW of RNCO)}} \times 100$

(VII) & (VIII)   X = $\dfrac{7.5 \times \text{OH \#} \times \text{NCO/OH}}{\text{\% NCO}}$

X = Grams of isocyanate product at certain percent free -NCO required to react with 100 grams of polyol with certain OH # at a certain Isocyanate Index (NCO/OH).

Derivation: Basic premise and chemical theory—when products react, equivalents of one component react with equivalents of the other component.

Basic Reaction

$$\underset{\text{(A)}}{\text{R-N=C=O}} + \underset{\text{(B)}}{\text{R'O-H}} \rightarrow \text{urethane}$$

Equivalents of R–N=C=O = Equivalents of R'–O–H

Mathematical expression of above equivalent of R–N=C=O is:

(1) An equivalent (R–NCO) = $\dfrac{\text{(A in grams)}}{\text{(eq. wt. RNCO in grams)}}$

Mathematical expression of above equivalent of R'–O–H is:

(2) An equivalent (R'OH) = $\dfrac{\text{(B in grams)}}{\text{(eq. wt. R'OH in grams)}}$

Mathematical expression of an equivalent weight is:

(3) Equivalent weight = $\dfrac{\text{(MW compd. in grams)}}{\text{(func. compd.)}}$

If (A) is X grams, and (B) is 100 grams in the following reaction:

$$\text{R–N=C=O} + \text{R'–O–H} = \text{urethane}$$
$$(X) \qquad\qquad (100)$$

Equivalents of (R–NCO) = $\dfrac{\text{(X in grams)}}{\text{(eq. wt. RNCO in grams)}}$

or = $\dfrac{\text{(X in grams)}}{\dfrac{\text{(MW RNCO in grams)}}{\text{(func. RNCO)}}}$

[from (3)]

Simplified:

(4) Eq. RNCO = $\dfrac{[\text{X (grams)}] \times \text{(func. RNCO)}}{\text{(MW RNCO in grams)}}$

Likewise,

Equivalents of R'OH = $\dfrac{\text{(100 grams)}}{\text{(eq. wt. R'OH in grams)}}$

or = $\dfrac{\text{(100 grams)}}{\dfrac{\text{(MW R'OH in grams)}}{\text{(func. R'OH)}}}$

[from (3)]

Simplified:

(5) Eq. R'OH = $\dfrac{100 \text{ grams} \times \text{(func. R'OH)}}{\text{(MW R'OH in grams)}}$

Since equivalents of RNCO react with equivalents of R'OH

Thermoset Polyurethanes   245

$$(4) = (5) = Eq.$$
$$Eq.\ RNCO = Eq.\ R'OH$$

Substitute values:

$$(6) \quad \frac{X\ \text{grams} \times (\text{func. RNCO})}{(\text{MW RNCO in grams})} = \frac{100\ \text{grams} \times (\text{func. R'OH})}{(\text{MW R'OH in grams})}$$

Isolate (X), the equation becomes:

$$(7) \quad (X\ \text{grams}) = \frac{100\ \text{grams} \times (\text{func. R'OH})}{(\text{MW R'OH in grams})} \times \frac{[\text{MW RNCO (grams)}]}{(\text{func. RNCO})}$$

Since $(\text{MW of R'OH}) = \dfrac{(\text{func. R'OH}) \times 56{,}100}{\text{OH \#}}$ [from (II)]

and substitute this into (7)

$$(X\ \text{grams}) = \frac{100\ \text{grams} \times (\text{func. R'OH})}{\dfrac{(\text{func. R'OH}) \times 56{,}100}{\text{OH \#}}} \times \frac{(\text{MW RNCO in grams})}{(\text{func. RNCO})}$$

Simplified, it becomes:

$$(8) \quad (X\ \text{grams}) = \frac{100\ \text{grams} \times \text{OH \#}}{56{,}100} \times \frac{(\text{MW RNCO in grams})}{(\text{func. RNCO})}$$

but, $\%\ \text{NCO} = \dfrac{(\text{MW -NCO in grams}) \times (\text{func. RNCO})}{(\text{MW RNCO in grams})} \times 100$ [from (VI)]

Isolate (MW RNCO) and the equation becomes:

$$(\text{MW RNCO in grams}) = \frac{(\text{MW -NCO in grams}) \times (\text{func. RNCO})}{(\%\ \text{NCO})} \times 100$$

Substitute this into the formulation in (8) and it becomes:

(9) $$X \text{ (grams)} = \frac{100 \text{ grams} \times OH\#}{56{,}100} \times \frac{(MW\ -NCO \text{ in grams}) \times (\text{func. RNCO}) \times 100}{\frac{\%\ NCO}{(\text{func. RNCO})}}$$

Simplified, it becomes:

(10) $$X \text{ (grams)} = \frac{OH\# \times (MW\ -NCO \text{ in grams})}{561 \times (\%\ NCO)} \times 100$$

Gram MW of NCO = N(14) + C(12) + O(16) = 42.02

$$= \frac{OH\# \times 42.02}{561 \times (\%\ NCO)} \times 100$$

(VII) $$X \text{ (grams)} = \frac{7.5 \times OH\#}{(\%\ NCO)} \text{ at stoichiometry.}$$

X = grams of % NCO product to react with 100 grams of polyol with certain OH # at stoichiometry.

However, urethane reactions require an experimentally determined Isocyanate Index (NCO/OH, equivalents ratio of isocyanate to polyol) of 0.95, 1.0, 1.05, or 1.10 of stoichiometry. Generally, a slight excess of isocyanate is used. Popular number is 1.05.

(VIII) $$X = \frac{7.5 \times OH\# \times NCO/OH}{(\%\ NCO)}$$

Again, X = grams of isocyanate compound with certain % NCO to react with 100 grams of polyol with certain OH # at a specified Isocyanate Index.

(IX) $$X' = 0.155 \times OH\# \times NCO/OH$$

X' = grams of TDI required to react with polyol with certain OH # and at a certain Isocyanate Index (NCO/OH).

Derivation: For pure toluene diisocyanate (TDI) the simplified calculation number becomes:

$$(\%\ NCO) \text{ of TDI} = \frac{(\text{func. -NCO}) \times (MW\ -NCO)}{(MW\ TDI)} \times 100$$

$$= \frac{(2 \times 42.02) \times 100}{174.2} = 48.22$$

Substitute 48.22 (% NCO) in (VIII) above.

$$X' = \frac{7.5 \times \text{OH \#} \times \text{Isocyanate Index}}{48.22}$$

Simplified:

(IX) $\qquad X' = 0.155 \times \text{OH \#} \times \text{NCO/OH}$

Again X' = grams of pure TDI required to react with 100 grams of polyol with certain OH # at certain Isocyanate Index (NCO/OH).

(X) $\qquad X'' = 0.22 \times \text{OH \#} \times \text{NCO/OH}$

X'' = grams of pure MDI required to react with 100 grams of polyol with certain OH # at certain Isocyanate Index.

Derivation: For pure methylene diisocyanate (MDI) the simplified calculation number becomes:

$$(\% \text{ NCO}) \text{ of MDI} = \frac{(\text{func. -NCO}) \times (\text{MW NCO})}{(\text{MW MDI})} \times 100$$

$$= \frac{(2) \times (42.02)}{250} \times 100 = 33.5$$

Substitute 33.5 (% NCO) in (VIII) above.

$$X'' = \frac{7.5 \times \text{OH \#} \times \text{NCO/OH}}{33.5}$$

Simplified:

(X) $\qquad X'' = 0.22 \times \text{OH \#} \times \text{NCO/OH}$

Again, X'' = grams of pure MDI required to react with 100 grams of polyol with certain OH # at certain Isocyanate Index (NCO/OH).

Above calculations are based on pure isocyanates. Allowances must be made for impure isocyanates by running % NCO and using these numbers in (VII) and (VIII) above.

Toluene Diisocyanate (TDI)

Isomers of TDI

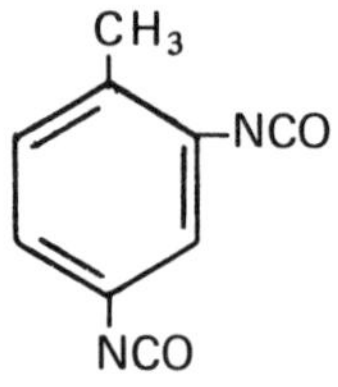

2,4-TDI

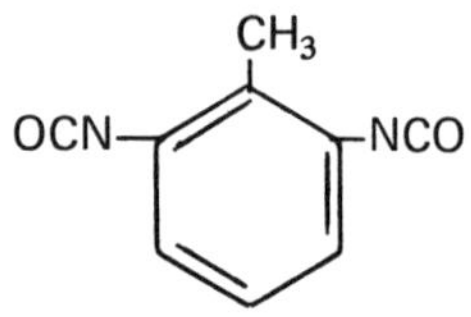

2,6-TDI

Gram molecular weight	=	174.2
Molecular weight -NCO	=	42.02 (N=14, C=12, O=16,)
% -NCO	=	48.22
Specific gravity	=	1.22 ± 0.01
Consists of isomers	=	2,4 and 2,6 normally at 80%/20% respectively
Boiling point (10 mm Hg)	=	248° ± 1.8°F (120° ± 1°C)
Flash point	=	270°F (132°C)
Freezing point	=	53.6° ± 1.8°F (12° ± 1°C)
Color	=	APHA 15

Derivation % NCO: The derivation of the % NCO is repeated here since it is widely used for calculations of reaction weights by the polyurethane industry. Molecular weight of pure TDI = 174.2. The molecular weight of the isocyanate group on the TDI is 84.04 (2 x 42.02).

$$\% \text{ free NCO} = \frac{(\text{func. NCO}) \times (\text{MW}-\text{NCO})}{(\text{MW TDI})} \times 100 = \frac{(2) \times (42.02)}{174.2} \times 100$$

$$= 48.22$$

$$X = \frac{7.5 \times \text{OH \#} \times \text{Isocyanate Index}}{\% \text{ free NCO}} \qquad [\text{from (VIII)}]$$

Substitute TDI's % NCO into above:

$$X = \frac{7.5 \times \text{OH \#} \times (\text{NCO/OH})}{48.22}$$

$$X = 0.155 \times \text{OH \#} \times \text{NCO/OH}$$

X = grams of TDI to react with 100 grams of polyol with certain OH # at certain Isocyanate Index (NCO/OH).

Discussion: Normal 80/20 TDI consists of two isomers. 80% 2,4-toluene diisocyanate and 20% 2,6-toluene diisocyanate. Furthermore, there are two versions of this 80/20 product available. A low acid-containing TDI (Type 1) used in one shot systems and a high acid-containing TDI (Type 2, i.e., Union Carbide TDI-P) used in making prepolymers for two-component systems. The high acid in the TDI contributes to the shelf stability of the prepared prepolymers. 65/35 isomer ratio TDI is available as well as the pure isomers themselves.

The significance of the position of the -NCO groups on the TDI molecule: The reactivity of the 2,6 positions on the aromatic ring is affected by the CH_3 group and is less reactive. The -NCO group at the 4 position is less hindered and is more reactive.

TDI is colorless when pure and fresh but it has a tendency to turn yellow with age and after exposure to the atmosphere.

TDI is normally a liquid at room temperatures and above, and it should be stored in a heated warm room at temperatures close to 80°F (27°C). When it is subjected to temperatures below 57°F (14°C), crystals of TDI will begin to precipitate out of the liquid. The isomers do not precipitate out at the ratio of the mixture. If subjected to cold temperatures long enough, the whole system will turn into crystals. If partial or total crystallization occurs, the product should be brought into a warm room and allowed to thaw and become liquid again. The reconstituted liquid should be thoroughly mixed to regain its original composition. Generally, there is no detrimental effect that can be detected after the TDI is properly thawed and mixed. Attaching heat bands or placing crystallized TDI containers on hot plates is certainly to be avoided as a means of reliquefaction.

Isocyanates undergo dimerization to form a uretidione in the presence of tertiary amines. This reaction is reversible at elevated temperatures and the dimer will dissociate and undergo the expected isocyanate reactions. These reactions are more thoroughly discussed in the prepolymer synthesis section.

Heating TDI along with phosphine oxides and phospholines has produced polymeric carbodiimides.

TDI based coating products turn yellow with age or upon exposure to sunlight. Historically, removing the unsaturation has eliminated the yellowing of the isocyanate based products. Where yellowing is not a problem, TDI is still the preferred diisocyanate of choice.

Hazards of Using TDI: The use of TDI requires proper ventilation to prevent inhaling the vapors, which causes difficulty in breathing in some individuals. Liquid TDI can cause burns on mucous membranes. Skin sensitization generally is *not* a problem with TDI. If TDI is not immediately removed with plenty of water and is allowed to remain on the skin, it will produce redness, swelling, and blistering. Even if TDI is taken off immediately, signs will remain on the skin that TDI had reacted with the skin to make it feel papery after a period of time.

Methylene Diphenyl Diisocyanate (MDI)

$$OCN-\phenyl-CH_2-\phenyl-NCO$$

4,4'-Methylene Diphenyl Diisocyanate (MDI)

Gram molecular weight	= 250.00
% -NCO	= 33.5
Room temperature state	= Solid
Density at 77°F (25°C) g/cm^3	= 1.22
Initial boiling point, 5 mm Hg, °F	= 374°F (190°C)
Melting point, °F (°C)	= 100 (38)
Decomposition temperature	= ~446°F (230°C)

Discussion: MDI is extremely sensitive to moisture degradation. Normally solid at room temperatures, it requires careful inventory control for longevity. Since the product slowly dimerizes in prolonged room temperature storage, this is not recommended. MDI should be stored at temperatures below 41°F (5°C). Stability for 3 months has been noted at these low temperatures.

MDI must be heated prior to use to 122°F (50°C) for 4 to 8 hours to liquefy and allow the dimers to precipitate. The precipitate is discarded and the supernatant portion is used for reacting. There will be very little noticeable change if MDI is kept at 104° to 122°F (40° to 50°C) for up to 14 days in the absence of moisture and sunlight. To minimize the dimer formation, MDI must not be heated above 158°F (70°C).

Heating MDI on a hot plate or by the use of heating bands is not recommended due to degradation by overheating at the point of heat application.

MDI is available in several modified forms from Upjohn Co. or from BASF Wyandotte Corp. Some modified MDIs contain relatively high levels of pure MDI but are mixtures of MDI and other aromatic diisocyanates and are generally liquid at room temperatures. Other modifications have high levels of pure MDI and are mixtures of MDI and polymeric aromatic diisocyanates and are also liquid at room temperatures.

Hazards of Using MDI: MDI is hazardous and is an irritant to the skin, eyes, and mucous membranes. It may cause allergic respiratory reaction if inhaled. Protective clothes, goggles, and fume trapping respirators are recommended when handling MDI. MDI on the skin should be removed immediately with soap and water.

Methylene Dicyclohexane-4,4'-Diisocyanate (Saturated MDI)

$$OCN-\bigcirc-CH_2-\bigcirc-NCO$$

Saturated MDI

Physical form	=	Water white, fused solid
Molecular weight	=	262.4
Equivalent weight	=	131–133
% NCO	=	31.8
Boiling point, °F	=	352°–363°F (178°–184°C)
Settling point, °F	=	108°–118°F (42°–48°C)

Discussion: Saturated MDI should be stored in air tight sealed containers at temperatures above 68°F (20°C). It should not be stored in glass containers due to its reaction with glass surfaces.

If crystallization occurs, heat to 100° to 122°F (37.7° to 50°C) in an appropriate heating environment to melt the crystals and stir well before use.

Saturated MDI is a very slow reacting diisocyanate, and unlike aromatic diisocyanates, a catalyst is required in the preparation of adducts or prepolymers. Cocure 30, a mercury catalyst from Cosan Chemical can be used.

The material cost of the product is significantly higher than MDI. Saturated MDI is popularly used as the basis for nonyellowing gloss retentive premium grade interior and exterior coatings. The use of saturated MDI in polyurethane structures prevents applied paint from being stained by color change. Because of this property, saturated MDI has lately become the diisocyanate of choice for substrate structures.

The saturation of MDI also contributes to higher service temperature use for the final product.

Hazards of Using Saturated MDI: Saturated MDI has a low odor level and this factor causes users to become careless with it. It should be used with adequate ventilation.

Saturated MDI, especially when it is handled at elevated temperatures, can cause severe irritation to the skin and mucous membranes. It is a strong skin sensitizer. Contact with eyes and skin, and inhalation should be avoided. Protective clothing, gloves, goggles, and respiratory protection should be used when handling saturated MDI.

Isophorone Diisocyanate (IPDI)

$$(CH_3)_2\text{—}\underset{\underset{CH_3}{|}}{\bigcirc}\underset{CH_2\text{—}N=C=O}{\overset{N=C=O}{|}}$$

3-Isocyanatomethyl-3,5,5-Trimethylcyclohexyl Isocyanate

Physical form	=	Colorless liquid
Molecular weight	=	222.3
Equivalent weight (NCO)	=	111.1
% –NCO	=	37.8
Specific gravity @ 68°F	=	1.058
Boiling point, °F	=	304°F (151°C)
Flash point (closed cup)	=	311°F (155°C)

Discussion: A low viscosity liquid with no tendency to crystallize when subjected to low temperatures. Due to its relatively high molecular weight the vapor pressure is very low which greatly simplifies handling this isocyanate.

IPDI is a slow reacting diisocyanate. The IPDI reactions with polyols can be influenced by most of the same catalysts that are used in toluene diisocyanate reactions (tin catalysts, DABCO, lead and zinc octoates, and mercury compounds).

IPDI is popularly used as the basis for nonyellowing polyurethane paints and varnishes for interior and exterior uses. Generally, when used with branched chain polyols, hard but flexible abrasion and chemical resistant coatings with good color retention and resistance to ultraviolet light, chalking and weathering are produced.

Hazards of Using Isophorone Diisocyanate (IPDI): The available data on the acute toxicity of isophorone diisocyanate indicate deleterious effects on health, which usually occur when organic isocyanates are handled carelessly. However, due to the relatively low vapor pressure of isophorone diisocyanate and the resulting relatively low concentrations of isocyanate vapor, the hazards are as a rule much reduced.

The use of protective clothing, gloves, goggles, and respiratory protection is required when handling this product.

(The above data was extracted from IPDI Technical Data Sheet. –Chemsche Werk Hüls AG, Germany.)

Determination of % Free –NCO

Reagents and Equipment:

(a) 2 N di-n-butylamine.

Preparation: Weigh 259 grams di-n-butylamine into a one liter volumetric flask. Dilute to mark with distilled toluene.
(b) 1.00 N HCl or known normality hydrochloric acid close to 1 Normal.
(c) Bromophenol blue indicator.
Preparation: Dissolve 0.1 grams powder in 1.5 ml 0.1 N NaOH.
Add 100 ml water by pipette.
(d) Isopropyl alcohol.
(e) 125 ml Erlenmeyer flasks.
(f) Hotplate.
(g) 50 ml burette.
(h) Tilt-up pipettes.

Procedure:

(1) Weigh 1 gram of the isocyanate material into a dry 125 ml conical Erlenmeyer flask. Add 25 ml distilled toluene. (Use tilt-up pipette dispenser.)
(2) Pipette 20.0 ml of amine reagent into the flask.
(3) Stopper the flask loosely. Swirl the flask to insure complete dissolving of the isocyanate compound and heat, while swirling, just to boiling on a hot plate. Allow to cool slightly. Add 100 ml isopropyl alcohol using tilt-up pipette.
(4) Add 0.5 ml indicator.
(5) Titrate to yellow end point with 1.00 N HCl. Record in ml.
(6) Run a blank.

Calculation:

$$\text{Amine equivalent} = \frac{(\text{wt. sample}) \times 1{,}000}{(\text{ml blank} - \text{ml sample}) \times N}$$

$$\% \text{ free NCO} = \frac{(\text{Gram MW of -NCO}) \times 100}{\text{Amine Equivalent}}$$

$$= \frac{42.02 \times 100}{\frac{(\text{wt sample} \times 1{,}000)}{(\text{ml blank} - \text{ml sample}) \times N}}$$

42.02 = gram MW of -NCO; N = 14; C = 12; O

$$\% \text{ free NCO} = \frac{42.02 \times (\text{ml blank} - \text{ml sample}) \times N}{\text{wt sample} \times 10}$$

Conversion factor:

$$\text{Amine equivalent} = \frac{4{,}200}{\% \text{ free NCO}}$$

Helpful Hints: Use dry Erlenmeyer flasks. Heat flask and flush inside surface with a blast of dry nitrogen to drive off moisture from the inside surface of the flask, and stopper the flask with a cork to protect it from intrusion of atmospheric moisture redepositing itself on the inside surface.

Weigh sample as closely to 1 gram as possible, but do it quickly. If many tests are going to be run, tilt-up pipette dispensers for toluene and isopropyl alcohol are a great convenience and time saver.

Titrate barely to the end point. Do not over titrate. Using this careful method, the flask can be emptied, and then rinsed three times with small amounts of clean acetone and blast dried with heated air and the flask can be used immediately for the next titration or it can be corked to be ready for the next free isocyanate test.

Theory: The isocyanate is reacted with di normal butylamine to form a urea. The excess amine is back titrated with standard HCl solution.

The amine equivalent is the equivalent weight of the molecule containing one equivalent of the isocyanate.

Determination of Hydroxyl Numbers

Equipment and Materials:

(a) 3 dried 500 ml Erlenmeyer Flasks with ground glass joints.
(b) 3 dried air condensers with ground glass joints.
(c) Reagent grade dry pyridine.
(d) Best commercial grade Butanol.
(e) Phenolphthalein in dry pyridine.
(f) Reagent grade acetic anhydride.
(g) Hot water bath.
(h) Standardized 0.5 or 1.0 N NaOH.
(i) 50 ml burette.
(j) Bath on hotplate.
(k) Fume hood.

Procedure:

(1) Weigh about 1 gram of hydroxy compound into each of two 500 ml flasks.
(2) Pipette into the two flasks with samples and into a third flask, 10.0 ml of a mixture of 3 volumes of dry pyridine and one volume of acetic anhydride.
(3) Attach an air condenser to each of the three flasks and submerge them into gently boiling water to the depth of the liquid level.

(4) Swirl the flasks after 10 minutes to thoroughly dissolve the hydroxyl compounds.
(5) Continue heating for 40 to 50 minutes, occasionally swirling the flasks.
(6) After this period, add 10 ml of distilled water through each of the three air condensers, carefully and thoroughly washing down the materials in the tubes.
(7) Heat for 2 to 5 minutes more.
(8) Cool thoroughly. Remove the air condensers and wash down the sides of the flask with 10 ml of butanol.
(9) Add at least 1 ml of phenolphthalein and titrate with 0.5 N or 1.0 N NaOH to a slightly pink endpoint.

Calculations:

$$\text{Hydroxyl number} = \frac{(\text{ml blank} - \text{ml sample}) \times N \times 56.1}{\text{grams of sample}}$$

Acetic anhydride is an esterifying agent which reacts with the hydroxyl groups to form esters and organic acids. By determining the quantity of the anhydride reacted with the sample, the titrations with NaOH produce results which are reported as mg KOH equivalent to 1 gram of sample. Add acid number to above hydroxyl number for true hydroxyl number.

Determination of Isocyanate Equivalents

If the compounded polyol is titratable using the following procedures, one can determine exactly the isocyanate equivalent of the system and also determine the Isocyanate Index. But this will not be the case in those components that contain strong bases. In such cases, the procedure will not produce desired results.

Materials and Equipment:

(a) Standardized 1.0 N HCl.
(b) 2.0 N di-n-butylamine in toluene.
(c) Dried toluene. (Dried by adding 8 mesh Molecular Sieves in appropriate amounts to the toluene reservoir and cap bottle.)
(d) Methanol.
(e) Catalyst such as N-methyl morpholine, triethyl amine or other tertiary amines.
(f) Bromophenol blue indicator.
(g) 250 ml dried Erlenmeyer flasks with ground glass joints.
(h) Dried (bore), reflux condensers.
(i) 2.25 N solution of phenyl isocyanate in toluene.
(j) 50 ml burette.
(k) Bath on hotplate.

Procedure:

(1) Weigh into each of two dry flasks about 0.03 equivalent (based on estimated active hydrogen atoms) of compounded polyol material.
(2) Add 10 ml of dry toluene into each flask and to a third dry flask.
(3) Add 6 drops of N-methyl morpholine catalyst into the three flasks.
(4) Attach reflux condensers and introduce through each condenser, 20 ml of a 2.25 N solution of phenyl isocyanate in toluene.
(5) Rinse down each condenser with 20 ml of dried toluene and attach drying tubes to the tops of the condensers.
(6) Submerge the flasks in gently boiling water to the level of the fluids and reflux for 1 hour.
(7) Raise the flasks out of the water and cool. Then introduce 25 ml of a 2 N toluene solution of di-n-butylamine through each condenser. Swirl flasks for complete dissolution.
(8) Place flasks again into gently boiling water to the level of the liquids and reflux additional 15 minutes. No longer!
(9) Remove flasks from the bath and allow them to cool.
(10) Rinse each condenser down with 100 ml of methanol.
(11) Add bromophenol blue and titrate to endpoint with standardized 1 N HCl.
(12) Run a blank.

Calculations:

$$\text{Isocyanate equivalent} = \frac{1{,}000 \times \text{wt. of sample}}{(\text{ml sample} - \text{ml blank} \times \text{N of HCl})}$$

Method for Preparation of TDI Prepolymers

(Excess isocyanate is reacted with the polyols and the resulting molecules have only –NCO end groups.)

Equipment:

(a) Laboratory. (The procedures should be carried out in a chemical exhaust hood. The hood should be equipped with heating and cooling water, and drain systems.)
(b) Glass or metal container to hold required amounts of reactants and attached with a bath facility for cooling or heating the reacting mass with water.
(c) Air stirrer or electric stirrer. (Air preferred)
(d) Nitrogen blanketing system.
(e) Compressed cylinder of dry nitrogen.

(f) Flow valve with meter.
(g) Tubings to inject nitrogen into the reactor environment.
(h) Thermometer with 250°F reading.
(i) Vacuuming system. (Bell jar, vacuum plate, vacuum pump)

Figure 7-16 illustrates the equipment.

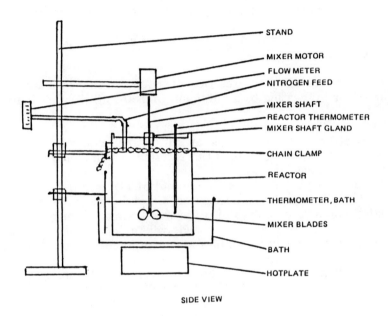

SIDE VIEW

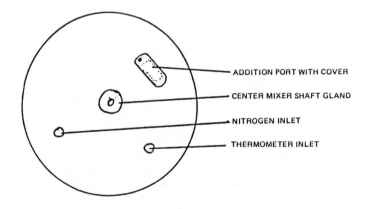

TOP VIEW OF REACTOR COVER

Figure 7-16: Reaction kettle diagram.

The simplest reactor is a closed tin can sitting in a pan of sufficient depth to hold water for the required heating or cooling. The tin can should be held in position.

Holes should be punched into the lid to accommodate:

(1) The pass through of the mixing blade from the mixer motor in the center of the lid.

(2) An inlet for dry nitrogen into the reactor (to maintain very slight positive internal pressure) at one edge.

(3) A closeable opening for adding materials into the reactor.

(4) A hole to pass through and hold the thermometer at one edge.

Procedure: (1) Dry the reaction vessel by storing in a dry environment or by heating to a minimum of 180°F (82°C) just before use.

(2) Set up the reactor as in the diagram.

(3) Start flow of dry nitrogen at 2 liters per minute for a gallon size vessel. Maintain flow until the reaction is completed and the container is removed and sealed.

(4) Carefully pour the required amount of TDI into the reactor.

(5) Start the mixer, at slow speed. Carefully pour the required calculated amounts of polyol into the vessel while the mixer is stirring. Continue stirring at slow speed, where there is just enough turbulence but not where bubbles are incorporated into the reacting mass.

(6) Exothermic heats. For those reactants which can exotherm by themselves to 180° to 185°F (82° to 85°C), cold water is used to control the temperature of the reacting mass, and initially a trial and error method is required to determine the precise level of the heat controlling medium in the jacket to maintain the reaction temperatures between 180° to 185°F (82° to 85°C). Care must be taken to prevent the exotherm from rising above 185°F (85°C).

Test sample for running "free NCO" is taken immediately after the exothermic temperature "peaks." The bulk of the reaction has occurred at this point and satisfactory adduction has taken place and the reactor can be shut down, although stirring should continue until the temperature drops below 150°F (65°C) (see Step 7 for handling of reactants). Free -NCO tests should be run quickly on the samples. Initially tests should be repeated at 5 minute intervals after peaking to ascertain that the reaction indeed has been satisfactorily completed. Repeat test results should be within 0.05% of each other and should coincide with theoretical free NCO of the reactants. The above describes the reaction of polyether polypropylene polyols of molecular weights below 3,000 with TDI.

For those reactions that require the addition of heat to initiate the reaction, the normal practice is to obtain reaction temperatures of 180° to 185°F (82° to 85°C) by the use of hot water in the bath to initiate and facilitate the reaction of the sluggishly reacting masses and to maintain these temperatures for at least one hour. Care must be taken to prevent the exothermic temperatures from rising beyond 185°F (85°C). At a crucial point (as determined by trial and error), hot water replacement by cold water may be required to control the exothermic temperatures from rising beyond 185°F (85°C). A free NCO test is taken after the reaction has been in progress for a minimum of one hour at 180° to 185°F (82° to 85°C). Shorter periods of "cooking" time can be determined by initiating the free NCO test earlier, after some peaking of the temperature is noted. Free NCO tests should be run until they are within 0.05% of each other and tests should coincide with the known theoretical free NCO of the reactants. Generally, polypropylene polyether polyols with a molecular weight above 3,000 reacted with TDI is described by the latter procedure. (A recording temperature chart is ideal for obtaining up-to-the-minute visual information of heat generation by the reactants.)

(7) After the test samples are taken and the results coincide and are close to theoretical values, remove the reacting vessel, with its contents still hot, from the reaction environment and wipe the outside dry and place the whole reaction vessel immediately in a vacuum bell jar.

(8) Vacuum until the bubbling subsides.

(9) Remove the vessel from the vacuuming chamber, place the lid partially onto the vessel and fill the space above the liquid with a blanket of dry nitrogen and seal the vessel for storage and later use.

Discussion: (Caution—A blanket of dry air or nitrogen should be maintained on the vessel during the entire process.)

The polyol should be charged into the reaction at a temperature between 70° to 80°F (21° to 27°C).

The type of TDI required for adducting is TDI Type II (high acid content TDI). Type II TDI will avoid a runaway reaction that occurs when low acid TDI is reacted with certain polyols. Furthermore the high acid TDI will contribute to shelf stability of the prepolymers.

Temperatures below 57°F (13°C) will cause uneven fractional crystallization of the isomers. The temperature of the TDI must be maintained well above its freezing point of 57°F (13°C). If crystallization does occur, warm the TDI to about 80°F (27°C) in a warm room and agitate before use. TDI must be added at temperatures above 68°F (20°C).

If reaction temperatures are allowed to exceed 200°F (93°C), there is the possibility of a runaway reaction occurring. Higher temperatures favor allophanate, urea, biuret type reactions, whereas, lower temperatures promote the linear urethane reactions. Below 122°F (50°C) -NCO does not ordinarily react.

In small batches, all the polyols can be added at the beginning. When reacting large batches that have tendencies to exotherm to temperatures above 200°F (93°C), the alternative method to the one mentioned above is to first add only half of the polyol. Let the exotherm occur. Then, allow the initial batch to cool and when the reaction is under control add the remainder of the required polyol.

Once the prepolymers are completed, extra high heat should not be applied to the prepolymer. The prepolymers are temporarily in an unstable condition and cool temperatures maintain this precarious state. Heat stresses alone, such as temperatures over 212°F (100°C) can lead to significant reductions of the isocyanate content and increases in the viscosities of the prepolymer. Even gelation of the prepolymer can occur, and ultimately, result in inferior physical properties. Upon excessive heating the biuret and allophanate reactions can occur within the polymer mass using up the free isocyanate radicals. Again, heat alone can contribute to the formation of these larger complicated polymers.

Allophanation can further be accomplished in the prepolymer system by:

(1) The second addition of TDI to the prepolymer system after sufficient linear polymerization has taken place.

(2) Heating and/or adding catalysts. (Optimum allophanation in prepolymer techniques are generally in the range of 212° to 248°F (100° to 120°C) and between 248° to 428°F (120° to 220°C). At these high temperatures, the reactions are hard to control. Above 428°F (220°C) a significant reverse reaction is noted.)

As described previously, adding the amine catalyst to the prepolymer is not desirable since amines will catalyze the reaction and may form dimers (diazides) and trimers (triazides). These are insoluble and may precipitate out as grains. Amines should not be added to the isocyanate adduct until just before application.

Water in the polyols—When water content is varied below 0.1% little effect is noted on viscosity build up in the preparation of prepolymers. However, above 0.1% the formation of multi-functional products leading to branching and crosslinking causes more rapid viscosity build up. Extra high water content in the polyol portion of the prepolymer is not desirable since this may cause unwanted crosslinking at the wrong time.

Method for Preparation of MDI Prepolymers

Use the equipment set-up and processes described previously for TDI, but make the following changes:

(A) Melt MDI at 122°F (50°C) for 4 to 8 hours prior to use in the reactor. At this length of time the dimerized MDI will crystallize out of the melted liquid MDI. Decant the liquid MDI and charge the liquids at 122°F (50°C) or higher into a heated reactor *and start agitation* and flow of dry nitrogen.

(B) Add heated dry polyols (hot enough to maintain the melted stage of MDI now in the reactor) to the system at a rate where the reaction temperatures will not surpass 158° to 176°F (70° to 80°C). Trial and error method will determine the amount of heat control agent required in the jacket.

(1) Obtain reaction temperatures of 180° to 185°F (82° to 85°C) and maintain these temperatures for one hour while stirring and continuing the flow of dry nitrogen. Care must be taken to prevent reaction temperatures from rising beyond 185°F (85°C).

(2) After one hour or after temperature peaking is noted, take a sample and run a free −NCO test. Initially, free −NCO tests should be run until they are within 0.05% of each other and they should coincide with the *known* theoretical free −NCO of the reactants. Shorter "cooking" time for the reactants can be determined in the same manner.

(3) Vacuum the hot mix and seal the container.

Discussion: (see TDI discussions.)

Raw Material Sources

Air Products & Chemicals, Inc., Allentown, Pennsylvania
American Cyanamid Co., Bound Brook, New Jersey
BASF Wyandotte Corp., Wyandotte, Michigan
Dow Chemical U.S.A., Midland, Michigan
Du Pont, Wilmington, Delaware
Mobay Chemical Corp., Pittsburgh, Pennsylvania
Olin Corporation, Stanford, Connecticut
Rubicon Chemicals Inc., Wilmington, Delaware
   (Importing technology from ICI Europe)
Hooker Chemical, Ruco Div., Hicksville, New York
Texaco Chemical Co., Bellaire, Texas
Union Carbide Corp., Canbury, Connecticut
Uniroyal Chemical, Naugatuck, Connecticut
Upjohn Polymer Chemicals Div., LaPorte, Texas

Tradenames	Source
Diisocyanates	
TDI, Toluene diisocyanate	
Mondur® TD 80	Mobay.
Niax®	Union Carbide.
Others	Olin, BASF Wyandotte, etc.
MDI, Methylene diphenyl-diisocyanate	
Lupranate® M	BASF Wyandotte.
Isonate® 143L "Liquid MDI"	The Upjohn Co.
MDI-PH	Mitsui-Nisso Corp.

Tradenames	Source
Diisocyanates	

Mixed isomers
   Isonate® 125M, 240, 181   The Upjohn Co.
   Lupranate® M-10, MP
      102, 103   BASF Wyandotte.
MDI-CR   Mitsui-Nisso Corp., Japan.
Saturated MDI   Uniroyal Chemical.
Isophorone diisocyanate   Chemische Werk Hüls AG,
      Germany.

Para-phenylene diisocyanate
   Elate® 160   Akzo Chemie America.
Trans-1,4-cyclohexane
   diisocyanate
      Elate® 166   Akzo Chemie America.
Naphthylene-1,5-diisocyanate   Mitsui-Nisso Corp., Japan.
Ortho-tolidine diisocyanate   Mitsui-Nisso Corp., Japan.

Aromatic Amines

MOCA® (4,4-bismethylene
   orthochloroaniline)
      Bis Amine A   Wakayama Seika Kogyo Co.,
          Japan.

MDA, Methylene dianiline
   Tonox®   Uniroyal Chemical.
Methylene bis-methyl anthra-
   nilate
      Poly-Cure® 1,000   P T M & W, Santa Fe Springs,
          California.

Meta xylelene diamine
   MXDA   Sherwin Williams Chemical.
1,3-Bis(aminomethyl)cyclo-
   hexane
      1,3-BAC   Sherwin Williams Chemical.
Trimethylene glycol di-para-
   aminobenzoate
      Polacure®   Polaroid Corp.
1,2-bis(2'-aminophenylthio)
   ethane
      Cyanacure®   American Cyanamid.

Aromatic Curing Agents

Hydroquinone di(beta-hydroxyethyl) ether
      HQEE   Eastman Chemical Products,
          Inc.

Tradenames	Source

Catalysts

Tradenames	Source
Triethylene diamine	
DABCO®	Air Products & Chemicals, Inc.
Tin catalysts	
Stannous octoate,	
M & T® T-9	M & T Chemicals.
Dibutyltin dilaurate,	
M & T® T-12	M & T Chemicals.
SUL-4	Witco Chemical Corp.
820	Ferro Corp.
DBTDL	Cincinnati Milacron.
Morpholines (morpholines, N methyl-, N ethyl-)	
Thancat®	Texaco.
Organo mercury compounds	
Co-Cure®	Cosan Chemical Corp., Clifton, New Jersey.
Lead octoate	
Nuodex®	Witco Chemical Corp.
Lead naphthanate	
Nuodex®	Witco Chemical Corp.
X8161, A tertiary amine	Air Products & Chemical Company.
Polycat® DBU, SA	Abbott.
Fomrez® UL1-1, UL1-28, UL1-29, Organo-tin	Witco Chemical Corp.
Fomrez® UL-6, UL-24, UL-32, etc.	Witco Chemical Corp.

Driers

Tradenames	Source
Molecular sieves	
Linde®	Union Carbide.
Baylith®	Mobay.
Drierite®	W A Hammond Drierite Co.

Diluents

Tradenames	Source
N-methyl pyrrolidone	
M-Pyrrol®	Monsanto.
Triethylene glycol, dicaprate and dicaprylate	
Plasticizer SC	Drew Chemical Corp.
Hallco® Plasticizer 4141	CP Hall Co.
Dipropylene glycol dibenzoate	
Benzoflex® 9-88S	Velsicol Chemical Corp.

Tradenames	Source
Polyhydroxyl Compounds	
Polypropylene polyether glycols	
Voranol®	Dow Chemical.
Niax®	Union Carbide.
Pluracol®, Quadrol®	BASF Wyandotte.
Poly-G®	Olin Chemicals.
Rucoflex®	Ruco Polymer Corporation.
Thanol®	Texaco.
Caprolactones	
Niax®	Union Carbide.
Polytetramethylene oxide diols	
Polymeg®	Q O Chemicals Incorporated.
Teracol®	Du Pont.

Pertinent ASTM Test Methods

Tensile strength, psi	ASTM D 412
Elongation, %	ASTM D 412
Modulus at 100% and 300%	ASTM D 412
Tear strength, pli notched die C	ASTM D 624
Hardness, Shore A and D Durometer	ASTM D 2240
Compression set, % constant deflection	ASTM D 395
Compressive properties	ASTM D 575
Abrasion resistance	TABER ASTM D 1044
Heat deflection	ASTM D 648
Hydrolytic stability	ASTM F 74
Immersion	ASTM D 471
Low temperature brittleness	ASTM D 736
Compression fatigue	ASTM D 623
Vertical rebound	ASTM D 2692

REFERENCES

1. Duzey, R.H., Weigel, J.E., Niax Polyether in Urethane Coatings, Union Carbide Corp.
2. Eagle, G., Ren Plastics, Polyurethane Sealants Challenge Vinyl Plastisols in Automotive Field, *Plastics Engineering*, p 29, (July 1980).
3. Wilson, J.B., Isocyanate Adhesives as Binders for Composition Board, *Adhesive Age*, p 41 (May 1981).
4. *Plastics Technology*, p 22 (August 1980).
5. Hirosawa, F.N. U.S. Patent 3,265,669; August 9, 1966; assigned to Furane Plastics Inc.
6. Hagen, E.L., MDI-Prepolymer Systems, Uniroyal Chemical.

7. Taller, R.A., Coe, J.A., Polylactone Polyols Expand Scope of Urethanes, Union Carbide Corp., *Modern Paint and Coatings*, p 58 (March 1975).
8. *Plastics Technology*, p 29 (January 1985).
9. von Hassell, A., For Urethanes, Flourishing R & D in Materials and Additives, *Plastics Technology*, p 75 (January 1985).
10. Wilwerth, L.O., K.J. Quinn & Co., Inc., Aliphatic Polyurethanes; Has Their Time Come? *Plastics Engineering*, p 25, (January 1984).
12. New Urethanes Take Hard Line, *Chemical Week*, p 45 (March 20, 1985).
13. Currier, V., Jefferson Chemicals, How to Compound High-Density Urethane Foams, *Plastics Technology*, p 35 (August 1966).
14. Anagnostou, T., Wyandotte, Synthesis of Blocked MDI Adducts, etc., *Journal of Coating Technology*, p 35 (February 1981).
15. Isophorone diisocyanate IPDI, Veba-Chemie AG.
16. Castor Oils and Chemical Derivatives, Baker Castor Oil Co. (1962).

8
Commercial Polyimides

Abraham L. Landis
Materials Science Department
Technology Support Division
Hughes Aircraft Company
El Segundo, California

After World War II, rapid expansion of the aerospace industry created a need for new materials for the new high technology which it generated. Many of these requirements could not be achieved with metallic and existing plastic materials. There was a need for materials which were lightweight, oxidatively and thermally stable, had good mechanical properties and could operate in space environments. During the last thirty years many new polymers were discovered which had unusual high temperature properties. However, relatively few achieved commercial success. There were many reasons but the main ones were because of their high potential cost and difficulty in being fabricated into useful forms. One class of polymers, the polyimides, was successfully introduced as commercial materials in the early sixties by E.I. du Pont de Nemours and Company.[1,2] The continued success of these polymers as commercial materials was due to the availability of inexpensive starting material and the ability to tailor these polymers to a variety of end uses.

The polyimides are characterized as having the imide structure (I) in the polymer backbone. This structure has exceptional thermal and oxidative stability.

I

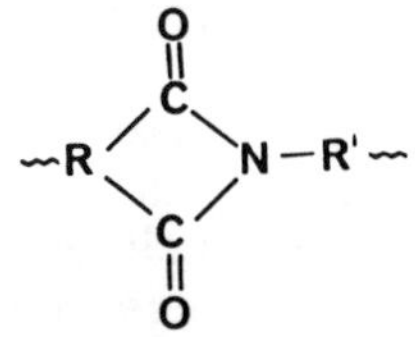

$$\sim R \diagup \substack{\displaystyle \overset{O}{\underset{\|}{C}} \\ \diagdown \\ \underset{\|}{\underset{O}{C}}} \diagdown N-R' \sim$$

These polymers have been prepared from a variety of starting monomers. However, the best thermooxidative properties have been achieved using aromatic structures for R and R'. Evidently it is necessary to have the imide structure along with the aromatic moieties because wholly aromatic polymers such as polyphenylenes are not as oxidatively resistant as the polyimides. The combination of the aromatic structure with the imide structure results in a polymer with a very high glass transition temperature (T_g) and good oxidative stability. The polyimides can be processed either as thermosetting or as thermoplastic resins depending on the polymer structure or the route chosen in processing the prepolymer into its final form or shape. The polyimides can be conveniently broken down into three categories depending upon their processing characteristics. One of these categories includes thermoplastic polyimides. However, the very high glass transition temperature of these materials overlap the glass transition temperature of some of the thermosetting polyimides. This category is included with the thermosetting polyimides because they represent an important advancement in polyimide processing.

The polyimides in the first category are processed via precursors which undergo condensation reactions to form the final product. The precursors are usually the amic acid or the amic ester intermediate. These precursors are soluble and tractible and are used in the form of lacquers. Upon conversion to the imide, the polymer is rendered insoluble and intractible. The scheme can be illustrated by the following generalized structures:

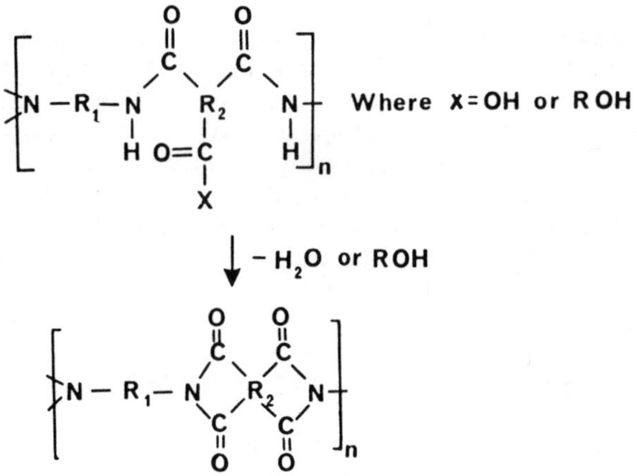

Where X = OH or ROH

$$\downarrow -H_2O \text{ or ROH}$$

The cyclodehydration step requires temperatures greater than 284°F (140°C). Generally temperatures up to 572°F (300°C) are used to ensure cyclization. It is also possible to effect cyclization by chemical dehydration with reagents such as aliphatic acid anhydrides, ketenes and strong Lewis acids such as phosphorous trichloride. The formation of the imide by these reagents can also proceed via the thermally unstable isoimide.

The polyimides in the second category owe their processibility to their thermoplasticity above their glass transition temperature. They are processed much in the same way as thermoplastics except at much higher temperatures. The thermoplasticity of polyimides is accomplished by modifying the polymer backbone or by the introduction of pendant side chains. In some cases these polyimides have sufficient solubility in select solvents to permit the formulation of lacquers so they can be used as coatings or laminating resins.

The third category owes their processibility to the use of soluble and fusible short chain polyimide oligomers which have homopolymerizable functional groups. These permit the polymer chain to grow preferably by reactions which do not evolve volatile by-products. The most successful prepolymers were based on bismaleimides (II) and acetylene terminated polyimides (III).

II

III $HC \equiv C-R_1-N \left[\begin{array}{c} O \\ \parallel \\ C \\ / \\ \diagdown \\ C \\ \parallel \\ O \end{array} R_2 \begin{array}{c} O \\ \parallel \\ C \\ \diagdown \\ / \\ C \\ \parallel \\ O \end{array} N-R_3 \right]_n N \begin{array}{c} O \\ \parallel \\ C \\ / \\ \diagdown \\ C \\ \parallel \\ O \end{array} R_2 \begin{array}{c} O \\ \parallel \\ C \\ \diagdown \\ / \\ C \\ \parallel \\ O \end{array} N-R_1-C \equiv CH$

These prepolymers undergo homopolymerization upon heating. In order to obtain a processible prepolymer it is necessary that they do not homopolymerize below their melting point and that they have reasonable solubility in a number of solvents that can be used as a lacquer for prepregs. Thus, crosslinked polyimides have been made by thermal polymerization of bismaleimides, preferably by heating them above their melting point in the presence of free radical catalysts such as dicumyl peroxide and also from acetylene-terminated polyimides by thermal polymerization.

POLYIMIDES FROM CONDENSATION REACTIONS

The most important polyimides of this first group are based on the monomer, pyromellitic dianhydride (PMDA), (IV)

IV

[PMDA structure: pyromellitic dianhydride]

This monomer is manufactured by Du Pont Company and obtained by the vapor phase oxidation of durene (1,2,4,5-tetramethylbenzene), using a supported vanadium oxide catalyst. Certain aromatic amines, namely m-phenylenediamine (MPD) (VII), benzidine and di-(4-aminophenyl)ether (ODA) (V) give polymers having a high degree of oxidative and thermal stability. The synthesis is carried out by a two stage process[3] which involves the preparation of the intermediate polyamic prepolymer and then conversion of the prepolymer to the imide form under a specific cure cycle. The imidization process has been studied using infrared spectroscopy. In the temperature range of 220°-250°C, complete imidization takes place. Higher temperatures appear to cause crosslinking by the formation of an interchain network. Russian workers report that at the higher temperatures the structurization depends on reversible cleavage of $-CO-N-$ bonds[4] and their recombination. The Du Pont Company has marketed a number of polyimides based on PMDA (IV) and the aromatic amine ODA (V) namely Kapton® (supplied as a film), Vespel® (supplied as sintered parts) and Pyralin® and Pyre ML® supplied as a lacquer of the polyamic prepolymer.

V H_2N—⟨⟩—O—⟨⟩—NH_2

The Kapton® film is supplied in several forms: Type H is the uncoated polyimide film. Figures 8-1 and 8-2 show two major applications of this film. Type V is similar to Type H but with superior dimensional stability and Type F which is a Type H film coated on one or both sides with Teflon fluorinated ethylene propylene (FEP) resin to provide a moisture barrier and to enhance chemical resistance. Some typical properties of these films are shown in Tables 8-1 through 8-6. Vespel® is supplied as sintered parts and formulated with various fillers. A description of the various Vespel® compositions commercially available is shown in Table 8-7. Some of the typical properties from these compositions are shown in Tables 8-8 through 8-12. These property data are for reference only. Since service conditions may differ from laboratory test conditions, users of Vespel® parts and shapes should independently evaluate the suitability of these parts using their own test procedures. The properties of Pyralin® are shown in Tables 8-13 and 8-14.

Figure 8-1: Kapton® cover on solar panel in the spaceship Discovery. (Courtesy: E.I. du Pont de Nemours and Co.)

Figure 8-2: Kapton® printed circuit in disk drive. (Courtesy: E.I. du Pont de Nemours and Co.).

Table 8-1: Kapton® Type H Film–Physical Properties[21] (1 mil)

Physical Properties	Typical Properties			Test Method
	−195°C	23°C	200°C	
Ultimate tensile strength, psi	35,000	25,000	17,000	ASTM D882-64T
Yield point at 3%, psi	–	10,000	6,000	ASTM D882-64T
Stress to produce 5% elongation, psi	–	13,000	8,500	ASTM D882-64T
Ultimate elongation, %	2	70	90	ASTM D882-64T
Tensile modulus, psi	510,000	430,000	260,000	ASTM D882-64T
Impact strength J/mm	–	23	–	Du Pont pneumatic impact test
Folding endurance MIT	–	10,000 cycles	–	ASTM D2176-63T
Tear strength–propagating (Elmendorf), g	–	8	–	ASTM D1922-61T
Tear strength–initial (Graver), g	–	510	–	ASTM D1004-61
Density, g/cm³	–	1.42	–	ASTM D1505-63T
Coef. of friction, Kinetic (Film-to-Film)	–	0.42	–	ASTM D1894-63
Refractive Index (Becke line)	–	1.78	–	Encyl. Dict. of Phys., Vol; Ave.3 samples,elongated at 5%, 7%, 10%
Poisson's ratio	–	0.34	–	–

Reprinted by permission from du Pont Co.

Table 8-2: Kapton® Type V Film—Typical Properties[21]

Property	Typical Values	Test Condition	Test Method
Dielectric Strength			
2 mil	5,400 V/mil	60 hertz	ASTM D149-64
3 mil	4,600 V/mil	1/4 inch electrode	
5 mil	3,600 V/mil		
Dielectric Constant			
2 mil	3.6	1 kilohertz	ASTM D150-64T
3 mil	3.7		
5 mil	3.7		
Dissipation Factor			
2 mil	0.0025	1 kilohertz	ASTM D150-64T
3 mil	0.0025		
5 mil	0.0027		
Volume Resistivity			
2 mil	8×10^{15} ohm.cm	125 Volts	ASTM D257
3 mil	5×10^{15} ohm.cm		
5 mil	1×10^{15} ohm.cm		

Reprinted by permission from du Pont Co.

Table 8-3: Kapton® Type H Film—Typical Electrical Properties[21]

Property	Typical Values	Test Conditions	Test Method
Dielectric Strength			
1 mil	7,000 V/mil	60 hertz	ASTM D149-61
2 mil	5,400 V/mil	1/4 inch electrode	
3 mil	4,600 V/mil		
5 mil	3,600 V/mil		
Dielectric Constant			
1 mil	3.5	1 kilohertz	ASTM D150-59T
2 mil	3.6		
3 mil	3.7		
5 mil	3.7		
Dissipation Factor			
1 mil	0.0025	1 kilohertz	ASTM D150-59T
2 mil	0.0025		
3 mil	0.0025		
5 mil	0.0027		
Volume Resistivity			
1 mil	1×10^{15} ohm·cm	125 Volts	ASTM D257-61
2 mil	8×10^{15} ohm·cm		
3 mil	5×10^{15} ohm·cm		
5 mil	1×10^{15} ohm·cm		

Reprinted by permission from du Pont Co.

Table 8-4: Kapton® Type F Film—Typical Electrical Properties[21]

Property	120F616	150F019	250F029
Dielectric Strength			
Total Volts	7,500	6,300	-
Volts/mil	6,800	4,200	4,000
Dielectric Constant	2.8	3.0	-
Dissipation Factor	0.0022	0.0014	-
Volume Resistivity			
ohm.cm@23°C	1.5×10^{16}	10^{18}	7×10^{17}
ohm.cm@200°C	5×10^{14}	10^{16}	-

Reprinted by permission from du Pont Co.

Table 8-5: Kapton® Type F Film—Properties[21]

Property	Typical Values Film Type*		
	120F616	150F019	250F029
Ultimate Tensile Strength, psi			
23°C	24,000	17,000	25,000
200°C	16,000	11,000	16,000
Yield Point at 3% psi			
23°C	9,000	7,300	10,000
200°C	5,500	4,000	8,000
Stress at 5% Elongation, psi			
23°C	12,500	9,000	-
200°C	7,500	5,500	-
Ultimate Elongation %			
23°C	65	75	80
200°C	95	85	-
Tensile Modulus, psi			
23°C	415,000	320,000	-
200°C	215,000	173,000	-
Impact Strength at 23°C			
Kg cm/mil	6.0	4.6	-
Tear Strength-Propagating Elmendorf			
g/mil	10	13.5	12
Tear Strength-Initial (graves)			
g/mil	750	435	-
Weight % Polyimide	80	57	73
Weight % FEP	20	43	27
Density gm/cm³	1.53	1.67	1.57

*120F616 = 1.2 mils nominal thickness; 0.1 mil Teflon® FEP; 1 mil Kapton® Type H, 0.1 mil Teflon FEP

150F019 = 1.5 mils nominal thickness, 0.0 mil Teflon® FEP; 1 mil Kapton® Type H, 0.5 mil Teflon FEP

250F029 = 25 mils nominal thickness, 0.0 mil Teflon® FEP; 2 mil Kapton® Type H; 0.5 mil Teflon FEP

Reprinted by permission from du Pont Co.

Table 8-6: Gas Permeability of Kapton® Type H Film[21]

Gas	(cc/100in^2) (24 hrs) (atm/mil)	Test Method
Carbon Dioxide	45	ASTM D1434@ 23°C
Hydrogen	250	
Nitrogen	6	
Oxygen	25	
Helium	415	
Water Vapor	g/(100in^2) (24 hrs)/min 5.4	ASTM E96-63

Reprinted by permission from du Pont Co.

Table 8-7: Vespel® Polyimide Compositions[22]

Resin Designation	Description	Characteristics
SP-1	Unfilled base resin	Provides maximum physical strength, elongation and toughness and best electrical and thermal insulation.
SP-21	15% by weight graphite filler	Graphite added to provide low wear and friction for bearings, thrust washers, and dynamic seals.
SP-22	40% by weight graphite filler	Same as SP-21 for wear and friction plus improved dimensional and oxidative stability. It has the lowest coefficient of thermal expansion.
SP-211	15% by weight graphite and 10% by weight Teflon® fluorocarbon resin fillers	Has lowest coefficient of friction over wide range of operating conditions. Also has lowest wear rate up to 300°F.
SP-3	15% by weight molybdenum disulfide	MoS$_2$ added to provide lubrication for seals and bearings in vacuum or dry environments.

Reprinted by permission from du Pont Co.

Table 8-8: Mechanical Properties* of Vespel® SP Polyimide Resins[22]

Property	Temp.°F	ASTM Method	Units	SP-1	SP-21	SP-22	SP-211	SP-3
Tensile Strength Ultimate	73	D1708 or E8	10^3×psi	12.5	9.5	7.5	6.5	8.2
	500			6.0	5.5	3.4	3.5	–
Elongation, Ultimate	73	D1708 or E8	%	7.5	4.5	3.0	3.5	4.0
	500			7.0	2.5	2.5	3.0	–
Flexural Strength Ultimate	73	D790	10^3×psi	19.0	16.0	14.0	10.0	–
	500			11.0	9.0	8.0	5.0	–
Flexural Modulus	73	D790	10^3×psi	500	500	700	500	–
				250	370	400	200	–
Compressive Stress								
@1% strain	73		10^3×psi	3.6	4.2	4.6	3.0	–
@10% strain				19.3	19.3	16.3	14.8	–
@0.1% offset				7.4	6.6	6.6	5.4	–
Compressive Modulus	73	D695	10^3×psi	350	420	475	300	–
Axial Fatigue Endurance Limit								
@10^3 cycles	73		10^3×psi	8.1	6.7	–	–	–
	500			3.8	3.3	–	–	–
@10^7 cycles	73		10^3×psi	6.1	4.7	–	–	–
	500			2.4	2.4	–	–	–
Flexural Fatigue Endurance Limit								
@10^3 cycles	73		10^3×psi	9.5	9.5	–	–	–
@10^7 cycles	73			6.5	6.5	–	–	–
Shear Strength	73	D732	10^3×psi	13.0	11.2	–	–	–
Impact Strength Izod, notched	73	D256	J/m	80.0	42.7	–	–	–
Impact Strength Izod, unnotched	73	D256	J/m	1601	427	–	–	–
Poisson's Ratio	73			0.41	0.41	–	–	–

*Properties are non-directional.

Reprinted by permission from du Pont Co.

Table 8-9: Wear and Friction Properties* of Vespel® SP Polyimide Resins[22]

Property	SP-1	SP-21	SP-22	SP-211	SP-3
Wear Rate - $M/10^5 s$	17-85	6.3	4.2	4.9	17-23
Friction Coefficient** PV=0.875 MPa m/s PV=3.5 MPa m/s	0.29 -	0.24 0.12	0.20 0.09	0.12 0.08	0.25 0.17
In Vacuum	-	-	-	-	0.03
Static in Air	0.35	0.30	0.27	0.20	-

* Properties are non-directional.
** Steady state, unlubricated in air.

Reprinted by permission from du Pont Co.

Table 8-10: Electrical Properties at 23°C of Vespel® SP Polyimide Resins[22]

Property	ASTM Method	Units	SP-1	SP-21
Dielectric Constant	D150			
@10^2Hz			3.62	13.53
@10^4Hz			3.64	13.28
@10^6Hz			3.55	13.41
Dissipation Factor	D150			
@10^2Hz			0.0018	0.0053
@10^4Hz			0.0036	0.0067
@10^6Hz			0.0034	0.0106
Dielectric Strength Short Time 0.002 m thick	D149	MV/m	22	9.84
Volume Resistivity	D257	ohm.cm	10^{14}-10^{15}	10^{12}-10^{13}
Surface Resistivity	D257	ohm	10^{15}-10^{16}	-

Reprinted by permission from du Pont Co.

Table 8-11: Thermal and Electrical Properties of Vespel® SP Polyimide Resins[22]

Property	Temp.°C	ASTM Method	Unit	SP-1	SP-21	SP-22	SP-211
Thermal Coefficient of Linear Expansion	23-300 -62-23	D696	μ in/in/°C	54 45	49 34	38 -	54 -
Thermal Conductivity	40		W/m/°K	0.35	0.87	1.73	0.76
Specific Heat	-	-	BTU/lb/°F J/Kg/°C	0.27 1130	-	-	-
Deformation under 2000psi Load	50	D621	%	0.14	0.10	-	-
Deflection Temperature		D648	°C	≳360	≳360		

Reprinted by permission from du Pont Co.

Table 8-12: Other Select Properties of Vespel® SP Polyimide Resins[22]

Property	ASTM Method	Units	SP-1	SP-21	SP-22	SP-211	SP-3
Water Absorption 24 hrs. @23°C 48 hrs. @50°C	D570	%	0.24 0.72	0.19 0.57	0.14 0.42	0.21 0.62	– –
Equilibrium—50%			1.0–1.3	0.8–1.1	–	–	–
Specific Gravity	D792		1.43	1.51	1.65	1.55	1.60
Hardness	D785	Rockwell "E" Rockwell "M"	45–58 92–102	32–44 82–94	15–40 68–78	5–25 69–79	40–55 –
Limiting Oxygen Index	D2863	%	53	49	–	–	–

Reprinted by permission from du Pont Co.

Table 8-13: Pyralin® Coatings, Lacquer Typical Properties*[23]

	PI-2540	PI-2545	PI-2550	PI-2555
Solids (2 gms, 2 hours @ 200°C)	14.5%	14%	25%	19%
Viscosity (LVF#3 @ 12 RPM)	50-70 poise 5-7 Pascal sec.	9-13 poise 0.9-1.3 Pascal sec.	50-70 poise 5-7 Pascal sec.	12-16 poise 1.2-1.6 Pascal sec.
Weight per gallon per liter	8.80 lbs. 1.06 kg	8.78 lbs. 1.06 kg	8.95 lbs. 1.08 kg	8.80 lbs. 1.06 kg
Solution Density	1.06 g/cc	1.04 g/cc	1.05 g/cc	1.06 g/cc
Solvent	NMP/aromatic hydrocarbon	NMP/aromatic hydrocarbon	NMP/acetone	NMP/aromatic hydrocarbon
Flash Point (Closed Cup)	64°C	64°C	-7°C	64°C
Filtration	0.1 micron nominal	0.2 micron absolute	0.1 micron nominal	0.2 micron absolute

*Typical properties; not to be used for specification purposes.

Reprinted by permission from du Pont Co.

Table 8-14: Pyralin® Polyimide Film Properties[23]

	PI-2540	(PI-2545)	PI-2550	(PI-2555)
Physical				
Tensile Strength (Ultimate)	17,000 psi	(1.17×10^8 Pascal)	19,000 psi	(1.31×10^8 Pascal)
Elongation	25%		10%	
Density	1.4 gms/cc		1.39 gms/cc	
Refractive Index (Becke Line)	1.78		1.70	
Flexibility	180° bend, no cracks		180° bend, no cracks	
Thermal				
Melting Point	None		None	
Weight Loss @ 316°C in air, after 300 hrs.	4%		4%	
Final Decomposition Temperature	560°C		560°C	
Coefficient of Thermal Expansion	2.0×10^{-5}/°C		4.0×10^{-5}/°C	
Coefficient of Thermal Conductivity	$37 \times 10^{-5} \frac{\text{cal}}{(\text{cm})(\text{sec})(°C)}$			
Flammability	Self-extinguishing		Self-extinguishing	
Specific Heat	0.26 cal/gm/°C		0.26 cal/gm/°C	
Electrical				
Dissipation Factor (1 KHz)	0.002		0.002	
Dielectric Strength	4000 volt/mil		4000 volt/mil	
Volume Resistivity	10^{16} ohm-cm		10^{16} ohm-cm	
Surface Resistivity	10^{15} ohm		10^{15} ohm	
Dielectric Constant (1 KHz)	3.5		3.5	

Reprinted by permission from du Pont Co.

Polyimides based on the acid dianhydride, 3,3',4,4'-benzophenonetetracarboxylic dianhydride (BTDA) (VI) and meta-phenylenediamine (MPD) (VII) have been marketed by Monsanto Company as Skybond 700 series and by American Cyanamid Company as FM-34 lacquers.

VI

VII

The Skybond® 700 lacquer consists of a solution of monomer reactants of the diester, diacid and diamine. Solvents such as ethanol, butanol, ethylene glycol and N-methylpyrrolidinone (NMP) are used as solvents for these monomers. Skybond® 700 is specifically designed for structural, electrical, and specialty applications where extended temperatures up to 700°F is required. These resins are useful for varnishes, coatings, and films. They are also useful for preparing prepregs for laminates that can be molded by either press or vacuum bag techniques. Mechanical properties of Skybond® 700 are shown in Table 8-15. The properties were obtained after a sixteen hour postcure cycle which included a final postcure for four hours at 700°F (372°C) for developing adequate initial hot strength at 700°F (372°C). Table 8-16 tabulates the electrical properties of Skybond® laminates. These properties were measured on laminates that were hot press molded so as to keep the void content low.

FM®-36 is a modified polyimide supplied as a supported film with a lightweight glass cloth. It is suitable for bonding metal-to-metal, composites and various sandwich structures; serviceable over a temperature range of -67°F (-55°C) to 550°F (287°C). Typical tensile shear properties of FM®-36 adhesive are shown in Table 8-17. FM®-34-18 adhesives are arsenic-free versions of FM®-34 polyimide adhesive film. This adhesive is available as a supportive and unsupported film and as a paste. These adhesives are noted for strength retention after long-term exposure to temperatures of -67°F (-55°C) to 700°F (370°C) in both metal and composite constructions. The mechanical properties of this adhesive are shown in Table 8-18.

Recently, General Electric Company introduced a polyimide[5] which also has the silicon-oxygen linkage in the polymer backbone. These are known as silicone polyimides (SiPI) and show promise for use as passivation coatings and as interlayer dielectrics for electronic applications. Unlike conventional polyimides, the SiPI possess excellent inherent adhesive properties which eliminate the necessity of primers. The adhesive properties seem to be insensitive to moisture.

Table 8-15: Mechanical Properties of Laminate Utilizing Skybond® 700[12]

The following results are representative of laboratory data on 1/8" thick laminates. Laminates were made using 181 glass cloth with A-1100 soft finish. The testing temperatures are the same as the exposure condition temperatures.

Property	High Temperature–High Pressure	Vacuum Bag
Flexural, Flatwise, PSI		
Standard conditions (75°F)	75–85,000	76–83,500
One-half hour at 700°F	45–60,000	22–32,000
100 hours at 700°F	20–35,000	20–24,000
Weight loss 100 hours at 700°F	3.0%	<5%
Modulus of Elasticity (x 10^6)		
Standard conditions (75°F)	3.12	2.8
335 hours at 570°F	3.12	–
100 hours at 700°F	–	1.8
Ultimate Tensile Strength, psi		
Standard conditions (75°F)	57,000	50,300
335 hours at 570°F	42,000	–
100 hours at 482°F (at R.T.)	–	48,800
100 hours at 572°F (at R.T.)	–	48,200
Barcol hardness	70	60
Flammability	Nonburning	Nonburning
Elongation		
Standard conditions (75°F), %	1.90%	2.0%
Tested at 75°F, 335 hours aging at 570°F	1.40%	–
100 hours at 482°F (at R.T.)	–	1.7%
100 hours at 572°F (at R.T.)	–	2.0%
Water Absorption		
24 hour immersion	0.70%	2.0%
24 hour immersion, coated	–	<1.0%

600°F Long-Term Aging Study (High Temperature–High Pressure Laminates)

Property	R.T.	500 Hours	860 Hours	1,850 Hours
Flexural strength (psi)	75,000	29,000	20,000	10,950
Flexural modulus (psi x 10^6)	–	2.61	2.59	2.08
Weight loss (%)	–	2.2	3.4	7.9

550°F Long-Term Aging Data (High Temperature–High Pressure Laminates)

Property	R.T.	2,300 Hours	4,500 Hours	9,000 Hours
Flexural strength (psi)	83,000	41,200	32,000	15,000
Flexural modulus (psi x 10^6)	3.2	2.63	2.95	2.00
Weight loss (%)	–	3.6	5.0	12.0

Reprinted by permission from Monsanto Co. Skybond is a registered trademark of Monsanto Co.

Table 8-16: Electrical Properties of Laminates Using Skybond® 700 (High Temperature-High Pressure Laminates)[12]

Property	As is	D 24/23	D 48/50	C 96/35/90
Dielectric Strength				
–Short time parallel to laminate (volts)	55,000	–	32,000	–
–Step-by-step parallel to laminate (volts)	38,000	–	16,000	–
–Short time (volts/mil)	179	–	–	–
–Stepwise (volts/mil)	140	–	–	–
Dielectric constant (1MC)	4.10	4.30	4.81	–
Dissipation factor (1MC)	.00445	.00639	.01650	–
Insulation resistance (megohms)	1.9×10^7	–	–	1.4×10^2
Volume resistivity (ohm-cms)	2.47×10^{15}	–	–	1.16×10^{11}
Surface resistivity (ohms)	3.35×10^{14}	–	–	2.90×10^{15}

X-BAND DATA (8.5 KMC)

Temperature	Dielectric Constant	Dissipation Factor
Room Temperature	3.74	0.016
50°C	3.74	0.015
100°C	3.74	0.014
150°C	3.74	0.018
200°C	3.74	0.013
250°C	3.74	0.010
300°C	3.70	0.015

Reprinted by permission from Monsanto Co.

Table 8-17: Typical Average Tensile Shear Properties of FM®36 Adhesive Tested with BR®36 Primer*[13]

Test Condition	Test Temp., °F	Aluminum ½ in. Large	Aluminum Blister Detection	Titanium ½ in. Lap
Tensile shear, psi	-67	3,025	2,895	–
Tensile shear, psi	75	2,640	2,670	2,550
Tensile shear, psi	350	2,730	2,425	–
Tensile shear, psi	550	2,750	2,390	1,850
Tensile shear, psi after 100 hrs. @ 550°F	75	–	–	2,000
Tensile shear, psi after 200 hrs. @ 550°F	75	–	–	2,000

*Cured two hours at 550°F.

Reprinted by permission from American Cyanamid Co.

Table 8-18: Mechanical Properties of FM® 34B-18 Adhesive Film with BR® 34B-18 Adhesive Primer[14]

Exposure Condition	Lap Shear Strength, psi 75°F (24°C)	Lap Shear Strength, psi 500°F (260°C)
Initial–as bonded	4,050	2,330
750 Hours @ 500°F (260°C)	3,361	2,290
2,000 Hours @ 500°F (260°C)	2,713	1,400
4,000 Hours @ 500°F (260°C)	1,134	1,700
192 Hours @ 600°F (315°C)	1,579	1,281
360 Hours @ 600°F (315°C)	364	699

Metal 0.050 inch. Type 301, half-hard stainless steel.
Cleaning Process: Prebond 700–acid pickle–acid etch.

Reprinted by permission from American Cyanamid Co.

The SiPI is a block copolymer prepared from BTDA (VI), methylenedianiline (MDA) (VIII), and propyltetramethyl disiloxane (GAPD) (IX).[5]

VIII $H_2N-\langle\!\!\!\bigcirc\!\!\!\rangle-CH_2-\langle\!\!\!\bigcirc\!\!\!\rangle-NH_2$

IX $\left(H_2NCH_2CH_2CH_2\underset{CH_3}{\overset{CH_3}{\underset{|}{\overset{|}{Si}}}}\right)_2 O$

The polyamic acid is prepared by first adding MDA to BTDA followed by the addition of GAPD. The scheme for the preparation of the 70/30 MDA/GAPD block copolymer SiPI is shown as follows.

[Chemical reaction scheme]

The electrical properties of SiPI are shown in Table 8-19 and the adhesion of SiPI and Pyralin® on various wafer surfaces are compared in Table 8-20.

Table 8-21 lists the various polymer designations and the corresponding manufacturer.

THERMOPLASTIC POLYIMIDES

Although polyimides are generally classified as thermoset resins because of the way they are processed and because of the very high melt temperatures, there is one class of polyimides which are thermoplastic. They are included in this discussion because they are very important materials and cannot be easily separated from the other polyimides. The thermoplastic polyimides can theoretically be processed through the polyamic acid precursor but are not generally processed in that manner. These processible polyimides were specifically developed so that they can be processed by methods analogous to the thermoplastics. Rather than being used as an amic acid precursor, they are used in the imidized form. This allows well formed moldings and composites with a very low void content since no volatiles are produced during the processing. Because the thermoplastic polyimides are not crosslinked they may have some solubility in select solvents. To make the polyimides thermoplastic it is necessary to modify the polyimide backbone. This is accomplished in a number of ways. One technique is to introduce pendent side chains such as phenyl groups. Soluble and fusible polyimides have been prepared by the introduction of flexible linking units into the

Table 8-19: Electrical Properties of SiPI (General Electric Company) Resin[15]

Property	Value
Dielectric Constant	3.0
Bulk Resistivity	10^{17} ohm-cm at 25°C
Surface Charge	1.8×10^{11} cm^{-2} (positive)
Dielectric Strength	5.5 MV/cm

Reprinted by permission from General Electric Co.

Table 8-20: Adhesion of SiPI (General Electric Company) Resin and Pyralin® (du Pont) on Various Wafer Surfaces[15]

Sample	Surface	Test Results*		
		No Boiling Water	1 Hour Boiling Water	3 Days Boiling Water
SiPI	Si	P	P	P
Pyralin® w/AP**	Si	P	F	F
Pyralin® wo/AP	Si	P	F	F
SiPI	SiO$_2$	P	P	P
Pyralin® w/AP	SiO$_2$	P	F	F
Pyralin® wo/AP	SiO$_2$	P	F	F
SiPI	Si$_3$N$_4$	P	P	P
Pyralin® w/AP	Si$_3$N$_4$	P	P	P
Pyralin® wo/AP	Si$_3$N$_4$	F	F	F
SiPI	Al	P	P	P
Pyralin® w/AP	Al	P	P	P
Pyralin® wo/AP	Al	P	P	P
SiPI	SiPI	P	P	P

* P-passed tape pull test; F-failed tape pull test.
** AP - Adhesion promoter.

Reprinted by permission from General Electric Co.

main polyimide chain. Thus, siloxane and phosphorous modified homo- and copolymers have been developed but because these systems contain aliphatic chains, the increase in processibility has been compromised by a decrease in thermooxidative stability. This limitation was solved by using fluoroalkylene rather than aliphatic groups in the polymer chain.[6] Aromatic polyimides containing the perfluoroisopropylidene group have been marketed by Du Pont Company.

Table 8-21: Some Polyimides and Their Manufacturers

Type	Company	Trade Name
Condensation	E.I. Du Pont de Nemours and Co.	Kapton®
		Vespel®
		Pryalin®
		PYRE ML®
	Monsanto Company	Skybond® 700 Series
	American Cyanamid Company	FM-34®
		FM-36®
	General Electric Company	SiPI
Thermoplastic	E.I. Du Pont de Nemours and Co.	NR-150®
	General Electric Company	Ultem®
	Amoco	Torlon®
	Ciba-Geigy Corporation	UX 218®
Addition	Rhone-Poulenc Company	Kerimid®
	NASA – Lewis	PMR
	NASA – Langley	LARC
	National Starch and Chemical Corp.	Thermid®
	Ciba-Geigy Corporation	XU-292

This group enhances solubility of polyimides in conventional solvents and also yields polyimides with relatively low glass transition temperature (T_g, 340°C). Thus, Du Pont introduced the NR-150 polymer prepared from the perfluoroisopropylidene containing acid dianhydride (6FDA) (X) and a number of aromatic diamines.[7] These resins at the present time have limited marketability due to the very high cost.

X

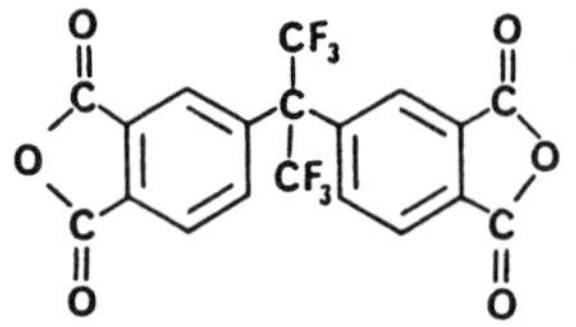

The NR-150 series of polyimides have some of the best long term thermooxidative stability, particularly at 700°F (371°C) of any of the commercially available polyimides. The excellent thermooxidative stability is in part attributed to the perfluoroisopropylidene moiety in the polymer backbone.

General Electric Company introduced the Ultem® polyetherimide resin (XI) system in 1982. This resin is an amorphous, high performance engineering thermoplastic. The ether linkage in the polyetherimide provides sufficient flexibility to allow good melt processibility yet retains the aromatic imide characteristics of excellent mechanical and thermal properties.

XI

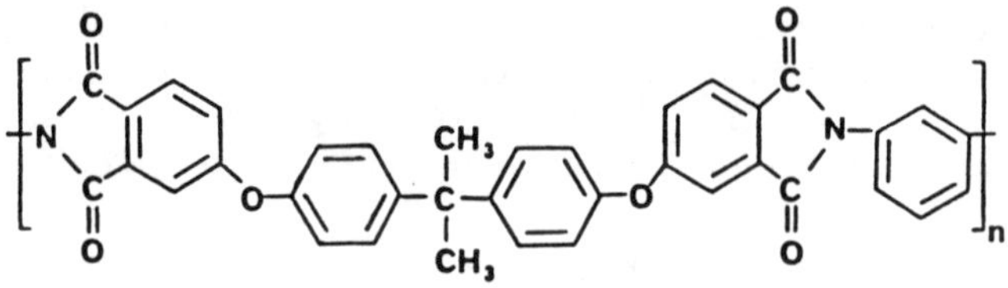

It is noteworthy that this polymer shows a balance of properties which is typical of an amorphous polymer and yet approaches the performance of some crystalline and thermoset resins. The synthesis of this class of resins is novel in that the synthesis route involves a cyclization reaction to form the imide rings and a displacement reaction to prepare the ether linkage and the polymer. This scheme is shown below.

Step I $O_2N\text{-(phthalic anhydride)} + H_2N-R-NH_2 \rightarrow O_2N\text{-(bisimide)}-NO_2 + 2H_2O$

$HO-Ar-OH + 2OH^- \rightleftharpoons {^-O-Ar-O^-} + 2H_2O$

Step II $O_2N\text{-(bisimide)}-NO_2 + {^-O-Ar-O^-} \rightarrow$

$[\text{-(imide-R-imide)-O-Ar-O-}]_n + 2NO_2^-$

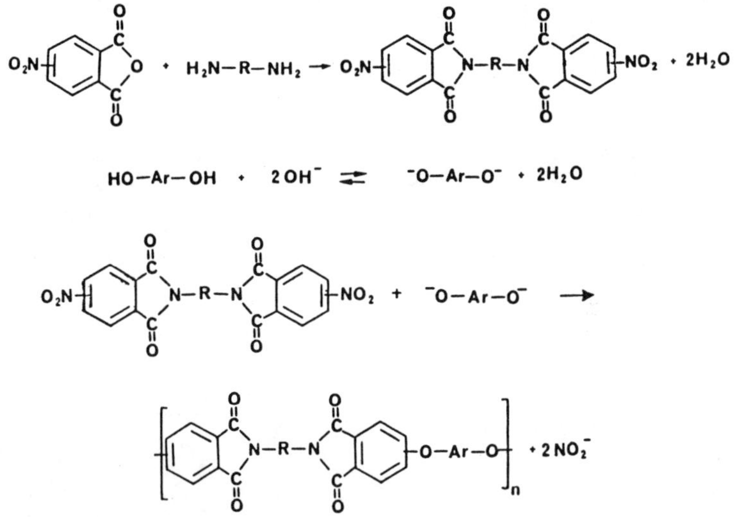

In Step I the bisimide monomer is formed by the reaction of nitrophthalic anhydride and a diamine. In Step II, the bisimide monomer is then reacted with a bisphenol dianion from the reaction of diphenol with two equivalents of base and removal of water.

A number of formulations of the Ultem® resins are marketed by General Electric Company to meet various engineering demands. Ultem® 1000 is an unreinforced grade offering a heat deflection temperature of 392°F (200°C) at 264 psi and a UL 94 flammability rating of V-O at 0.016 inch thickness. The Ultem® 2000 series are glass reinforced resins to provide greater stability and improved dimensional stability while retaining excellent processibility. The Ultem® 6000 series are a new family of materials exhibiting even greater heat capabilities than the 1000 and 2000 series. These are particularly suitable for military electrical components which must survive 392°F (200°C) testing. All Ultem® 1000 and 2000 series are available with two options: low viscosity resin, and mold release agent. Either option increases the flow length of the product by approximately 10-25% and will facilitate molding of complex thin-walled parts. Optimum flow length is obtained by combining both options which typically leads to flow improvements of 20-50%. Typical properties of the Ultem® resins are shown in Tables 8-22, 8-23, 8-24 and 8-25. Figure 8-3 shows a novel use for Ultem®, a nonplanar circuit board.

Table 8-22: Typical Mechanical Properties of Ultem® (General Electric Company) Resins[16]

Mechanical	ASTM Test	Units	ULTEM 1000	ULTEM 2100	ULTEM 2200	ULTEM 2300
Tensile strength, yield	D638	10^3 x psi	15.2	16.6	20.1	24.5
Tensile modulus, 1% secant	D638	10^3 x psi	430	650	1000	1300
Tensile elongation, yield	D638	%	7 - 8	5	—	—
Tensile elongation, ultimate	D638	%	60	6	3	3
Flexural strength	D790	10^3 x psi	21	28	30	33
Flexural modulus, tangent	D790	10^3 x psi	480	650	900	1200
Compressive strength	D695	10^3 x psi	20.3	22.5	24.5	23.5
Compressive modulus	D695	10^3 x psi	420	450	515	550
Gardner Impact	—	in - lb	320	—	—	—
Izod Impact	D256					
notched (1/8")	—	ft - lb/in	1.0	1.1	1.6	2.0
unnotched (1/8")	—	ft - lb/in	25	9.0	9.0	8.0
Shear strength, ultimate	—	10^3 x psi	15	13	13.5	14
Rockwell hardness	D785	—	M109	M114	M118	M125
Taber abrasion (CS17, 1 Kg)	D1044	mg wt loss/ 1000 cycles	10	—	—	—

Reprinted by permission from General Electric Co.

Table 8-23: Typical Thermal and Flammability Properties of Ultem® (General Electric Company) Resins[16]

	ASTM Test	Units	ULTEM 1000	ULTEM 21000	ULTEM 2200	ULTEM 2300
Thermal						
Deflection temperature, unannealed	D648					
@ 264 psi (1/4")		°F	392	405	408	410
@ 66 psi (1/4")		°F	410	410	410	414
Vicat softening point, method B	D1525	°F	426	434	438	442
Continuous service temperature index (UL Bulletin 746B)		°F	338	338	338	356
Coefficient of thermal expansion (to 300°F), mold direction	D696	10^{-5} x in / in · F	3.1	1.8	1.4	1.1
Thermal conductivity	C177	BTU-in / h-ft² °F	1.5	—	—	—
Flammability						
Oxygen Index (0.060")	D2863	%	47	47	50	50
Vertical burn (UL Bulletin 94)		—	V-0 @ 0.016" 5V @ 0.075	V-0 @ 0.016"	V-0 @ 0.016"	V-0 @ 0.01
MBS smoke, flame mode (0.060")	E662					
D_s @ 4 min			0.7	—	1.3	—
D_{MAX} @ 20 min			30	—	27	—

Reprinted by permission from General Electric Co.

Table 8-24: Typical Electrical and Other Properties of Ultem® (General Electric Company) Resins[16]

	ASTM Test	Units	ULTEM 1000	ULTEM 2100	ULTEM 2200	ULTEM 2300
Electrical						
Dielectric strength (1/16")	D149					
in oil		V/mil	710	700	670	630
in air		V/mil	830	–	–	770
Dielectric constant @ 1 kHz, 40% RH	D150	–	3.15	3.5	3.5	3.7
Dissipation factor						
@ 1 kHz, 50% RH, 73°F			0.0013	0.0014	0.0015	0.0015
@ 2450 MHz, 50% RH, 73°F			0.0025	0.0046	0.0049	0.0053
Volume resistivity (1/16")	D257	ohm–cm	6.7×10^{17}	1.0×10^{17}	7.0×10^{16}	
Arc resistance	D495	sec	128			85
Other Properties						
Specific gravity	D729	–	1.27	1.34	1.42	1.51
Mold shrinkage		in/in	0.007	0.005–0.006	0.003–0.005	0.002–0.004
Water absorption	D570					
@ 24 hrs., 73°F		%	0.25	0.28	0.26	0.18
@ equilibrium		%	1.25	1.0	1.0	0.9

Reprinted by permission from General Electric Co.

Table 8-25: Typical Properties of Ultem® (General Electric Co.) Resins[16]

Mechanical	ASTM Test	Units	6000 Series High Temperature Copolymer Grades			
			Ultem 6000	Ultem 6100	Ultem 6200	Ultem 6202
Tensile strength, yield @ 73°F	D638	psi	15,000	17,000	21,000	14,000
@ 392°F	D638	psi	5,000	*	*	5,000
Tensile elongation, yield @ 73°F	D638	%	7–8	—	—	—
Tensile elongation, ultimate, @ 73°F	D638	%	30	6	4	6
Flexural strength, @ 73°F	D790	psi	21,000	28,000	30,000	21,000
@ 392°F	D790	psi	9,000	**	**	9,000
Flexural modulus, tangent, @ 73°F	D790	psi	440,000	670,000	950,000	550,000
@ 392°F	D790	psi	315,000	*	**	410,000
Compressive strength	D695	psi	20,000	*	25,000	20,000
Izod impact	D256					
notched (1/8″)		ft-lb/in	0.8	1.0	1.6	0.8
unnotched (1/8″)		ft-lb/in	25	8	8	7
Shear strength, ultimate	—	psi	15,000	14,000	14,000	14,000
Thermal						
Deflection temperature, unannealed	D648					
@ 264 psi (1/4″)		°F	420	430	433	420
@ 66 psi (1/4″)		°F	430	432	437	*
Vicat softening point, method B	D1525	°F	453	465	470	459
Continuous service temperature index (UL Bulletin 746B)	—	°F	*	*	*	*
Coefficient of thermal expansion (0° to 300°F), mold direction	D696	in/in·°F	2.9×10^{-5}	*	1.4×10^{-5}	2.5×10^{-5}

(Continued)

Table 8-25: (Continued)

........6000 Series High Temperature Copolymer Grades........

Flammability	ASTM Test	Units	Ultem 6000	Ultem 6100	Ultem 6200	Ultem 6202
Oxygen index	D2863	%	44	44	44	48
Vertical burn (UL Bulletin 94)**	—	—	V-O @ $^1/_{16}$"****	V-O @ $^1/_{16}$"****	V-O @ $^1/_{16}$"****	V-O @ $^1/_{16}$"****
NBS smoke, flaming mode (0.060")	E662					
D_S @ 4 min		—	5	—	4	—
D_{MAX} @ 20 min		—	70	—	70	—
Electrical						
Dielectric strength ($^1/_{16}$"), in oil	D149	V/mil	750	*	580	530
Dielectric constant @ 1kHz, 50% RH	D150	—	3.0	3.1	3.1	3.1
Dissipation factor @ 1 kHz, 50% RH, 73°F	D150	—	0.001	0.001	0.001	0.001
Volume resistivity ($^1/_{16}$")	D257	ohm-cm	1.0×10^{17}	1.0×10^{17}	1.0×10^{17}	1.0×10^{17}
Arc resistance	D495	seconds	127	—	—	140
Other						
Specific gravity	D792	—	1.29	1.35	1.43	1.42
Mold shrinkage	D955	in/in	0.005–0.007	*	*	0.005–0.007
Water absorption @ 24 hours, 73°F	D570	%	0.28	0.24	0.22	0.22

*Testing in progress.
**This rating is not intended to reflect hazards presented by this or any other material under actual fire conditions.
***General Electric Company test data.

Reprinted by permission from General Electric Co.

Figure 8-3: An Ultem® single-sided MINT-PAC™ board with three-dimensional features on its unplated side. All MINT-PAC exterior board dimensions, through-holes, and three-dimensional features are molded in a single step eliminating a great deal of costly machining. (Courtesy: General Electric Company).

Another route to fusible and processible polyimides is to introduce the amide moiety into the polyimide chain. These poly(amide-imide) polymers are commercially available as Amoco's Torlon®. Chemically, Torlon® is a polymer from the reaction of Amoco's trimellitic anhydride and aromatic diamines. The polymer is depicted by structure XII.

XII

The polymer chain comprises amide alternating with imide linkage which results in a thermoplastic polymer. The versatility and processibility has made these resins attractive materials. Torlon® has even been used to make a plastic gasoline engine (see Figure 8-4).

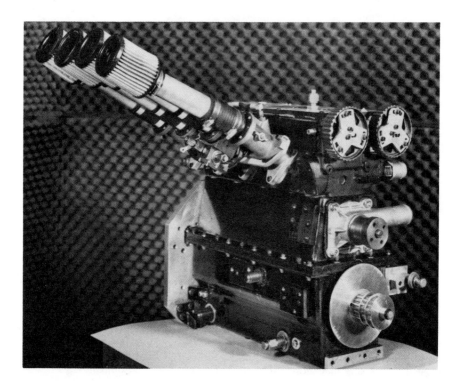

Figure 8-4: The 2.0 liter, dual overhead cam plastic engine developed by Polimotor Research, Inc. utilizes Torlon®, an injection moldable thermoplastic made by Amoco Chemicals Corporation. Weighing in at 175 pounds and developing 318 brake horsepower at 9,500 rpm, the engine is 200 pounds lighter than its steel counterpart. (Courtesy: Amoco Chemicals Corporation).

The resulting plastic engine delivers surprising performance and reduced the weight of the 2-liter, four-cylinder engine to about 168 pounds, about 200 pounds less than an all metal engine. The Torlon® polymer was chosen because of its strength, toughness and temperature resistance. It would not be surprising if the plastic engine parts used in this demonstration engine may appear in passenger cars before long.

Various grades of Torlon® are available for a number of engineering applications. The grades and their application are described in Table 8-26. The mechanical properties are shown in Table 8-27, electrical properties in Table 8-28, and thermal and general properties in Table 8-29. Figure 8-5 shows parts fabricated from Torlon® which take advantage of this material's excellent wear and temperature resistance.

Table 8-26: Grades of Torlon® Commercially Available[17]

Grade Code Number	Description
TORLON 4000T	This is a powdered form of the resin that is used for compression molding. It is the base resin used to produce most other grades of TORLON.
TORLON 4203L	This grade contains TORLON 4000T plus 3 percent TiO_2 and 1/2 percent pTFE, compounded and pelletized. It is used primarily for injection molding. PTFE is added to enhance its moldability and mold release. 4203L is used where elongation and impact resistance are important considerations.
TORLON 4301	This grade contains TORLON plus 12 percent graphite powder and 3 percent pTEF, compounded and pelletized. It was designed to reduce the coefficient of friction and to increase its wear resistance in lubricated and non-lubricated bearing applications.
TORLON 4275	This grade contains TORLON 4000T plus 20 percent graphite powder and 3 percent pTFE, compounded and pelletized. It was specifically designed to provide maximum wear resistance and a lower coefficient of friction.
TORLON 4347	This grade contains 12 percent graphite powder and 8 percent pTFE, compounded and pelletized. This new material is designed for the lowest coefficient of friction and should be useful for bearings under high load or surfaces that are subjected to reciprocating motion.
TORLON 5030	This grade contains TORLON 4000T plus 30 percent glass fiber and one percent pTFE, compounded and pelletized. It is used primarily in application requiring high modulus, low shrinkage, high creep resistance, high fatigue resistance, improved resistance to deformation at high temperatures, or good dielectric properties.
TORLON 7130	This grade contains TORLON 4000T plus 30 percent graphite fibers and one percent pTFE, compounded and pelletized. The most metal-like of all TORLON resins, this grade was specifically developed as a structural metal replacement for use in the aerospace industry. The incorporation of graphite fibers results in a material with a high modulus that is highly resistant to creep, and has the highest fatigue resistance available in a TORLON resin.
TORLON 9040	This grade is the first in a new series of TORLON resins made possible by developments in the basic polymer chemistry. This version contains 40 percent glass and carries a volume tag of $5.40 per pound. Even higher heat resistance has been achieved with a minor loss of impact strength. Ground forms of 4203, 5030, 4301, 4275 are also available for use by compression molders.
TORLON 5430	This grade is compounded with 30 percent glass fibers. It offers superior resistance to creep and excellent stiffness at elevated temperatures as well as improved processibility.

Reprinted by permission from Amoco Chemicals Corp.

Table 8-27: Torlon® Resin Compositions—Mechanical Properties[17]

	Testing Temperature °F	4203 and 4203L	4301	4275	4347	5030	5430	7130	9040
Tensile Strength 10³ psi, D1708	-321	31.5	-	1818	-	29.5	-	22.8	-
	73	27.8	23.7	22.0	29.7	29.7	27.7	29.4	25.5
	275	16.9	16.3	16.3	15.1	23.1	19.0	22.8	24.0
	450	9.5	10.6	8.1	7.8	16.3	12.9	15.7	19.8
Tensile Elongation %, D1708	-321	6	-	3	-	4	-	3	-
	73	15	7	7	9	7	12	6	4
	275	21	20	15	21	15	10	14	8.3
	450	22	17	17	15	12	8	11	7.7
Tensile Modulus 10⁶ psi, D1708	73	7.0	9.5	11.3	8.7	15.6	15.8	3.2	-
Flexural Strength 10³ psi D790	-321	41.0	-	29.0	-	54.4	-	45.0	-
	73	34.9	31.2	30.2	27.0	48.3	39.7	50.7	49.0
	275	24.8	23.5	22.4	20.5	35.9	29.7	37.6	39.5
	450	17.1	16.2	15.8	14.3	26.2	20.7	25.2	22.8
Flexural Modulus 10⁵ psi, D790	-321	11.4	-	13.9	-	20.4	-	35.7	-
	73	7.3	10.0	10.6	9.1	17.0	16.7	22.8	21.1
	275	5.6	7.9	8.1	6.4	15.5	14.2	27.2	20.9
	450	5.2	7.2	7.4	6.2	14.3	13.2	22.8	18.3
Compressive Strength 10⁶ psi, D695	73	32.1	21.7	17.8	18.3	34.8	34.1	32.7	-
Shear Strength 10³ psi, D732	73	18.5	16.1	11.1	11.5	20.1	19.1	17.3	-
Izod Impact ft lbs/in, D256 notched	73	2.7	1.2	1.6	1.3	1.5	1.3	0.9	1.5
unnotched	-	20.0	7.6	4.7	-	9.5	-	6.4	1.8

Reprinted by permission from Amoco Chemicals Corp.

Table 8-28: Torlon® Resin Compositions—Electrical Properties[17]

Property	4203L	4301	4275	4347	5030	5430
Dielectric Constant, D150						
@ 10^3 Hz	4.2	6.0	7.3	6.8	4.4	4.2
@ 10^6 Hz	3.9	5.4	6.6	6.0	6.5	4.0
Dissipation Factor, D150						
@ 10^3 Hz	0.026	0.037	0.059	0.037	0.022	0.020
@ 10^6 Hz	0.031	0.042	0.063	0.071	0.023	0.036
Volume Resistivity, ohm – in, D257	8×10^{16}	3×10^{15}	3×10^{15}	3×10^{15}	6×10^{16}	2×10^{16}
Surface Resistivity, ohms, D257	5×10^{18}	8×10^{17}	4×10^{17}	1×10^{18}	1×10^{18}	3×10^{17}
Dielectric Strength, volts/mil	580	–	–	–	835	650

Reprinted by permission from Amoco Chemicals Corp.

Table 8-29: Torlon® Resin Compositions—Thermal and General Properties[17]

	4203 and 4203L	4301	4275	4347	5030	5430	7130	9040
Deflection Temperature °F @ 264 psi, D646	532	534	536	532	539	534	540	536
Coefficient of Linear Thermal Expansion 10^{-6} in/in, °F, D696	17	14	14	15	9	10	6	7
Thermal Conductivity BTU in/hr, Ft^2, °F, C177	1.8	3.7	–	–	2.5	–	–	–
Flammability Underwriters Lab 94	94VO	94VO	94VO	94VO	94VO	94VO	94VO	94VO
Limiting Oxygen Index %, D2863	45	44	45	46	51	48	48	50
Density g/cc, D792	1.38	1.43	1.44	1.44	1.56	1.62	1.42	1.66
Hardness Rockwell E, D792	86	72	70	66	94	93	94	107
Water Absorption, % D57	0.33	0.28	0.33	0.17	0.24	0.31	0.26	0.21

Reprinted by permission from Amoco Chemicals Corp.

Figure 8-5: Torlon® parts are long lasting in wear-intensive applications. (Courtesy: Amoco Chemicals Corporation).

Ciba-Geigy Corporation markets a fully-imidized polyimide called XU218, which is depicted by structure (XIII).

XIII

The polymer is supplied as a powder in the fully imidized form. It has high solubility in volatile solvents such as methylene chloride, tetrahydrofuran as well as in solvents such as N-methylpyrrolinone. It is an excellent film former with a relatively high glass transition temperature [T_g, 320°C (608°F)]. Films made from this polyimide have excellent chemical resistance. Exposure in 5 N sodium hydroxide for 48 hours produced only surface pitting. Under similar conditions, Kapton® film completely lost its integrity. Select properties are shown in Tables 8-30, 8-31 and 8-32. A comparison of some properties of the XU218 film with Kapton® H film is shown in Table 8-33. This polyimide may be useful for various electronic applications requiring films and conformal coatings.

Table 8-30: XU 218 Polyimide Resin, Thermal Stability—Aging of 1 Mil Films[18]

Temperature, °C	Environment	Hours to Embrillement
200	air	>2000
225	air	925
250	air	250
250	nitrogen	>2500
300	vacuum	>2500

Reprinted by permission from Ciba-Geigy Corporation.

Table 8-31: XU-218 Polyimide Resin—Electrical Properties[18]

Property	Temp. °C	Units	Value
Dielectric Constant			
100 Hz	25	-	3.4
1000 Hz	25	-	3.3
1000 Hz	60	-	3.0
1000 Hz	100	-	3.0
1000 Hz	150	-	3.0
1×10^6 Hz	25	-	3.0
Dissipation Factor			
100 Hz	25	-	0.0061
1000 Hz	25	-	0.0026
1000 Hz	60	-	0.0039
1000 Hz	100	-	0.0061
1000 Hz	150	-	0.0073
1×10^6 Hz	25	-	0.0091
Dielectric Strength	-	KV/mil	5.6
Volume Resistivity	25	ohm-cm	4.4×10^{16}

Reprinted by permission from Ciba-Geigy Corporation.

Table 8-32: XU-218 (Ciba-Geigy) Polyimide Film, Tensile Properties*[18]

Property	Temp. °C	Tensile Strength (Kpsi) Yield	Tensile Strength (Kpsi) Break	Tensile Modulus (Kpsi)	% Elongation Yield	% Elongation Break
Tensile Strength (Machine direction)	25	–	22.5	560	–	17
	100	14.4	18.4	370	8.8	30
	150	11.9	17.8	330	7.7	41
	204	9.1	14.4	363	6.4	47
	260	5.9	8.8	270	3.9	38
(Cross-machine direction)	25	–	14.6	460	–	10
	100	11.7	11.1	340	8.0	32
	150	9.0	8.3	273	6.8	49
	204	7.0	6.4	275	5.4	51
	260	4.7	4.3	260	3.4	69

*0.5 mil film – uniaxially stretched from 1.0 mil film oriented in machine direction.

Reprinted by permission from Ciba-Geigy Corp.

Table 8-33: Comparison of Select Properties of H Film (Du Pont) and XU-218 (Ciba-Geigy)

Room Temperature Properties	Du Pont H Film	Ciba-Geigy XU-218
Ultimate Tensile Strength, psi	25,000	22,500
Tensile Modulus, psi	430,000	560,000
Dielectric Constant at 1000 Hertz	3.5	3.3
Dissipation Factor at 1000 Hertz	0.0027	0.0026

ADDITION POLYIMIDES

The polyimides which are processed by conversion to the imidized structure by a condensation reaction present some formidable processing problems. They are difficult to use as molding compounds and for the fabrication of composite structures, very high molding pressures and a carefully controlled post-cure process are required to allow volatile by-products from the condensation process to escape. To overcome these problems the addition polyimides were conceived. These polyimides are formed by homopolymerization of fusible and soluble short chain polyimide oligomers having reactive functional end groups. Ideally, these functional groups homopolymerize into thermo and oxidatively resistant structures without the evolution of volatile by-products.

One group of polymers using this approach is based on N-substituted bismaleimides. By choosing the appropriate moieties in the polymeric chain it is possible to prepare very soluble and fusible oligomers. These prepolymers can be heated above their melting point in the presence of a free radical catalyst, such as dicumyl peroxide, to polymerize into a crosslinked polyimide resin. Rhone-Poulenc Company introduced the Kerimid® resins which are based on mixtures of bismaleimides and aromatic diamines. This results in linear chain extension via a Michael addition reaction. Improved high temperature properties could be obtained from these resins if nonstoichiometric mixtures of aromatic diamine and bismaleimide were used. Two types of reactions are postulated.[8] First, Michael addition to form linear polymers as depicted below,

$$\underset{O}{\overset{O}{\|}}\text{N-R-N}\underset{O}{\overset{O}{\|}} + H_2N-R'-NH_2 \longrightarrow \left[\underset{O}{\overset{O}{\|}}\text{N-R-N}\underset{O}{\overset{O}{\|}}-\underset{H}{\overset{H}{\text{N}}}-R'-\underset{}{\text{N}}\right]_n$$

and secondly, the double bond is opened and followed by a free radical cure leading to crosslinking.

Table 8-34 shows typical molding parameters of these resins. Thus Kerimid® FE70003 coated on graphite or glass cloth fibers can be molded using vacuum bag techniques. Properties of the neat resin are shown in Table 8-35. The low viscosity and reasonable gel times are very important parameters which make these resins easy to process. Tables 8-36, 8-37 and 8-38 show typical thermal mechanical and electrical properties obtained for Kerimid® 601/181E glass cloth laminates.

Table 8-34: Kermid® FE 70003 Modified Bismaleimide Resin[24]

Processing

Impregnation

Reinforcement from glass cloth, graphite, etc., can be coated from melted resin (90°C) or from a lacquer in methylene chloride or methylethyl ketone.

Molding

Can be molded by a vacuum bag technique. The cure takes place at 200-250°C at pressures of 75 to 150 psi. It is advisable to postcure the parts 12 to 24 hours at 250°C.

Properties of Composite

Seven plies graphite fabric 215 g/m² with 42% resin content and molded and cured as indicated above.

	Flexural Strength 10^3 x psi	Flexural Modulus 10^6 x psi
68°F	98.6	0.77

Reprinted by permission from Rhone-Poulenc Inc.

Table 8-35: Kerimid® FE 70003 Modified Bismaleimide Resin— Select Properties of Neat Resin[24]

Property	Units	Value
Density	g/cc	1.25
Viscosity	cps	
60°C	"	27000
70°C	"	4500
80°C	"	1200
90°C	"	500
100°C	"	450
Gel Time	minutes	
150°C	"	65
180°C	"	20
Solubility		

Solutions of 50% in weight can be obtained with methylene chloride or methylethyl ketone.

Reprinted by permission from Rhone-Poulenc Inc.

Table 8-36: Kermid® 601/181E Glass Cloth 18 Ply Laminate— Thermal Aging and Mechanical Properties[25]

Temp.°F of Aging	Hours	Flexural Strength psi x 10^3		Flexural Modulus psi x 10^3		Weight Loss %
		77°F	390°F	77°F	390°F	
180°F	Initial	71	57	3850	3150	-
	1000	67	59	3600	3350	0.25
	2000	67	59	3800	3350	0.30
	5000	67	57	3850	3400	0.5
	8000	59	52	3500	3300	0.9
	10000	59	47	3850	3350	1.4
390°F	Initial	71	57	3850	3150	-
	1000	66	56	3650	3400	0.4
	2000	66	54	3650	3400	0.5
	5000	57	54	3650	3350	1.1
	8000	46	37	3300	3150	1.9
	10000	37	30	3600	3000	2.7
430°F	Initial	71	57	3850	3150	-
	1000	64	56	3800	3400	0.7
	2000	64	54	3700	3200	0.9
	3000	59	51	3600	3200	1.4
	5000	44	37	3300	3050	2.8
	8000	19	16	2800	2600	5.4
	10000	17	12	1850	2450	7.7
480°F	Initial	71	57	3850	3150	-
	1000	57	47	3300	3150	4.0
	2000	50	44	3100	2950	5.7
	3000	26	23	2600	2600	813

The specimens were taken from laminates prepared in the following conditions; impregnation bath; solution of KERIMID 601 at 45% in NMP; fiberglass fabric: continuous filament yarn, satin weave of the 181E type with aminosilane finish; resin content in the prepreg: 30 to 35%; rate of flow of the prepreg: 30 to 40%; stack of 18 plies; curing under 210 psi pressure at 480°F and postcuring for 48 hours at 390°F, or curing under 210 psi at 390°F and postcuring for 24 hours at 480°F. Resin content 22 to 24%, sp. gr. 1.94 and Barcol Hardness, ca 70.

Reprinted by permission from Rhone-Poulenc Inc.

Table 8-37: Kermid® 601/181E Glass Cloth Laminate—Electrical Properties as a Result of Aging[25]

Condition	Dielectric Strength KV/mm	Volume Resistivity ohm x cm	Dielectric Constant at 1 kHz	Dissipation Factor at 1 kHz
ASTM Method	D149	D257	D159	D150
Initial	25	6×10^{14}	4.5	0.012
After 24 hrs. Immersion in Water	20	1.5×10^{13}	5.4	0.016
1000 hrs. at 355°F	>16.5	-	-	-
1000 hrs. at 390°F	>16.5	-	-	-
1000 hrs. at 430°F	12	-	-	-
2000 hrs. at 480°F	-	2.2×10^{15}	-	-
10000 hrs. at 355°F	-	-	5.5	-
10000 hrs. at 390°F	-	-	5.5	-
10000 hrs. at 430°F	-	-	4.7	-

Reprinted by permission from Rhone-Poulenc Inc.

Table 8-38: Select Properties of Kerimid® 601/181E Glass Cloth Laminates[25]

Property	ASTM	Unit	Value	
Flexural Strength	D790	psi x 10^3		
77°F (25°C)			ca	7
390°F (200°C)			ca	70
480°F (250°C)			ca	60
			ca	50
Flexural Modulus	D790	psi x 10^3		
77°F (25°C)			ca	4000
390°F (200°C)			ca	3800
480°F (250°C)			ca	3200
Tensile Strength	D638	psi x 10^3		
77°F (25°C)			ca	50
Compressive Strength	D695	psi x 10^3		
77°F (25°C)			ca	50
Delaminating Strength	D2355	psi		
77°F (25°c)			ca	2150
Izod Impact Strength	D256	ft x lb/in		
77°F (25°C)				
notched			ca	13
notched			ca	15

Reprinted by permission from Rhone-Poulenc Inc.

In the nineteen-sixties, TRW developed bismaleimide polymers marketed by Ciba-Geigy Corporation known as P13N polyimides. They were difficult to process and thereby developed only a small market. They were eventually discontinued. The P13N polyimides were based on norbornenyl-terminated imide oligomers that underwent a reverse Diels-Alder reaction at elevated temperatures of 527°-662°F (275°-350°C) to form macromolecules. The polymerization of the P13N polyimide oligomers is depicted as follows.

Under the direction of Dr. Serafini of NASA Lewis, a novel class of addition polyimides known as PMR (for in-situ polymerization of monomer reactant) polyimide[9] was developed to replace the P13N. The PMR system is characterized by good solubility of the reacting monomers, good reactivity at elevated temperatures and no volatiles produced during the cure cycle.

In the PMR approach, the reinforcing fibers are impregnated with a solution of a mixture of monomers in low boiling alcohols. At room temperature the monomers are essentially unreactive but at elevated temperatures react to form a thermooxidatively stable polyimide resin. The solution of monomers consists of a high solid content (70-90%) of a dialkyl ester of an aromatic tetracarboxylic acid, an aromatic diamine, and a monoalkyl ester of 5-norbornene-2,3-dicarboxylic acid (NE) (XIV).

XIV

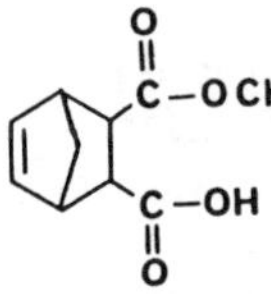

Various modifications of the PMR polyimides were prepared. The first generation of PMR resin was designated PMR-15 and the second generation PMRII. Thus, PMR-15 used (NE), the dimethylester of 3,3',4,4'-benzophenonetetracarboxylic acid (BTDE) (XV) and 4,4'-methylenedianiline (MDA) (VIII).

XV

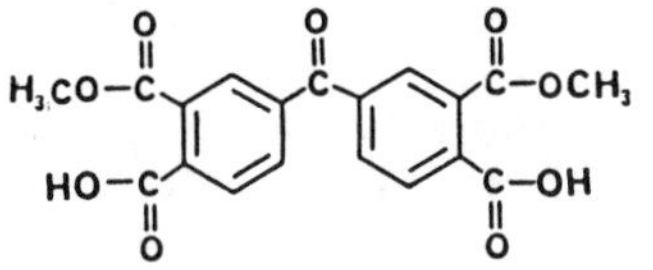

The second generation of PMR resins used the perfluorinated isopropylidene analogue of BTDE known as 6FDE (XVI) and p-phenylenediamine.

XVI

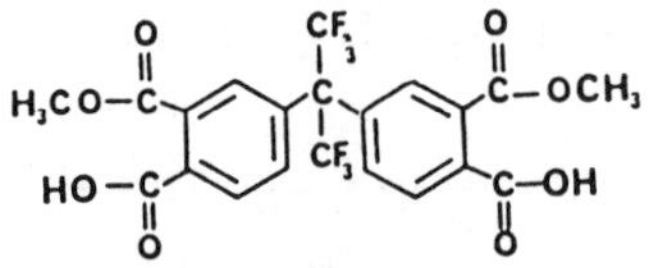

The second generation of PMR resins should become more widely used since American Hoechst Corporation recently announced that it intends to market the precursor to 6FDE, namely the 6F dianhydride. A number of systems have been investigated and modifications of the reactants as well as changes in the stoichiometry of the monomeric reactants have shown the ability of the PMR approach to produce "tailor-made" resins with varying degrees of flow.

The PMR system developed at the NASA-Langely Research

Center were given the name LARC. The LARC 160 which is a modified PMR-15 uses the monomers BTDE, NE and a polyaromatic amine known as Jeffamine AP-22 depicted by structure (XVII) with an average molecular weight of 234.

XVII

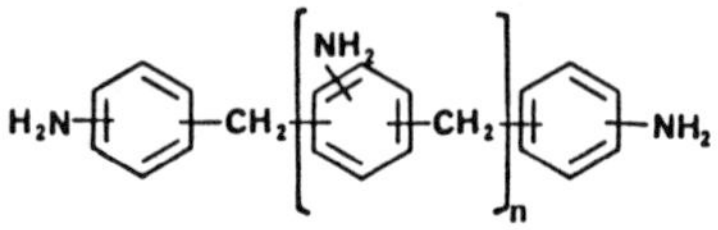

The LARC 160 resin is readily processed and gives excellent mechanical properties, but it is a brittle system and has marginal thermooxidative stability at 600°F (316°C). The LARC-13 system was developed as a high temperature adhesive for bonding large areas. Previous adhesive systems frequently yielded poor results because for bonding large areas a large volume of volatiles is given off during the cure process. The LARC-13 system is an addition type cure which results in the evolution of few volatiles. Its unique polymer chain incorporates meta-linkages in the oligomeric chains. This is shown by structure (XVIII).

XVIII

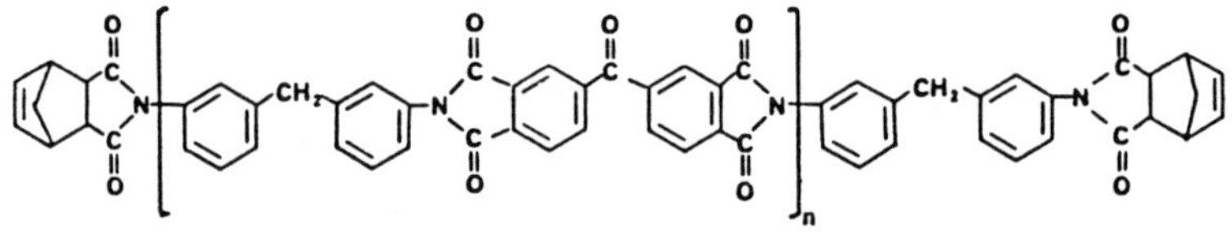

The oligomer has a relatively low softening temperature and can be processed at 512°-600°F (266°-316°C). This makes it attractive for bonding metals. The resin can also be used for preparing composites.

The room temperature properties of PMR-15 neat resin are shown in Table 8-39. Some typical properties for graphite PMR composites are shown in Table 8-40. The excellent retention of interlaminar shear strength of graphite fiber/PMR-15 composites as a function of air aging and testing at 600°F is shown in Table 8-41.

Table 8-39: Room Temperature Properties of PMR-15 Neat Resin[11]

Property	Units	Value
Tensile Strength	10^3 x psi	8.1
Tensile Modulus	10^6 x psi	0.47
Compressive Yield Strength	10^3 x psi	16.5
Compressive Strength	10^3 x psi	27.2
Thermal Coefficient of Expansion	10^{-6} x in/in/°F	28.0

Reprinted by permission of Van Nostrand Reinhold Co.

Table 8-40: Properties of HT-S/PMR-15 Composites*[11]

Property	Units	6(0)	6(90)	4(+10-10)	8(0,+10,0-10,-10,0,+10,0)	13(+40,9(0)+40)
Tensile Strength	10^3 x psi	180	9.74	150	191	124
Tensile Modulus	10^6 x psi	21.7	1.15	21.1	19.0	14.0
Compressive Strength	10^6 x psi	135	34	–	–	–
Flexural Strength	10^3 x psi	206	16.35	230	197	145
Flexural Modulus	10^6 x psi	17.6	1.07	16.8	19.0	9.3
Short-beam Interlaminar Shear	10^3 x psi	16.0	–	–	–	–
Miniature Izod Impact Energy	in-lb	15.2	1.8	–	–	–
Coefficient of Thermal Expansion	10^{-6} x in/in/°F	0	14.5	–	–	–

*Fiber Volume, 55% by volume; tested along 0° ply direction.

Reprinted by permission from Van Nostrand Reinhold Co.

Table 8-41: Interlaminar Shear Strength of Graphite Fiber/PMR-15 Composites as a Function of Air Aging and Testing at 600°F[9]

Time, Hours	Interlaminar Shear Strength, 10^3 × psi, Fiber Reinf. and			
	Celion 6000(58%)	HTS-2(62%)	Fortafil 3	Thornel
0	7.3	6.5	5.4	6.5
400	7.7	6.5	6.1	6.5
700	-	-	6.5	-
800	7.7	6.2	6.3	6.1
1200	7.3	5.6	5.7	5.5
1400	7.0	4.8	4.8	4.8

Reprinted by permission from Dr. T.T. Serafini—NASA Lewis

A NASA sponsored study by Boeing Aerospace Company on the effects surface treatment of titanium on the lap shear strengths of titanium bonded with LARC-13 shows a wide range of values with the best values being obtained with a chromic acid anodized treatment [2,980 psi at room temperature and 220 psi at 450°F (232°C)] and the poorest with Pasa-Jell 107 [850 psi at room temperature and 580 psi at 450°F (232°C)].[10]

Several years ago Ciba-Geigy introduced a two-component bismaleimide system "XU292" which when combined and cured is suitable for high temperature advanced composites and adhesives applications. The system is based on 4,4'-bismaleimide-diphenylmethane (XIX) and o,o'-diallyl bisphenol A (XX).[26]

XIX

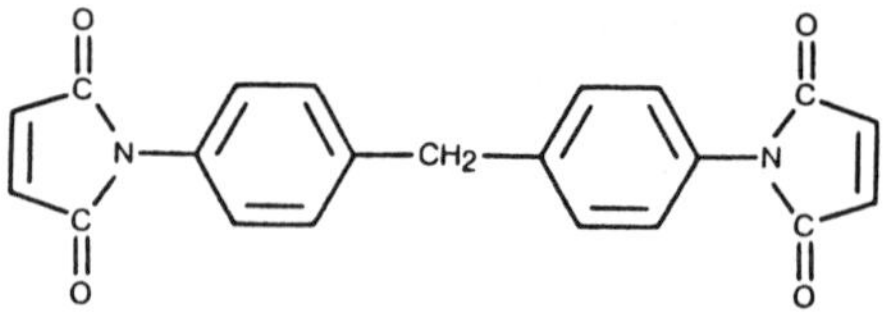

XX

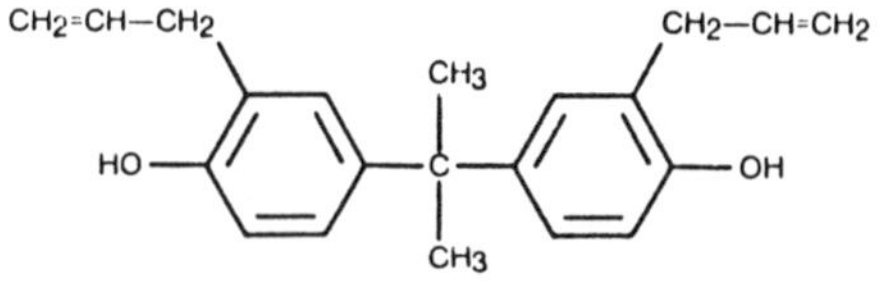

Upon combining the two components with continuous stirring to 120°-150°C (248°-302°F), a clear homogeneous solution is obtained.

The resulting liquid can be used either as a casting resin for the preparation of prepregs or as an adhesive. The two-component system provides flexibility to provide the optimum formulation for the prepreger. One formulation of the neat resin reported by Ciba-Geigy[27] reports a room temperature tensile strength of 13.6 ksi, a modulus of 564 ksi, an elongation of 3.0%, a flexural strength of 26.8 ksi and a flexural modulus of 580 ksi. At 400°F, these values drop to 10.4 ksi for tensile strength and 394 ksi for the modulus. The T_g for TMA penetration method is 282°C (540°F).

In the late nineteen-sixties and early nineteen-seventies, chemists at Hughes Aircraft Company with AFWAL sponsorship developed unique thermosetting polyimides based on the homopolymerization of terminal ethynyl (acetylene) groups. They developed fusible and tractable acetylene-terminated polyimides which homopolymerized upon the application of heat. The monomer system used for these oligomers was designed so that the prepolymer would melt before undergoing homopolymerization. The prepolymer was based on three monomers, 3-aminophenylacetylene (APA) (XXI), benzophenone-tetracarboxylic dianhydride (BTDA) (X) and 1,3-bis(3-aminophenoxy)benzene (APB) (XXII).

XXI HC≡C—⌬—NH₂   XXII H₂N—⌬—O—⌬—O—⌬—NH₂

The prepolymer is shown by structure (XXIII) where n is generally one, two or three.

XXIII

Higher values of n give prepolymers which have limited solubility in solvents such as N-methylpyrrolidone (NMP). Their higher melting temperature is too close to the homopolymerization temperature and therefore narrows the processing temperature. To improve the processibility of these acetylene-terminated oligomers, Hughes Aircraft Company chemists developed acetylene-terminated isoimides using the same monomers as for the imide form but differing only in that the polymer backbone contained the isoimide moiety. This structure isomerizes to the imide form upon application of heat. In the next chapter, Dr. Lau develops the chemistry of this system more fully. The polyimides based on acetylene-terminated polyimides are now marketed by National Starch and Chemical Company. These oligomers were previously available from Gulf Oil Chemical Company. A number of formulations are available from National Starch and Chemical Company. Thermid® IP-600 is a polyisoimide powder,

Thermid® MC-600, a preimidized molding powder, Thermid® LR-600, a amic acid form with 50% solids in N-methylpyrrolidone (NMP) and Thermid® AL-600, an amine and ester monomeric mixture with 75% solids in ethanol. Developmental quantities of higher molecular weight polyimides (IP 603 to IP 650) are also available. The Thermid® IP-600 series is characterized by lower melting temperature, longer gel time and good solubility in low-boiling solvents, retention of strength following high-temperature exposure and superior processibility in vacuum bag/autoclave composite structure fabrication. These key attributes solve many processing problems encountered with conventional polyimides. The higher values for the degree of polymerization (DP) for the Thermid® IP-600 series give resins which have excellent film-forming capabilities, lowered glass transition temperature, and improved strain capability, all with little change in strength.

The properties of Thermid® IP-600 neat resin are shown in Table 8-42. The properties of Thermid® MC-600 neat resin are shown in Table 8-43. Thermid® LR-600 is an excellent adhesive for titanium. The University of Dayton Research Institute generated data using heat cleaned 112 glass cloth as an adhesive carrier at 70% level of impregnation. The adhesive was formulated with 5% hydroquinone plus primer of a 10% solids solution of Thermid® LR-600 in methyl ethyl ketone. 500 hours at 550°F (288°C) gave a lap shear value of 2,600 psi and 500 hours at 600°F (316°C), a value of 2,500 psi. All the specimens were heat soaked at test temperatures for 30 minutes before testing.

Table 8-42: Thermid® IP-600 (National Starch and Chemical Co.)— Neat Resin Properties[19]

Property	Temp.°F	Units	Value
Tensile Strength	78	10^3 psi	8.8
	600	10^3 psi	4.5
Modulus	78	10^6 psi	0.75
	600	10^6 psi	0.20
Elongation	78	%	1.4
	600	%	4.0
T_g after postcure in air			
8 hrs. @ 700°F		°F	572
15 hrs. @ 750°F			662
4 hrs. @ 750°F			626
8 hrs. @ 750°F			669

Reprinted by permission from Hughes Aircraft Company.

Table 8-43: Thermid® MC-600 Neat Resin Properties[20]

Physical

Flexural Strength, psi	19,000
Flexural Modulus, psi	650,000
Tensile Strength, psi	12,000
Tensile Modulus, psi	570,000
Tensile Elongation at Failure, %	2
Compressive Strength, psi	25,000
Hardness, Barcol	53
Shore D	91
Water Absorption, % wt. gain (1000 hrs. @ 50°C @ 95% R.H.)	1

Electrical

Dielectric Constant	@ 10.0 MHz	5.38
	@ 9.0 GHz	3.13
	@ 12.0 GHz	3.12
Loss Tangent	@ 10.0 MHz	0.0006
	@ 9.0 GHz	0.0068
	@ 12.0 GHz	0.0048

Thermal

Thermal Expansion Coefficient 10^{-5} in/in °F (73°F to 572°F)	4.42
Effect of 600°F Aging	
% wt. Loss - After 500 hrs.	2.89
– After 1000 hrs.	4.04
Flexural Strength (psi)	
Initial	
70°F	18,600
600°F	4,200
After 1000 hrs.	
70°F (% Retention)	13,400 (72)
600°F (% Retention)	2,600 (62)

Reprinted by permission from National Starch and Chemical Corp.

The acetylene-terminated polyimides homopolymerize at about 392°F (200°C) to 482°F (250°C). There has been much speculation as to the type of bonds generated during the cure. Although low molecular weight model compounds were found to undergo some trimerization to form aromatic rings, the homopolymerization of the higher molecular weight acetylene-terminated polyimides is more complicated. There is some evidence that ene-yne type structures are formed.

The various types of polyimides discussed in this chapter and their sources are listed in Table 8-21.

REFERENCES

1. E.I. du Pont de Nemours Company, French Patent 1,239,491 (1960); Australian Patent 58,424 (1960).
2. Kreuz, J.A., U.S. Patent 3,413,267 (1968); Kreuz, J.A., U.S. Patent 3,541,057 (1970); Kreuz, J.A., Endrey, A.L., Gray, F.P. and Sroog, C.E., *J. Poly. Sci.* Al, 2607 (1966); Endrey, A.L., U.S. Patent 3,179,630 (1965); Hoegger, E.R., U.S. Patents 3,282,898 (1966) and 3,345,342 (1967).
3. Jones, J., Ochnyski, J. and Rockly, F., *Chem. Ind.* 1686 (1962).
4. Boldyrev, A.G., Adrova, N.A., Bessonov, M.I., Koton, M.M., Kuvshinskii, E.V., Rudakov, A.P. and Florinskii, F.S., *Dolk. Akad. Nauk. SSSR* 163, 1143 (1965); Rudekov, A.P., Bessonov, M.I., Koton, M.M., Pokrovskii, E.I., Fedorova, E.F., *Dolk. Akad. Nauk. SSSR* 161, 617 (1965); Rudakov, A.P., Florinskii, F.S., Bessonov, M.I., Viasova, K.N., Koton, M.M. and Tanunine, P.M., *Plast. Massig* No. 9, 26 (1967).
5. Davis, G.C., Heath, B.A. and Gildenblat, G., General Electric Report N82-CRD331, Polyimide Siloxane: Properties and Characterization for Thin Film Electronic Applications.
6. Critchley, J.P., Grattan, P.A., White, M.A. and Pippett, J.S., *J. Polym. Sci.* Al, 10, 1789 (1972); Webster, J.A., Butler, J.M. and Morrow, T.J., *Polym. Preprints* 13, 612 (1972).
7. Gibbs, H.H., Proc. 28th Ann. Tech. Conf. Reinf. Plast./Comp. Inst. Soc. Plas. Ind. (1973); Gibbs, H.H. and Breder, C.V., *Polym. Preprints* 15, 775 (1974); Gibbs, H.H., 7th Nat. SAMPE Tech. Conf. 7, 244 (1975); 21st Nat. SAMPE Symp. 21, 592 (1976).
8. Mallet, M.A.J. and Darmory, F.P., Polyaminobismaleimides, *New Industrial Polymers*, D. Desnin (ed), Marcel Dekker, New York, ch. 9, p. 112 (1975); Dermory, F.P., Processible Polyimides, *New Industrial Polymers*, D. Desnin (ed), Marcel Dekker, New York, ch. 10, p. 124 (1975).
9. Serafini, T.T., Delvigs, P. and Lightsey, G.R., *J. Appl. Polym. Sci.* 16, 905 (1972); Serafini,T.T., Delvigs, P. and Lightsey, G.R., U.S. Patent 3,745,149 (July 10, 1973); Serafini, T.T., *Proceedings of the 1975 International Conference on Composite Materials* (edited by E. Scala), Vol. 1, *AIME*, New York (1976), p. 202; Delvigs, P., Serafini, T.T. and Lightsey, G.R., NASA TN D-6877 (1972).
10. Boeing Aerospace Company, Evaluation of High Temperature Structure Adhesive for Extended Service, Contract NAS1-15605, Semi-Annual Progress Report (June 1979).
11. Serafini, T.T., High Temperature Resins, ch. 6, Handbook of Composites, edited by George Lubin, Van Nostrand Reinhold, New York (1982).
12. Monsanto Co., Skybond® 700, Technical Bulletin No. 5042C.
13. American Cyanamid Company, Aerospace Adhesives, FM® 36 Adhesive Film, BR® 36 Primer, and FM® 30 Adhesive Foam, Bulletin BPT-640.
14. American Cyanamid Company, FM® 34B-18 and 18U Adhesive Films, and BR® 34B-18 Adhesive Paste, Bulletin 4-2117 (May 1984).
15. General Electric Co., Report 82CRD331 (Dec. 1982).
16. General Electric Co., Report VLT-301B.
17. Amoco Chemicals Corporation, Bulletin TAT-35 (Dec. 1984).

18. Ciba-Geigy Corp., Polyimide UX-218, Report JB:HD 9601/A10.
19. Hughes Aircraft Company, IR&D Reports 1980-1984, Material Sciences Department, Technology Support Division.
20. National Starch and Chemical Corporation, Thermal Materials, Thermid® Polyimides (1985).
21. E.I. du Pont de Nemours and Co., Kapton®, Summary of Properties, Report E-50533, 8/82.
22. E.I. du Pont de Nemours and Co., The Properties of du Pont Vespel® Parts, Report E-26800.
23. E.I. du Pont de Nemours and Co., Preliminary Process Bulletin PC-1.
24. Rhone-Poulenc, Inc., Kerimid® FE 70003, Professional Leaflet (Nov. 1983).
25. Rhone-Poulenc, Inc., Kerimid® High Temperature Composite Polyimide Resin Report K/IE/783 (July 1983).
26. Sheik Zahir and Renner, Alfred, U.S. Patent 4,100,140 (Jan. 1978).
27. Ciba-Geigy Corporation, Developmental System XU292, Report CR9213-C93 (1983).

9
Silicones

Bernard Schneier

Reliability Engineering Department
Space and Strategic Engineering Division
Hughes Aircraft Company
El Segundo, California

Silicones are a class of synthetic compounds consisting of a polymer chain of silicon-oxygen-silicon atoms linked together. They are found in a variety of applications ranging from requirements for long life at elevated temperatures to fluidity at low temperatures. The basic straight chain Si-O-Si structure, which is similar to the repeating inorganic structure found naturally in silicate minerals, might be expected to impart thermal stability to the polymers. Organic side groups that are attached to the silicon atom confer upon the silicone polymers their characteristic properties. Groups that have been attached include alkyl, aryl, and vinyl moieties.

Silicon appears below carbon in the Periodic Table and, like carbon, has a normal valency of four. However, silicon can expand its shell and, in some instances, shows a coordination number of six. Silicon is also less electronegative than carbon and its bonds with carbon and oxygen are partly ionic. The larger size of the silicon atom and the polar nature of the bonds may account, in part, for the great freedom of motion and flexibility of the Si-O-Si bond. For example, it was found that very free rotation of methyl and bulkier alkyl and aryl groups about the silicon-carbon bond persisted even at low temperatures. The very free rotation about bonds attached to silicon, including the Si-O bond, is a contributing factor to the unusual properties of the Si-O-Si chains. The premise of weak intermolecular forces between the polysiloxane chains is used[1] as a partial explanation for the viscosity characteristics of silicone fluids as well as their low freezing points, remarkably low second-order transition temperatures and very low boiling points. On the other hand, it is suggested[2] that polysiloxane molecules readily form coils and that these coils are the ultimate structure of the liquids. The theory of helix formation assumes that relatively large forces are satisfied in forming attachments between different parts of the same molecule so that attractions between the coiled molecules themselves are diminished; controversy on the subject still exists.

Broadly speaking, polysiloxane chains, Si-O-Si, are linear long chain polymers whose properties can be varied by introducing organic side chains on the silicon atom. Silicones can be produced in the form of liquids, greases, rubbers and hard resins. Silicone fluids are the lowest in molecular weight amongst the silicone polymers.

SILICONE FLUIDS

Commercially-available silicone fluids are principally linear dimethylpolysiloxanes and methylphenylpolysiloxanes. Table 9-1 shows some approximate physical properties of methylpolysiloxanes.[1] The fluids differ in the degree of condensation; for molecular weights greater than 2,500, viscosity (η, in centistokes) at 25°C is given by the expression $\log \eta = 1.00 + 0.0123 M$, where M is molecular weight. Examination of Table 9-1 shows that viscosity can vary between 5 and 2,500,000 cs. Except for viscosity, the physical properties of the fluids exhibiting a viscosity of 300 cs, or greater, vary only slightly, if at all. Fluids with a viscosity of 20 cs or greater exert a vapor pressure of about 0.02 mm mercury at 200°C. In addition, their flash point is greater than 200°C. The methylpolysiloxane fluids are characterized by low freezing points and low viscosity-temperature coefficient constants.

Table 9-2 compares the effect of fluid type on the change of viscosity with temperature. Over the temperature range of -25°C to 120°C, the silicone fluid changes sixteen fold and the mineral oil changes a thousand fold. It should be noted (Table 9-1) that the viscosity-temperature coefficient increases slightly with increase in the viscosity of the starting fluids; the less viscous product exhibits a smaller change than the more viscous product.

The rheological behavior of the silicone fluids (η = 1,000 cs) is characterized as Newtonian for shear rates up to about 10,000 sec^{-1}. For fluids with a viscosity greater than about 1,000 cs, Newtonian behavior is exhibited below a certain level of shear rate; the higher the initial viscosity of the fluid, the lower the level of shear rate for the onset of pseudoplastic behavior. Beyond the critical value of shear rate, the behavior is described as "pseudoplastic" (apparent viscosity is less than initial viscosity extrapolated to a zero gradient).

Table 9-1: Approximate Physical Properties at 25°C of Methylpolysiloxane Fluids (Rhodorsil Oil 47V)

VISCOSITY cSt	VTC [1]	SPECIFIC GRAVITY	FLASH POINT °C	FREEZING POINT °C	SURFACE TENSION DYNES/CM	VAPOR [2] PRESSURE mm.Hg	VCE [3] $\frac{cm^3}{cm^3}$ °C	DIELECTRIC CONSTANT [4]	DIELECTRIC STRENGTH kV/mm
5	0.55	0.910	136	-65	19.7	-	1.05×10^{-3}	2.59	
10	0.57	0.930	162	-65	20.1	-	1.08×10^{-3}	2.63	13
20	0.59	0.950	230	-60	20.6	1×10^{-2}	1.07×10^{-3}	2.68	
50	0.59	0.959	280	-55	20.7	1×10^{-2}	1.05×10^{-3}	2.8	15
100	0.60	0.965	>300	-55	20.9	1×10^{-2}	0.95×10^{-3}	2.8	16
300	0.62	0.970	>300	-50	21.1	1×10^{-2}	0.95×10^{-3}	2.8	16
500	0.62	0.970	>300	-50	21.1	1×10^{-2}	0.95×10^{-3}	2.8	16
1000	0.62	0.970	>300	-50	21.1	1×10^{-2}	0.95×10^{-3}	2.8	16
5000 to 2,500,000	0.62	0.973	>300	-45	21.1	1×10^{-2}	0.95×10^{-3}	2.8	18

(1) Viscosity/Temperature Coefficient = 1- (viscosity at 99°C/viscosity at 38°C)
(2) At 200°C
(3) Volume Coefficient of Expansion between 25°C and 100°C
(4) Between 0.5 and 100 kHz

Data taken from Reference 1, reprinted with permission from Rhone-Poulenc Inc.

Table 9-2: Effect of Fluid Type and Temperature on Viscosity

OIL TYPE	VISCOSITY, cSt. at		
	$-25°C$	$25°C$	$120°C$
Rhodorsil Oil 47V 100	350	100	22
Mineral Oil	5000	100	5

Data taken from Reference 1, reprinted with permission from Rhone-Poulenc Inc.

This change is reversible. Below the critical rate of shear, viscosity exhibits Newtonian behavior once again. Viscosity is unaffected by intense shearing of long duration. The resistance of dimethylsilicone fluids to intense and prolonged shearing finds application for these materials as hydraulic and damping fluids.[1]

Silicone fluids may be blended in order to obtain a fluid of some nonstandard viscosity. The fluids selected for blending should be those with viscosities closest to the desired intermediate viscosity.[1] Since the blend will have a broader molecular weight distribution than the starting materials, the physical properties of the blend may be different, except for viscosity at a low rate of shear.

An unusual feature of dimethylsilicone fluids is their high compressibility. Table 9-3 shows that a 100 cs fluid may be compressed about 15% at 3,500 kg/cm^2 without becoming a solid. Compressibility decreases as the viscosity of the fluid increases. The compressibility of silicone fluids is greater than the degree of compression of mineral oil. This is probably related to the freedom of rotation of the substituents about the silicon atom.[1]

Table 9-3: Compressibility of Dimethylpolysiloxanes

APPLIED PRESSURE, kg/cm^2	REDUCTION IN VOLUME, %		
	OIL 47V 100 cSt	OIL 47V 1000 cSt	MINERAL OIL
500	4.5	3.8	3.1
1,000	7.3	6.5	5.2
2,000	11.2	10.7	6.8
3,500	15.1	14.4	-

Data taken from Reference 1, reprinted with permission from Rhone-Poulenc Inc.

Dimethyl Types

Dimethylsilicone fluids can be used over a temperature range of -60°C to 200°C at atmospheric pressure without risk of gelling. Moreover, in the absence of air, these products withstand exposure at 250°C for several hundreds of hours.[1]

Dimethylsilicone fluids are insoluble in water, low molecular weight alcohols and glycols, and higher molecular weight hydrocarbons such as petroleum, vegetable oils and fatty acids. They are soluble in hydrocarbons (hexane, heptane, benzene, xylene) including chlorinated hydrocarbons, and higher molecular weight ketones, such as methyl ethyl ketone. Solubility is, in fact, a function of the viscosity of the products. Low viscosity fluids may provide limited solubility in solvents in which high viscosity oils are completely insoluble.

Methylphenyl Types

Partial replacement of the methyl by bulkier phenyl groups imparts better protection of the Si-O-Si chain through steric hindrance. This results in less susceptibility to attack of the backbone by oxidizing agents and improvements in the thermal stability of the polymer as well as compatibility with slightly polar organic groups. The bulkier groups interfere with the freedom of rotation of the C-Si bond about the Si-O-Si bond. Table 9-4 shows that these effects are reflected in increased viscosity-temperature coefficients and higher freezing points for methylphenylpolysiloxanes as compared to the methyl derivatives of the same viscosity. Compressibility of the phenyl derivatives is also less than the methylpolysiloxanes. For example, at an applied pressure of 1,000 kg/cm^2, the 125 cs and the 500 cs fluids compress about 5.2 volume percent and 4.2 volume percent, respectively.[1]

The 125 cs fluid is stable to oxidation and radiation.[1] The fluid is suggested for use between -50°C to 250°C, and, is not affected after 1,000 hours heating at 250°C in air, except that contact with lead should be avoided. Lead is a catalyst which would alter the physical characteristics of the silicone fluid should contact occur. In contrast to the methylpolysiloxane fluids which offer relatively poor resistance to radiation, the methylphenylpolysiloxane materials withstand irradiation of 150 megarads at ambient temperature. They are used as a heat transfer medium for metal treatment baths, a dielectric coolant, and lubricant for plastic gear systems, among others.

Silicone fluids can be prepared containing substituents which are attached to the phenyl group and impart reactivity to the fluid. For example, methylchlorophenyl polysiloxanes are used in lubrication applications under severe conditions.[1] The chlorine group enables a chemical bond to be effected between the lubricant and the metal surfaces. The chemical bond is stable at high temperatures and the phenyl group confers resistance to high temperature degradation. The fluid is recommended for use in steel/steel lubrication and in hydraulic systems operated at high or low temperatures.[1]

Table 9-4: Approximate Physical Properties at 25°C of Methylphenylpolysiloxane Fluids (Rhodorsil Oils)

VISCOSITY cSt	VTC[1]	SPECIFIC GRAVITY	FLASH POINT °C	FREEZING POINT °C	SURFACE TENSION DYNES/CM	VAPOR[2] PRESSURE mm.Hg	VCE[3] $\frac{cm^3}{cm^3 \cdot °C}$	DIELECTRIC CONSTANT[4]	DIELECTRIC STRENGTH kV/mm
125	0.76	1.065	300	-50	24.5	1×10^{-2}	0.75×10^{-3}	2.9	14
500	0.86	1.103	300	-22	28.5	40×10^{-2}	0.77×10^{-3}	2.95	14

(1) Viscosity/Temperature Coefficient = 1- (viscosity at 99°C/viscosity at 38°C)
(2) At 200°C
(3) Volume Coefficient of Expansion between 25°C and 100°C
(4) Between 0.5 and 100 kHz

Data taken from Reference 1, reprinted with permission from Rhone-Poulenc Inc.

Controlled chemical reactions can provide copolymers of polysiloxanes with organic intermediates. Copolymers may be obtained that are polycondensates of ethylene oxide and/or propylene oxide and polysiloxanes (Figure 9-1).

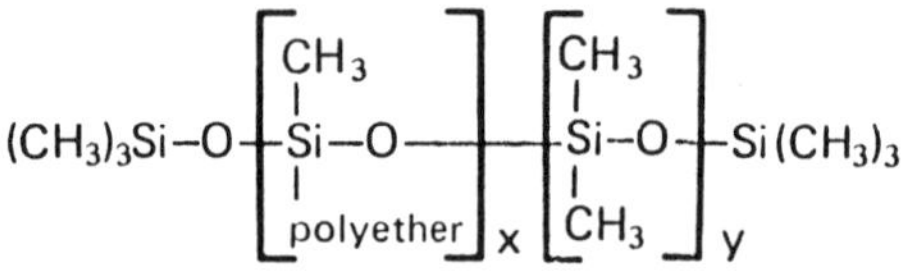

$$(CH_3)_3Si-O-\left[\begin{array}{c}CH_3\\|\\Si-O\\|\\polyether\end{array}\right]_x-\left[\begin{array}{c}CH_3\\|\\Si-O\\|\\CH_3\end{array}\right]_y-Si(CH_3)_3$$

Figure 9-1: Schematic representation of polysiloxanes and polycondensates of ethylene oxide and/or propylene oxide. (Data taken from Reference 1, reprinted with permission from Rhone-Poulenc Inc.)

This part-silicone, part-organic structure exhibits very unusual surface tension properties in different media and is used, for example, to control the cell structure of urethane foams. The low-surface-tension characteristic of silicone fluids is translated into consumer markets such as car and furniture polishes and into such industrial applications as a mold release agent for plastics, metals and elastomers.[1]

SILICONE RUBBERS

Silicone rubbers are elastomers based on high molecular weight linear polymers, generally polydimethylsiloxanes, which also may be modified with functional groups. They are available in the form of liquid or paste consistencies as room-temperature-vulcanizate (RTV) sealants and adhesives, gums, bases, and compounded stocks for fabricating heat-cured rubber products. The RTV silicone rubbers are supplied as either one-component or two-component systems, which differ in their cure chemistries.[3]

Room Temperature Vulcanizate

One Component Systems: One-component silicone rubbers use moisture in the air to hydrolyze a functional group and provide sites for the formation of a network structure of Si–O–Si bonds. Generally the one-component silicone rubbers utilize a moisture vapor cure chemistry in one of two different variations. The two cure chemistry variations are nominally called acetoxy and alkoxy.[3] These designations refer to the cure by-products generated during the curing process, namely acetic acid and methyl alcohol, respectively. Acetoxy and alkoxy cure chemistries, while similar in that each requires atmospheric moisture to effect a cure, differ in cure speed, rate of evaporation of cure by-products, and in other ways such as, but not limited to, odor, corrosion potential, tack-free time (work life) and adhesion.[3] Table 9-5 presents some key differences between the two cure chemistries.

Table 9-5: Effect of Hydrolyzable Type on some Properties of One Component Silicone Rubbers

	Acetoxy	Alkoxy
Tack free time, hrs., 25°C (77°F), 50% RH	¼–½	2–4
Cure time, hrs., 25°C (77°F), 50% RH	12–24	24–72
Adhesion		
Glass and ceramic	Excellent	Excellent
Aluminum	Excellent	Excellent
Iron, steel	Good	Excellent
Copper	Not Appl.	Excellent
Plastics		
Polycarbonate	Good	Good
Acrylic	Good	Good
PVC	Fair	Fair
Odor during cure	Acetic acid, pungent	Methanol, mild
Corrosion potential	Moderate during cure	Very low to none

Data taken from Reference 3, reprinted with permission from General Electric Corp.

The one-component system is limited to end-uses exhibiting a thin cross section since the cure depends on the diffusion of moisture into the system and on the release of acetic acid or alcohol for products using the acetoxy or alkoxy cure system, respectively. The cure process[3] begins with the formation of a skin on the exposed surface and progresses inward through the material. At 25°C and 50% relative humidity, a surface skin will form which is tack-free to the touch in 10 to 45 minutes. Table 9-5 shows that alkoxy products will form a tack-free skin in 2 to 4 hours. High temperatures and high humidity will accelerate the cure process; low temperatures and low humidity will slow the cure rate. Products using the acetoxy or alkoxy cure systems will usually be free of acetic acid or alcohol, respectively, within one to three days of application.

One-component RTV silicone rubbers are used in applications such as a formed-in-place gasketing, sealants and other uses involving bonding to a variety of substrates.[4] Table 9-6 shows some typical properties of a silicone rubber that gives off methanol during the cure process at room temperature. The physical properties of the formulation, including bonding, are adequate for handling after three days cure at 25°C and 50% relative humidity. Physical properties develop as the RTV cure progresses which may take several weeks. The electrical properties after seven days cure are also shown in Table 9-6.

Table 9-6: Typical Properties of an RTV Silicone Rubber

As Supplied	
Color	Black, white
Flow, sag or slump, inches	0.1
Specific gravity	1.50
Extrusion rate ($1/8$" orifice, 90 psi), grams/minute	300
Cure Characteristics—exposed to air, 77°F (25°C) and 50% RH	
Skin-over time, minutes	20 to 30
Working time, minutes	30
Tack-free time, hours	2 to 3
Cure time ($1/8$" thickness), hours	24
Physical Properties—cured 3 days at 77°F (25°C) and 50% RH	
Durometer hardness, shore A, points	28
Tensile strength, psi	150
Elongation, percent	550
Tear strength, die B, ppi	27
Adhesion, lap shear (to glass and aluminum), psi	75
Adhesion, peel strength,* lb/inch	20
Electrical Properties—cured 7 days at 77°F (25°C) and 50% RH	
Volume resistivity, ohm-cm	4.7×10^{14}
Dielectric strength, volts/mil	400
Dielectric constant at 100 Hz	3.6
Dielectric constant at 100 KHz	3.6
Dissipation factor at 100 Hz	0.0021
Dissipation factor at 100 KHz	0.0010

*Laboratory tests and market tests have demonstrated superior adhesion to many substrates, including:
- Plastic surfaces—acrylic, polycarbonate, polyvinylidene fluoride, polyvinyl chloride, polystyrene, acrylonitrile-butadiene-styrene
- Metals—milled aluminum, anodized aluminum, steel, galvanized steel, stainless steel
- Other—glass, wood, cement, painted surfaces

Note: Test placement prior to general use is recommended.

Data taken from Reference 4, reprinted with permission from Dow Corning Corp.

Tables 9-5 and 9-6 show that superior adhesion is obtained between the RTV silicone system and a number of plastic, metal and other substrates. In many instances, preparation of the surfaces for bonding involves wiping of the surface with a clean, oil-free rag wet with solvent such as mineral spirits, naphtha or ketones, and then, allowing the surface to dry thoroughly before applying the adhesive/sealant. In other instances, the surface may need to be abraded with the aid of a wire brush or a sandblast; after the residual dust is removed, a solvent wipe is recommended. When bonding is difficult, priming is required. Some primers are organofunctional silane derivatives which react with the substrate leaving a modified surface which can bond to the silicone rubber.[4]

Since cure time increases with thickness, use of one-component

RTV rubbers should be limited to section thicknesses of 6 mm (¼ in.) or less, as measured from the exposed surface or edge inward.[3] Thus, where broad surfaces are to be mated, sealant should be applied in a thin, less than 6 mm (¼ in.) diameter, bead or ribbon around the edge of the surface to be bonded. For applications where section depth exceeding 6 mm is required, two-component RTV silicone rubber compounds are recommended for use.

Two Component Systems: Two-component RTV silicone rubber compounds consist of silanol-terminated polymers that need to be mixed with a crosslinking agent to effect a cure. They utilize either of two different cure system chemistries, referred to as condensation cure and addition cure.[5] The condensation cure system consists of a base compound which may be mixed with any one of several interchangeable curing agents and in varying proportions depending on mixing methods employed (hand or machine) and application and cure time required for the production cycle. After addition of the curing agent, cure takes place at room temperature to form a moderate strength, durable, resilient silicone rubber. Cured section depth of virtually any thickness is possible with proper selection of curing agent. Condensation cure RTV silicone rubber compounds are recommended for applications that are not completely sealed prior to full cure because moisture and an escape path for cure by-products are required to complete the cure. Addition cure products are generally suggested for completely sealed assemblies.

Condensation Cure: Separate interchangeable curing agents for condensation cure products are normally metallic soaps. Dibutyl tin dilaurate (DBT) is the standard curing agent and is generally preferred for most applications. DBT cures at moderate speeds. It is an essentially colorless liquid curing agent which is added in a very small amount (medicine dropper proportion) to the base compound. It is easy to use, requires no sophisticated measuring equipment, and is recommended for hand mixing operations and small volume applications.

For hand or machine mixing, greater accuracy and therefore improved control over the curing rate can be achieved by use of paste curing agents which are based on tin soaps as the active ingredients. Paste curing agents offer color contrast with base compounds for visual determination of thorough mixing. They are designed for a 10:1 (base compound to curing agent) mix ratio, suitable for automatic mixing equipment. Paste curing agents used with automatic mixing equipment will minimize pot life problems, reduce material waste, save time and manpower, and are ideally suited for large volume applications.

Stannous tin octoate (STO), a clear, colorless liquid curing agent provides a faster cure than DBT and is recommended for use in small volume applications and short production cycles where rapid production of cured parts is required. STO liquid curing agent will provide RTV silicone rubber cure in approximately 30 minutes. While careful measuring is necessary to ensure proper cure characteristics, only a small portion of curing agent is added to the base compound and no sophisticated measuring equipment is needed. Because of the fast cure rate of STO, however, work time is reduced considerably, and the catalyzed material must be poured or applied immediately after curing agent addition and thorough mixing. Deep section cure greater than 25 mm (1 in.) is possible with STO curing agent but a specially-formulated paste curing agent is usually preferable because it permits longer working time.

Condensation cure products will cure in contact with virtually all types of materials without cure inhibition. When adhesion to nonsilicones is desired, however, a primer should be used.

Condensation cure RTV silicone rubbers are available in a wide range of viscosities characterized as an easily-pourable product of 12,000 cp to a pastelike product of 600,000 cp. Three examples of methylphenyl products that are offered for extremely low temperatures (-175°F) and high temperature (400°F to 500°F) range applications are given in Table 9-7. Some suggested uses for the products include potting materials for airborne electronic assemblies, aerospace materials for mechanical applications, thermal insulation ablative material, and sealants.[5]

Addition Cure: The addition cure systems[5] consist of two-component RTV silicone rubber compounds supplied in prepackaged, premeasured kits (typically 10:1 base compound to curing agent ratio). Each base compound (A portion) has its own specific curing agent (B portion). Use of addition cure products thus eliminates the need for curing agent selection, assures uniform quality, and allows simplified inventory control. Some addition cure RTV silicone rubber products offer high-strength properties, and all of them offer reliable noncorrosive, deep-section cure because there are no cure by-products. They readily cure in sections of unlimited depth, even in completely enclosed assemblies. Cure may be achieved at room temperature without exotherm or may be accelerated with heat. More rapid cure with elevated temperatures allows higher unit production in shorter cycle time and reduces storage space required during cure. Addition cure products, are, however, susceptible to cure inhibition caused by the surface contaminants present in some materials. For example, a primer coating may be needed to minimize possible cure inhibition. Use of a primer is also recommended to obtain adhesion to nonsilicone materials.

Addition cure silicone rubber products are available as low viscosity systems which will flow freely in and around complex parts providing electrical insulation and shock resistance. Unfilled systems can cure to a soft transparent gel or to a tough transparent rubber (Table 9-8). They are useful where clear rubber is required such as solar cell potting (Figure 9-2), optical instrument applications, and windshield interlayers.

Table 9-7: Typical Properties of Condensation Cure Methylphenyl RTV Silicone Rubber Products Cured at Room Temperature

VISCOSITY cps	SPECIFIC GRAVITY	HARDNESS SHORE A	TENSILE STRENGTH psi	STRAIN %	USEFUL TEMPERATURE RANGE °F	DIELECTRIC STRENGTH volts/mil	DIELECTRIC CONSTANT @ 100 Hz	DISSIPATION FACTOR @ 100 Hz
20,000	1.18	40	300	130	-175 to 400	500	4.1	0.005
30,000	1.42	55	600	110	-175 to 500	500	4.4	0.006
600,000	1.35	40	400	130	-175 to 400	500	4.1	0.006

Data taken from Reference 5, reprinted with permission from General Electric Corp.

Table 9-8: Typical Properties of Addition Cure Clear RTV Silicone Rubber Products (Heat Accelerated Cure)

VISCOSITY cps	SPECIFIC GRAVITY	HARDNESS SHORE A	TENSILE STRENGTH psi	STRAIN %	USEFUL TEMPERATURE RANGE °F	DIELECTRIC STRENGTH volts/mil	DIELECTRIC CONSTANT @ 100 Hz	DISSIPATION FACTOR @ 100 Hz
600	0.97	-	-	-	-65 to 400	500	3.0	0.001
4,000	1.02	45	900	150	-65 to 400	500	3.0	0.001
5,000	1.06	45	900	150	-175 to 400	500	3.0	0.001

Data taken from Reference 5, reprinted with permission from General Electric Corp.

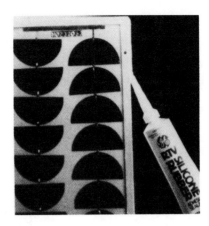

Figure 9-2: Potting of solar cells using RTV 655. (Reprinted by permission from General Electric Corp.)

Heat Cured Systems

Raw or unvulcanized silicone rubber is supplied to parts manufacturers in a state ranging from a soft to a relatively stiff doughlike consistency.[6] Before fabrication, the gum stock generally needs to be compounded on a two-roll mill or in a Banbury mixer with vulcanizing agents, extending and reinforcing fillers and special additives to tailor the properties of the finished product or to fabricate products that meet various industry specifications. Silicone rubber parts can be produced in a broad range of sizes and shapes using rubber processing techniques such as compression, transfer or injection molding, extrusion and calendering.

Silicone rubbers are grouped by polymer type and performance characteristics. Polymer classifications are based on the organic group side chains attached to the silicon-oxygen chain, methyl groups either alone or in combination with vinyl, phenyl, and fluorine-containing groups.[7] When classified by performance characteristics,[8] silicone rubber is available in four basic types: general purpose (methyl or methyl and vinyl), high performance (methyl and vinyl), extreme low temperature service (phenyl and methyl or phenyl, methyl and vinyl), and, solvent-resistant rubbers (vinyl, methyl and fluorine-containing groups).

Compounding: Compounding provides a means of producing high volume rubber products in an efficient manner.[9] The approach can be divided into three categories: extending, blending, and modifying. Extending is the process of adding semi-reinforcing fillers, such as diatomaceous earth or ground quartz, to a silicone rubber product to reduce cost, increase hardness or increase fluid resistance. Extending fillers are often used together with reinforcing fillers. Reinforcing fillers are employed to make silicone rubber products with optimum physical properties. Specific reinforcing fillers are often chosen for special applications: fume process (pyrogenic) silicas for the strongest

Silicones 331

vulcanizates and the best retention of electrical properties under wet conditions, wet process silicas for a low tendency to creep-harden, and carbon black for electrically conductive stocks.

The enhancement of specific features of a formulation is carried out by blending and/or modifying.[9] Two or more silicone rubber gum stocks may be blended to provide a rubber with different processing characteristics or properties, such as "green" strength or lower compression set. Modifying is the process of improving an aspect of a formulation by incorporating small portions of the modifier into a rubber. Depending on the type used, the modifier can enhance specific traits such as flame retardancy, high temperature stability, internal mold release, or improved shelf life, among others.

Curing: The curing of the compounded silicone rubber usually occurs in the presence of peroxide vulcanizing agents at elevated temperatures.[10] Table 9-9 lists some typical curing agents that are recommended for various uses. Molding temperatures vary from about 100°C to 180°C depending on the method of processing used and the physical dimensions of the vulcanized product. In addition, the selection of a curing agent is related to the polymer type, and desired properties of the finished product, among others. Specific applications will require the use of air oven post cures.

Table 9-9: Typical Peroxide Curing Agents

Peroxides	Commercial Grades	Form	%	Typical Molding Temperatures	Recommended Use
Bis (2,4 Di-chlorobenzoyl) Peroxide	Cadox TS-50[1] or Luperco CST[2]	50% Active Paste	1.2	104-132C (220-270F)	Hot Air Vulcanization
Benzoyl Peroxide	Cadox BS[1] or Luperco AST[2]	50% Active Paste	0.8	116-138C (240-280F)	Molding Steam Curing
DiCumyl Peroxide	DiCup 40C[3]	40% Active Powder	1.0	154-177C (310-360F)	Molding Thick Sections, Bonding, Steam Curing
2,5 DiMethyl-2,5 Di(t-butyl Peroxy) Hexane	Varox[4] or Luperco 101-XL[2]	50% Active Powder	0.8	166-182C (330-360F)	Molding Thick Sections, Bonding,
		100% Active Liquid	0.4		Steam Curing

CURING AGENT SUPPLIERS
[1]Trademark of and available from Noury Chemical Corporation, Route 78, Burt, New York 14028
[2]Trademark of and available from Lucidol Division, Pennwalt Corporation, 1740 Military Road, Buffalo, New York 14240
[3]Trademark of and available from Hercules Powder Company, 910 Market Street, Wilmington, Delaware 19899
[4]Trademark of and available from R.T. Vanderbilt Company, 30 Winfield Street, E. Norwalk, Connecticut 06855

Data taken from Reference 10, reprinted with permission from General Electric Corp.

Silicone Laminates

The properties of the finished silicone rubber depend on the type of gum, filler, curative, modifiers and solvents used, if any. Solutions

332 Handbook of Thermoset Plastics

or solvent dispersions of silicone rubber are used in the fabrication of laminates comprised of sheets of silicone solid rubber reinforced with glass cloth; normally, both sides of the fiber glass are coated.[11] The silicone coated fabrics are thin and tough, dimensionally stable and flexible. Their many applications include belting, vacuum blankets, press pads and diaphragms (Figure 9-3). If three plies of fiber glass are used, the laminates show the extra rigidity and breaking strength needed in extremely high pressure applications. The properties of the laminate may be modified by varying the glass style and rubber formulation; special constructions include one side coated and alternate base fabrics.

Figure 9-3: A multilayer layup showing COHRlastic® silicone rubber press pad material on both top and bottom. (Reprinted by permission from CHR Industries, Inc.)

The most outstanding property of silicone rubber is its great resistance to temperature extremes.[12] When compared with many popular organic rubbers at room temperature, silicone rubber is relatively weak. However, under normal operating conditions, at temperatures as high as 500°F and as low as -150°F, silicone rubbers stay elastomeric and flexible. The estimated useful life of silicone rubber at elevated temperatures is shown in Table 9-10. Useful life is defined as the period of time during which the rubber retains an elongation of 50% or more. The results indicate that parts are expected to be serviceable for 10 to 20 years at 250°F. Extrapolation to normal operating temperatures indicates a very long life for silicone rubbers. The aging resistance of silicone rubber is superior to organic-based rubbers.

Table 9-10: Estimated Useful Life of Silicone Rubber at Elevated Temperatures

Service Temperature	Useful Life*
250°F	10 to 20 years
300°F	5 to 10 years
400°F	2 to 5 years
500°F	3 months to 2 years
500°-600°F	1 week to 2 months
600°-700°F	6 hours to 1 week
700°-800°F	10 minutes to 2 hours
800°-900°F	2 to 10 minutes

*Retention of 50% elongation.

Data taken from Reference 12, reprinted with permission from General Electric Corp.

Silicone polymers have inherently good electric insulating properties. They are nonconductive because of their chemical nature and, when compounded with proper fillers and additives, are used for a wide range of electrical insulating applications. As a rule, rubber compounded for optimum retention of physical properties after heat aging will also show optimum retention of electrical properties after heat aging.[6]

Silicone rubbers swell when immersed in various liquids but solvent resistance usually improves as curing time or temperature increases.[6] The results indicate that the degree of swelling and the degree of crosslinking are interrelated. Undoubtedly, swelling is related also to the difference between the cohesive energy densities of the rubber and the liquid.

TRADE NAMES

Trade Name	Company
COHRlastic®	CHR Industries, Inc.
Dow Corning®	Dow Corning Corp.
Eccosil®	Emerson & Cumings
Green-Sil	Perma-Flex® Mold Co.
Rhodorsil	Rhone-Poulenc Inc.
RTV	General Electric Corp.
SC	Thermoset Plastics, Inc.
Silastic®	Dow Corning Corp.
Silicone Elastic Adhesive (SEA)	General Electric Corp.
Silite®	Devcon Corp.
Sylgard®	Dow Corning Corp.
Tufel®	General Electric Corp.

REFERENCES

1. Rhone-Poulenc Inc., Technical Bulletin, *Rhodorsil Oils* X03-04 (April 1979).
2. *Silicones*, S. Fordham Editor, William Clowes and Sons, Ltd., London, p 82 (1960).
3. General Electric Technical Bulletin, *The Sealers* S-2H Rev 7/83.
4. Dow Corning Technical Bulletin, *Silastic® 739 RTV Plastic Adhesive.*
5. General Electric Technical Bulletin, *The Versatiles* S-35C Rev 6/83.
6. Dow Corning Technical Bulletin, *Designing with Silastic® Silicone Rubber* 17-158A-79.
7. American Society for Testing Materials ASTM D1418-81, *Rubber and Rubber Latices—Nomenclature.*
8. Dow Corning Technical Bulletin, *Information About Silicone Elastomers* 17-80A dated 4/76.
9. Dow Corning Technical Bulletin, *The Silastic® Compounding System* 17-264-79.
10. General Electric Technical Bulletin, *Silplus™ Elastomeric Systems SE6035, SE6075.*
11. CHR Industries Technical Bulletin, *COHRlastic® Silicone Rubber Products* SR-2-7/83.
12. General Electric Technical Bulletin, *Silicones* S-1E.

10

Crosslinked Thermoplastics

Bernard Schneier

*Reliability Engineering Department
Space and Strategic Engineering Division
Hughes Aircraft Company
El Segundo, California*

Enhancement of properties is an underlying incentive for the commercial development of crosslinked thermoplastics. Crosslinking of polymers provides improved resistance to thermal degradation of physical properties and improved resistance to cracking effects by liquids and other harsh environments, as well as to creep, among other effects. This chapter deals with the crosslinking of primarily aliphatic polymers. High intensity radiation from electron beams or ultraviolet sources have been used to initiate polymerization in systems of oligomers capped with reactive methacrylate (acrylic) groups or isocyanates. Using these techniques, films with low shrinkage and high adhesion properties have been employed in applications such as pressure sensitive adhesives, glass coatings and dental enamels. On the other hand, for thermoplastics such as polyethylene, crosslinking by chemical or irradiation techniques is recognized as a technology for manufacturing industrial materials such as cable coverings, cellular materials, rotationally molded articles and piping.

For crystalline olefin polymers, the structure of the crystallite and the connected noncrystalline regions contribute to the polymer properties, including density. For example, polyethylenes with densities in the range of 0.94 to 0.96 g/cc may contain 65 to 90% crystalline material while polyethylene with a density of about 0.92 g/cc commonly contains 50 to 60% crystalline material.[1] The proportion of amorphous material is increased by high energy radiation, an effect which is evidently due to crosslinking of the molecules; exposure of low density polyethylene sufficient to cause crosslinking of about 10% of the carbon atoms gives a product which is entirely amorphous at room temperature.

Polyethylene, like other crystalline macromolecules, melts over a temperature range; as the temperature rises, the proportion of amorphous material increases until all the crystalline regions are melted. High density polyethylenes melt finally at 125°-131°C while lower density polyethylenes melt at 110°-115°C (Figure 10-1). These results are as expected. The presence of structural or stereochemical irregularities in macromolecules lowers the melting point and decreases the degree of crystallinity. The spread of molecular weight in a specimen is not expected to influence the final melting point or the melting range unless there is an appreciable proportion of molecules of molecular weight below 1,500 which is not expected in commercial processes.[1]

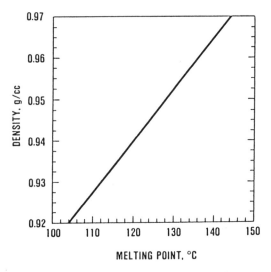

Figure 10-1: Change of melting point with density of polyethylene homopolymers. (Data taken from Reference 5a, reprinted with permission from Phillips Chemical Co.)

CROSSLINKING OF THERMOPLASTICS

Thermoplastics may be crosslinked using irradiation techniques and chemical agents. Radiation chemical studies[2a] on macromolecules provided the first evidence for the formation of crosslinks, formation of insoluble gel (due to crosslinking), production of main chain unsaturation, peroxide formation and changes in physical properties, among other effects.

Initially, the radiation chemical studies on crystalline olefin polymers were carried out mostly on polyethylene.[2a] When polyethylene is irradiated with x-rays, gamma-rays, high energy electrons, or pile irradiation, a considerable change in physical properties can result from a limited degree of chemical change. Hydrogen gas is liberated together with paraffins such as methane, ethane, and propane in smaller amounts. The polymer becomes increasingly insoluble; at first it becomes more flexible and transparent but after protracted radiation it hardens and becomes brittle. The structural changes involved in this transformation occur in the absence of heat and chemicals. These crosslinking reactions, induced by radiation, can be closely controlled and may cover an extremely wide range.

Although some property studies have been made on the effect of irradiation at very high doses, most studies have concerned themselves with irradiation doses in the range of 0 to 150 Mrad. The reason for this is simply that the beneficial changes in properties occur at relatively low doses, and increasing doses serve to degrade some of the properties of olefin polymers. From a technological standpoint, therefore, the dose range which is of most importance is in the range of 0 to 60 Mrad. The radiation allows the crosslinking reactions to proceed in the solid state so that an object molded into its final shape can be irradiated subsequently and modified to give it the required properties.[1]

The two fundamental actions of radiation are to ionize and to excite, and both ions and excited molecules can give free radicals.[3] The first step[4] in the production of chemical effects by high energy radiation is the interaction between the electromagnetic field of the gamma ray and the electrons of the polymer to produce ionization or electronic excitation. The gamma rays give rise to fast electrons which have appreciable energy. As the energy required to produce chemical change is only a few electron volts per molecule, a fast electron is capable of altering several thousand molecules and possibly producing a polymeric segment in an energy rich state. The gamma ray photons lose part of their energy by ejecting electrons from sites along the macromolecule (Compton scattering). Interaction with outer electrons is the principal process which produces chemical change. As a result of photon absorption, the outer electron is excited to higher energy levels or is ejected to leave ions. The positive ions, being electron deficient, possess the properties of free radicals as well as of ions. The ionization should be considered as the removal of an electron from the molecule as a whole, rather than just from one part. Consequently, the positive charges can be present anywhere in the molecule, although it will tend to be localized at certain positions more than at others, as for example, at electron deficient sites such as double bonds. Positive ions may be neutralized by combination with an electron or a negative ion. Considerable energy will usually be evolved so that the neutral product will be highly excited and will react further.

Excited molecules can also be formed by the high-energy particles themselves.[4] Fast particles excite molecules to the same levels as ultraviolet light, mainly the lowest allowed excited state, but slow electrons can excite to any level. The excitation also is of the whole molecule, not just one part of it, and in large molecules reaction can occur at a site remote from the track of the particle. The same phenomenon occurs in photochemistry where energy is absorbed by the chromophore and yet reaction or emission of fluorescence can occur elsewhere in the molecule. Excited molecules, if they do not phosphoresce or fluoresce, can lose their energy by internal conversion to give a strongly vibrating lower electronic state. The energy of vibration may then be removed by collisions. Excited molecules may also decompose into free radicals, but unless the decomposition is an energetic one there is, in condensed phases at any rate, the possibility that the radicals may recombine within the solvent cage (Franck-Rabiniwitch effect) leading to no net reaction. This effect should be especially marked with large molecules. Another mode of decomposition is to yield molecular products directly either by a unimolecular process, or by reaction with other excited molecules. Finally, some excited molecules may react chemically like free radicals.

The role of ions and excited molecules in radiation chemistry is reasonably well understood in general terms, but in specific instances, especially in the irradiation of liquids and solids, it is very difficult to sort out what is happening. However, both ions and excited molecules can give free radicals as well as stable molecular products, and the nature and reactions of the free radicals can be studied even though it is rarely possible to decide exactly how they were formed.[2a]

As mentioned, hydrogen gas is liberated by irradiation. In order to produce hydrogen from polyethylene, the C–H bond must be broken. Since the chemical bond strength of the C–H bond is greater than the strength of the C–C bond, chain scission might be expected to occur more frequently than scission of the C–H bond. Despite this, experimental evidence[3] shows that chain scission is low for low density polyethylene. If a C–C bond is broken, the sigma bonding electrons are separated and free radicals are formed. It would be expected that the two long chain fragments that are produced are held firmly in the solid matrix, and, as a result, recombination of the free radicals is likely (cage effect).

In polybutadiene, no chain scission on irradiation can be detected.[3] When the polymer chain contains carbon atoms connected to four carbon atoms as in polyisobutylene or to three as in polypropylene, then considerable chain scissions are produced by the irradiation. Substances containing chlorine yield HCl, but chlorine cannot be liberated so long as there are hydrogen atoms present. Polyvinyl chloride is among the least resistant of all plastics to irradiation.

On breaking the C–H bond in polyethylene, an alkyl free radical of the structure $-CH_2\dot{C}HCH_2-$ results.[3] If two free radicals are formed on neighboring chains, a very probable reaction would be crosslinking by recombination of the free radicals. The evidence is that the crosslinks occur primarily in the amorphous phase of the polyethylene as well as in the amorphous surface layers of single crystals of polyethylene. The liberation of hydrogen from polyethylene also produces unsaturation in the molecule.[2a] There are three types of unsaturation observed: vinylidene, terminal and trans-vinylene. While trans-vinylene unsaturation is produced in the molecule during irradiation there is concommitant decrease of vinyl and vinylidene unsaturation. A further change which occurs on irradiation is the reduction of pendant methylene groups present in the original polymer. When irradiation of polyethylene is carried out in air, crosslinking is inhibited and oxidation takes place mostly on or near the surfaces. The oxygen effect is attributed to the formation of peroxides and hydroperoxides. The mechanism may involve the combination of oxygen with a free radical to form a free radical of the type $RO_2\cdot$; this free radical can then abstract a hydrogen to form a hydroperoxide which, in turn, can decompose into two free radicals. Other mechanisms are possible. It should be emphasized[3] that oxygen gas dissolves only in the amorphous regions of polymers, and can, therefore, oxidize the material only within the amorphous phase or in the surface. Also the extent of oxidation per unit of radiation dose is very dependent on the dose rate. The lower the dose rate the more time the oxygen has to diffuse into the plastic per unit of dose so that the net amount of oxidation for the same dose is much greater at low rates.

Oxidation and crosslinking subsequent to irradiation have been observed.[3] The phenomena have been attributed to the persistence of free radicals in the polymers. Not only does oxidation occur during irradiation but it may continue for weeks after irradiation if the alkyl free radicals are not annealed out by heating, in the case of polyethylene, to 100°C or higher after irradiation. The post-irradiation oxidation of polyethylene has been followed, using as a basis, the increase in the carbonyl infrared absorption band at 1725 cm^{-1} even after sixty days exposure to air at room temperature. Carbonyl formation may be accompanied by chain scission. Also, a delayed crosslinking reaction was observed when a high density polyethylene was irradiated and then immediately heated above its crystalline melting point. The increased crosslinking was attributed to the migration of free-radical centers through the polyethylene by random jumps of a hydrogen atom from an adjacent site to the free-radical site.

EFFECTS OF CROSSLINKING ON POLYMER

Polyethylene

Table 10-1 gives the effects of gamma and beta irradiation on the properties of high density polyethylene.[5a] The data indicate that polymer crosslinking is accompanied by an increase in tensile strength and hardness and a decrease in solubility. Beta irradiation produces significant changes in tensile properties at 132°C, near the ultimate melting point of the crosslinked polymer (Figure 10-1). Table 10-1 shows that, at 132°C, unirradiated polyethylene exhibits insignificant properties. Beta irradiation may provide material with properties that are adequate for an application even at 132°C. Irradiation also increases resistance to environmental stress cracking.

Table 10-1: Effects of Gamma and Beta Irradiation on Properties of Marlex® High Density Polyethylene

Typical Properties	Temperature, °F (°C)	Beta Irradiation Dosage (Megarads)				
		0	5	10	15	50
Tensile Strength, psi (MPa)	82 (28)	4110 (28.3)	4217 (29.1)	4293 (30)	4400 (30.3)	4560 (31.4)
	200 (93)	1303 (8.98)	1567 (10.8)	1640 (11.3)	1120 (7.7)	1477 (10.8)
	270 (132)	——	180 (1.2)	212 (1.46)	455 (3.13)	745 (5.13)
Elongation, %	82 (28)	20	18	22	20	20
	200 (93)	167	375	520	505	133
	270 (132)	——	510	445	385	110
Hardness, Shore D		64	67	67	68	70
Density, g/cm³		0.96	0.96	0.96	0.96	0.96
Solubility, Tetralin, 266° (130°C)		Soluble	Insoluble	Insoluble	Insoluble	Insoluble
Color		White	White	Ivory	Ivory	Tan

Typical Properties	Temperature, °F (°C)	Gamma Irradiation Dosage (Megarads)			
		0	1	10	100
Tensile Strength, psi (MPa)	82 (28)	5840 (40.2)*	7007 (51.7)	7120 (49.1)	8360 (57.6)
Elongation, %	82 (28)	13	15	15	1
Hardness, Shore D		64	68	70	70
Density, g/cm³		0.952	0.955	0.955	0.967
Solubility, Tetralin 266°F (130°C)		Soluble	Insoluble	Insoluble	Insoluble

Type of Irradiation Dosage, rads	F_{50} Values, h**	
	Gamma	Beta
None	20	20
1 x 10⁶	20	—
3 x 10⁶	24	—
6 x 10⁶	110	40
1 x 10⁷	700	350
3 x 10⁷	350	350
1 x 10⁸	1	—

*Measured by different laboratories.   **Environmental Stress Cracking in Igepal CO-630 at 122°F (50°C).

Data taken from Reference 5a, reprinted with permission from Phillips Chemical Company.

Table 10-2 shows typical properties of a series of radiation crosslinked closed cell polyethylene foams.[6a] The foams, ranging in density from 1.5 to 12 pounds per cubic foot, are characterized by excellent mechanical, thermal and chemical properties, together with a fine cell structure and an exceptionally smooth surface; they are available in thicknesses from 1/32 of an inch to more than an inch. A crosslinked polyethylene foam sheet with an integral skin is available also in the same range of densities and thicknesses. The skin offers increased abrasion resistance without reducing the foam's flexibility. Table 10-2 shows a Type E foam which is described as a crosslinked polyethylene copolymer foam especially formulated to provide more flexibility and resilience than the standard grade Type A. Polypropylene foam is also available.

Table 10-2: Typical Properties of Closed-Cell, Radiation Crosslinked Polyethylene Foam

Foam Product	Nominal Density (pcf)	Compressive Strength (psi) @ 50% Deflection	Compression Set (% Orig Thick.) ASTM D-395	Tensile Strength (psi) ASTM D-1564 M / C	Elongation (% to Break) ASTM D-1564 M / C	Tear Resistant (lb/in) ASTM D-624 M / C	Thermal Stability % Linear Shrink After 3 Hours 180° / 215°	K-Factor	Water Absorption lb/ft³ Cut Surface Max. ASTM D-1667
VOLARA 1.5A	1.5	11-14	15 max.	38 / 25	121 / 101	8 / 6	2.5 / 8.5		
VOLARA 2A*	2	12-16	16 max.	50 / 41	138 / 114	11 / 8	1.5 / 3.0	0.25	0.04
VOLARA 2MF	2	12-16	30 max.	48 / 36	78 / 62	13 / 8	1.5 / 2.2		
VOLARA 2E	2	11-15	21 max.	60 / 48	250 / 250	11 / 10	3.6 / 20.0	0.25	0.04
VOLARA 2EE	2	10-15	25 max.	35 / 29	190 / 200	6 / 5	14.0 / 50.0		
VOLARA 4A	4	19-24	12 max.	100 / 82	174 / 148	22 / 18	1.2 / 2.8	0.30	0.04
VOLARA 4E	4	17-21	13 max.	129 / 111	302 / 300	23 / 20	2.6 / 15.0	0.30	0.04
VOLARA 6A	6	25-31	9 max.	148 / 124	220 / 178	35 / 28	1.0 / 2.2	0.32	0.04
VOLARA 6E	6	22-27	8 max.	200 / 172	350 / 348	35 / 31	2.1 / 9.9	0.32	0.04
VOLARA 8M	8	60-80	14 max.	250 / 200	165 / 120	65 / 50	0.2 / 0.5		
VOLARA 12A	12	75-100	15 max.	294 / 226	284 / 255	76 / 71	2.1 / 3.8		
VOLASTA 3A	3	18-23	20 max.	68 / 60	97 / 98	13 / 11	2.0 / 5.8		

*AVAILABLE IN FORMULATION CONTAINING FIRE RETARDANT ADDITIVES

Data taken from Reference 6a, reprinted with permission from Voltek, Inc.

The crosslinked polyethylene and polypropylene foams are resilient cushioning materials. Most applications are based on this property together with some other quality such as buoyancy, shock absorption, thermal or electrical insulation, vibration damping and moisture protection. Automotive applications include gasketing, sun visors, an insulating liner for air-conditioner housing and carpet backing. Recreation and sport uses are based on protection against repeated shock from relatively high stresses (Figure 10-2). The foams also find use in medical products since they add comfort to orthopedic braces and cervical collars, for example.

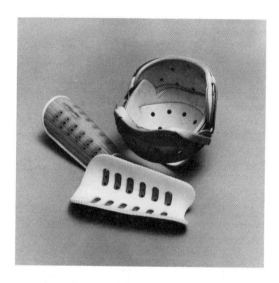

Figure 10-2: Crosslinked polyolefin foam in sport applications. (Reprinted by permission from Voltek, Inc.)

The applications of the crosslinked polyolefin foams are very diverse because these materials can be combined with others using proven plastics industry methods but with variations because of the foamed materials.[6b] The foams are produced from polyolefin resins which results in a low energy surface that is not easily wet by laminating adhesives. The most common way to improve wettability and adhesion is corona treatment. A similar effect can be obtained by flame treating the polyolefin foam prior to lamination. Using heat and pressure, foam may be laminated to itself, to urethane foam, to polyolefin films and to fabrics made from natural fibers. Each substrate is preheated using a gas flame or electric heaters just before they are joined at the laminating nip. A heat reactivation method can be used to bond to a fabric, film or other substrate. This technique employs a thermoplastic film or preapplied coating which, after heating and applying pressure, bonds to another substrate without the need to evaporate water or carrier solvent. Examples of adhesive films include low density urethane foam, solid polyethylene or ethylene-vinyl acetate films. These materials melt completely and become a solid continuous layer in the final laminate. For substrates that cannot be bonded using heat and/or pressure, adhesives may be used. According to the manufacturer, adhesives are available which will bond polyolefin foam without pretreatment of the foam. Other adhesives require pretreatment.

The radiation crosslinked polyolefin foams can be compression molded alone or in combination with other foams of different color and density, fabrics, films, foils, nonwovens, solid plastics and cellulosics.[6c] By combining the crosslinked foams with other materials, a wide variety of products with desired properties and appearances can be tailored to specific end-use requirements. While the combining is usually done by lamination prior to molding, materials can be bonded together in the mold. The choice is based on the cost and ease of laminating various materials. For example, a two-way stretch nylon fabric should be laminated before molding whereas a rigid high density polyethylene insert would be heated and then placed in the mold between layers of hot foam.

Polypropylene

The use of stabilizers that permit the sterilization of biomedical products by gamma irradiation may broaden the range of disposable products that are made from polypropylene. The sterilization of polypropylene biomedical products by gamma irradiation results in severe resin degradation.[7] Table 10-3 shows the changes in molecular weight of polypropylene and polystyrene samples after gamma irradiation was carried out by a Cobalt-60 source at dosages to 2.5 Mrad at a 1/2 Mrad-per-hour rate. The results show that, for polystyrene, weight average and number average molecular weights and molecular weight distribution were nearly unchanged. For polypropylene, however, the data show the shift from high to low molecular weights for unirradiated and irradiated polypropylene. The results clearly indicate that severe chain scission has occurred.

During gamma irradiation of polypropylene in air, random free radical formation is believed to occur followed by oxidative propagation of these radicals.[8] The incorporation[7] of stabilizers into polypropylene, such as a hindered amine or phenol, suppresses oxidation and inhibits the deterioration of the properties of the polypropylene rendering it more resistant to irradiation. The reduction in amine stabilizer concentration during irradiation and the subsequent storage period was attributed to stabilizer mechanisms of the additive (radical scavenger, chain transfer agent, or, peroxide decomposer). The conclusion was that regardless of the mechanism, stabilizers play a sacrificial role in protecting polypropylene from further degradation following irradiation.

Table 10-3: Changes in Molecular Weight of Polypropylene and Polystyrene After Gamma Irradiation of 2.5 Megarads

Samples	Unirradiated	Irradiated
Polypropylene*		
Weight average molecular weight	4.0×10^5	9.4×10^4
Number average molecular weight	1.1×10^5	1.2×10^4
Molecular weight distribution	3.7	8.0
Polystyrene**		
Weight average molecular weight	2.6×10^5	2.6×10^5
Number average molecular weight	1.5×10^5	1.6×10^5
Molecular weight distribution	1.8	1.7

*Hercules Prefax 6501.
**Dow Styron 685D.

Data taken from Reference 7, reprinted with permission from Plastics Engineering.

The progressive deterioration during storage of polypropylene articles following gamma irradiation has been attributed to the presence of hydroperoxide groups.[8] Polypropylene hydroperoxide groups are known to be thermally unstable and are considered a potential source for storage oxidation together with "trapped" peroxyl radicals. "Trapped" peroxyl radicals are thought to be in solvent accessible, crosslinked (amorphous) domains. It has been suggested[9] that incorporation of the additive increases the internal free volume in the amorphous phase of the polymer, thus enhancing main chain mobility and accelerating main chain recombination; this holds true both during and following irradiation.

Polyvinyl Chloride

In polyvinyl chloride (PVC), carbon-chlorine and carbon-hydrogen bonds are susceptible to cleavage by ionizing radiation, producing free radical sites on the polymer backbone.[10] During irradiation, these radicals would initiate dehydrochlorination and small amounts of main chain scission and crosslinking. This dehydrochlorination proceeds via a chain reaction to produce conjugated double bonds. These unsaturated structures absorb in the ultraviolet-visible region, thereby producing discoloration in the polyvinyl chloride.

A number of insulations and coatings used in high temperature environments are produced by the radiation crosslinking of polyvinyl chloride resin.[10] A typical formulation for a radiation-curable coating includes a base resin, a crosslinking sensitizer (e.g., polyfunctional monomer), and a plasticizer. A typical resin sensitizer must be capable of being crosslinked by irradiation as well as incorporating the resin molecules into the network via grafting reactions. Polyfunctional monomers that have been shown to act as such crosslinking sensitizers for polyvinyl chloride include allyl esters, dimethacrylates, trimethacrylates, triallyl isocyanurate, divinyl benzene, and triacrylates. The polyfunctional methacrylates and acrylates were found to possess the greatest sensitivity. The plasticizers are used to obtain the required physical properties for a specific coating application. In one study, the system of polyvinyl chloride blended with trimethylolpropane trimethacrylate (TMPTMA) and diundecyl phthalate (DUP) was selected as a representative example of such radiation-curable coatings.

In the absence of plasticizer, there was an initial preference for TMPTMA homopolymerization after which PVC molecules were bound into the network.[10] Increasing the irradiation temperature primarily increased all the reaction rates equally. However, dehydrochlorination of the PVC did begin to compete with the grafting and crosslinking mechanisms at the high temperatures (80°C). Post-irradiation thermal treatment was shown to markedly alter the physical properties of the irradiated blends. This was caused by the reactions of residual monomer molecules and unreacted double bonds in the crosslinked network. TMPTMA does not use all of the available double bonds to form the network.

Over a wide range of blend compositions the crosslinking rate was found to be proportional to the TMPTMA concentration.[10] As the TMPTMA concentration decreased, soluble graft copolymers were produced in addition to insoluble networks. In the blends, the free radical sites are scavenged by TMPTMA, which grafts to PVC and initiates incorporation into the network via structures like PVC–(TMPTMA)$_x$, PVC–(TMPTMA)$_x$–PVC, etc. Blends with no TMPTMA (pure PVC or PVC/DUP blends) showed no significant gel formation.

The introduction of DUP into the mixtures enhanced (1) TMPTMA homopolymerization, (2) TMPTMA grafting, (3) PVC crosslinking (at low doses), and (4) reactivity of double bonds.[10] These effects are interrelated. For example, TMPTMA grafting produced precursors for PVC crosslinking. Among the parameters that determine the chemical kinetics of the system are the concentration and mobility of the reactants. The competition between reactions (1), (2) and (3) above are determined by the reactivity and mobility of the double bonds. The ease with which PVC, TMPTMA and free radicals can diffuse through the matrix and react together will be constantly changing throughout the reaction. With PVC/TMPTMA blends, the medium changes from a flexible PVC resin (plasticized by the monomer) to a 3-dimensional network (strong and brittle). With DUP present, the blends remain flexible after irradiation. On a molecular level, this means that the mobility of the reactive species remains high. Until high conversions were reached, the mobility of the reactive sites was enhanced by the presence of DUP and the double bonds were accessible for reaction.

These results show that in the manufacture of crosslinked coatings, the inclusion of DUP would result in energy efficiency, double bond efficiency, and a more stable product.[10] The increase in mobility (due to DUP) produced faster crosslinking, and therefore required less irradiation (less energy). A typical dose required was 4 Mrad.

CHEMICAL CROSSLINKING

Chemical crosslinking of saturated polymers such as polyethylene, polypropylene and polyvinyl chloride consists of forming bonds between chains using organic peroxides, in general, as a source of free radicals. Crosslinking with free radicals involves hydrogen abstraction to produce a free radical initiator site on the polymer molecule. It would be expected that recombination of free radicals formed on adjacent chains would produce a crosslinked material with enhanced thermal stability. Polyethylenes, modified with organic peroxides, are used in applications such as rotational molding[5b] and piping for hot water use.[11]

Organic peroxides are useful as free radical initiators because they are stable compounds until heated and their decomposition rate is temperature dependent.[12] Since, in many instances, peroxide decomposition follows first order kinetics, decomposition rate at a particular temperature is usually expressed as half-life. It is defined as the time necessary to decompose one-half the amount of peroxide originally present. For example, dicumyl peroxide, which can be used to chemically crosslink polyethylene, has a half-life of 12 hours at 115°C, 1.8 hours at 130°C, and 0.3 hour at 145°C; the temperature for a half-life of 10 hours is 117°C. The melting point of low density polyethylene ranges from 110° to 115°C. After mixing or milling of the polymer and peroxide, the mixture may be shaped and then heated to induce the decomposition of the initiator with consequent crosslinking of the polymer in the molten state. Bearing in mind the oxygen effect leading to carbon groups in an oxidized state, the structure of the chemically crosslinked polymer most probably is more complex than simply a crosslinked paraffin.

Polyethylene

Polyethylene crosslinked with dicumyl peroxide exhibits two separate but overlapping regions of dielectric loss in the region of -150° to -50°C. Polyethylene shows three regions of dielectric loss centered around 60°C, 0°C, and -100°C, which are labeled alpha, beta, and gamma-loss regions, respectively.[13] These regions essentially result from carbonyl dipoles produced by adventitious or deliberate oxidation. Whereas both the beta and gamma losses originate from the amorphous phase, the alpha region is associated with the crystalline phase. Crosslinked polyethylene is now widely used as insulation for underground electric cable and is also gaining acceptance in foamed form as microwave insulation. When dry-cured, the polyethylene contains approximately equal amounts of the by-products of the dicumyl peroxide initiator, namely acetophenone and 2-phenyl-2-propanol. Both of these molecules are polar and would be expected to be located in the amorphous phase and hence to give rise to beta or gamma losses but with activation energies different from those of oxidized polyethylene. The major loss effects that occur at the lower temperatures are attributed to the individual loss peaks of the two major by-products of the crosslinking agent. Analog materials prepared by Yang et al.[13] from linear low density polyethylene by blending in either acetophenone or 2-phenyl-2-propanol confirm the hypothesis.

The concentration of dicumyl peroxide used to modify a low density polyethylene (number-average molecular weight = 32,000) affects the static and dynamic mechanical properties of the chemically crosslinked polymer. Kunert[14] reported on studies that were carried out at levels of 0.5 to 2.5 weight percent to determine the changes in structure of crosslinked polyethylene. The results of dynamic mechanical property testing in shear, using a torsional pendulum, showed that the maximum value of the logarithmic decrement occurred in the vicinity of 47°C (the so-called alpha relaxation point) in the uncrosslinked and the crosslinked polyethylene. At 27°C, values for the storage modulus of crosslinked polyethylene were below the values of the unmodified polymer; storage modulus decreased slightly as the peroxide concentration decreased to about 2%. The same correlation was observed at 80°C. At 87°C, however, the storage modulus of the crosslinked polymer was greater than for the noncrosslinked polyethylene; storage modulus decreased slightly with increasing peroxide concentration. Kunert concluded that a storage modulus at 27°C is probably affected mainly by crystallinity since this modulus exhibited higher values for the uncrosslinked polyethylene than for the crosslinked material. However, at 87°C, which is near the melting point of polyethylene crystallites, the storage modulus of the uncrosslinked polyethylene shows lower values than the crosslinked material. This is mainly due to an increase in stiffness of the polyethylene network at this temperature.

The Weissenberg rheogoniometer was used to test the crosslinked polyethylene as a function of frequency over four decades at room temperature.[14] The storage modulus increases as a function of frequency between 5×10^{-3} and 5 Hz. A plot of storage modulus as a function of peroxide concentration showed that, at constant frequency, the characteristic feature of the plot was the appearance of two maxima for storage modulus occurring at peroxide concentrations of 0.5 and 2% (Figure 10-3). A resonance method in the frequency range of 100 to 900 Hz gave similar results.

The absence of any maximum of storage modulus plotted against the peroxide concentration when testing with the torsion pendulum was attributed to the effect of a large deformation that probably exceeded the range of linear viscoelasticity of this material. Thus the

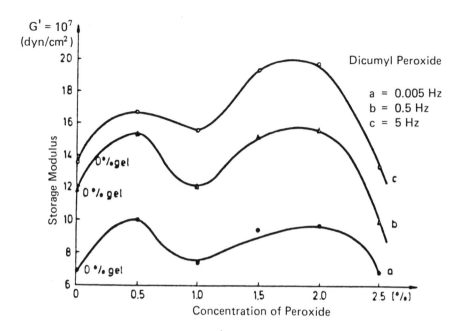

Figure 10-3: The storage modulus G' of XLPE as a function of dicumyl peroxide concentration measured in a Weissenberg rheogoniometer. (Data taken from Reference 14, reprinted with permission from John Wiley & Sons, Inc.)

storage modulus represents the averaged resultant properties of different phases in the crosslinked polyethylene specimen. On the other hand, the appearance of some maxima with the Weissenberg and resonance methods was taken to indicate that the deformations are at least half those of the torsional method and are probably within the range of linear viscoelasticity. Thus the mechanical properties of different phases are not averaged. Two rigid structures in crosslinked polyethylene are proposed. The first maximum is presumably related to the appearance of crystallites (excessive crystallinity) and the second maximum is probably caused by a uniform rigid polyethylene network. The structure attributed to excessive crystallinity is presumed to occur at low concentration of peroxide ($\sim 2\%$); the very regular and perfect polyethylene network is presumed to occur at the high peroxide concentration.

The results of static mechanical property testing showed that, as the peroxide content increased, Young's modulus and stress at yield point decreased while percent elongation at the yield point increased. A plot of ultimate tensile strength exhibited a maximum in the vicinity of 0.5% peroxide concentration. Elongation at break exhibited a similar maximum. Kunert's conclusion was that Young's modulus depends on the amount of rigid amorphous phase in crosslinked polyethylene. As the amount of peroxide is increased, the amount of soft phase increases. Kunert considers the soft phase may act as a plasticizer for the hard phase. He postulates that ultimate tensile strength depends on the amount of crystallites. Except for this, static mechanical properties, and especially Young's modulus, depend on the amount of hard, amorphous structures.

At the other end of the spectrum, Lem and Han[15] reported on studies in which the highest peroxide concentration was 0.1 weight percent. In this range, little, if any, insoluble gel particles were formed between the particular virgin resins and dicumyl peroxide. A comparison of the molecular weight distribution curves of a low density polyethylene (number-average molecular weight = 7,500) before and after reacting with peroxide showed that the modified material has more high molecular weight portions compared to the starting resin; both number-average and weight-average molecular weight increase with increasing amounts of dicumyl peroxide. In addition, the modified low density polyethylene has more low-molecular-weight portions compared to the virgin resin. The results show that, from a mechanistic viewpoint, dicumyl peroxide added to the polymer and extended the polymer chains (especially, the long chain molecules) giving rise to larger molecules. There was evidence also that the degree of long-chain branching is increased in the presence of dicumyl peroxide free radicals.

The molecular weight distribution curve of a high density polyethylene (number-average molecular weight = 15,300) showed that, after reacting with dicumyl peroxide, the entire curve of the modified resin was shifted toward the right-hand side, indicating that polymer chains are extended in the presence of peroxide free radicals and little dissociation of C–C bonds has taken place.[15] The little degradation of high density polyethylene observed, compared to low density polyethylene, was explained by the fact that, when subjected to dicumyl peroxide free radicals, the tertiary carbon atoms present in the long- and short-chain branching of low density polyethylene are very susceptible to degradation compared to those in the straight chains.

The rheological properties of the modified polyethylene were determined with a Weissenberg rheogoniometer. Plots of first normal stress difference versus shear stress for virgin and modified low density and high density polyethylene show that the resin becomes more elastic as the amount of dicumyl peroxide is increased. The results showed that if the concentration of dicumyl peroxide is low enough not to yield insoluble gel particles, modified resins are obtained which may be considered to have resulted from chain extension rather than crosslinking.[15]

Polypropylene

Radicals produced by peroxide decomposition may abstract hydrogen at any site along the chain. If unsaturation is present, hydrogen abstraction will likely occur from a carbon group alpha to a double bond. In the absence of an allylic group, abstraction is expected at the site of tertiary hydrogen. Tertiary hydrogen atoms characterize polypropylene and are found at branch points in polyethylene. Hydrogen abstraction of a tertiary hydrogen forms free radicals that tend to undergo chain scission at the expense of crosslinking. In contrast to polyethylene, radical formation in polypropylene is accompanied by degradation and crosslinking. The mechanism of degradation reaction of polypropylene is generally explained by fragmentation of tertiary alkyl macroradicals, while the crosslinking is explained by a combination of secondary alkyl macroradicals.[16] The decreased efficiency of peroxide crosslinking of copolymers of ethylene and propylene was accounted for in the same way.

In order to counteract this inherent low crosslinking tendency of polypropylene, monomers containing several olefinic bonds may be incorporated into the polypropylene so that the overall system is provided with a very much increased number of crosslink sites. In one study,[16] pentaerythritol triallyl ether was evaluated as a "coagent" in chemical crosslinking of isotactic polypropylene with the aid of organic peroxide. Looking at the structure of the peroxides used in this study, it seems that the most efficient crosslinking initiators formed benzoyloxy or phenyl radicals upon decomposition. On the other hand, initiators giving alkyl radicals were inefficient.

ROTATIONAL MOLDING

Powder process techniques for the making of hollow parts of polyethylene evolved into rotational molding.[5b] In this method, a split metal mold is used. A cold mold is filled with a powdered resin, the mold is placed in an oven and rotated simultaneously about two perpendicular axes. During this stage, a uniform layer of resin is deposited on the mold. After sufficient time has elapsed to properly fuse the resin, and while it is still rotating, the mold is cooled. After cooling, the part is removed from the mold and more resin is added to start the cycle again.

The absence of positive pressure in the powder molding process places certain limits on the type of resin that can be used. The higher molecular weight resins used for blow molding and sometimes injection molding cannot be rotationally molded because they will not flow out in the absence of pressure to form a homogeneous, void-free part. The polyethylene resins used in powder processes generally range from as low as 3 to 5 to as high as 70 melt index. (If molecular weight distribution is considered constant, the lower the molecular weight of the resin, the higher the melt index.) The resins with the 3 melt index have better stress cracking resistance, impact strength, and resistance to creep than the high melt index resins but are also more difficult to mold. Parts with hard-to-fill areas would require a slightly higher melt index for good moldability. Because of the somewhat limited stress cracking resistance and creep resistance of the powdered polyethylene resins, parts fabricated from these resins may be limited to the severity of service to which they can be subjected.

On the other hand, tanks made in the rotational molding process using crosslinkable high density resins have been used in a variety of applications including the handling of corrosive chemicals. Tanks varying in sizes from a few to thousands of gallons are being used in agriculture and industrial applications.[17] Crosslinking gives products that have excellent resistance to stress cracking and chemical attack, excellent impact strength, weathering characteristics and overall toughness. Although the crosslinked product goes through a crystalline melt point at a temperature similar to uncrosslinked resin, it has sufficient melt strength to support itself at temperatures up to about 210°C. Smaller molded products that are not subjected to a load will not deform even at temperatures at which uncrosslinked high density polyethylene will melt and flow.

For crosslinkable resins, Marlex® CL-50 and CL-100, generally require lower oven temperatures and slightly longer heating cycle times than normally used with conventional high density polyethylene resins.[5b] In most cases, an oven temperature between 550°F and 625°F should be used. Temperatures below 550°F can be used quite satisfactorily but longer cycle times are required. Temperatures above 625°F tend to generate too rapid a decomposition of the crosslinking agent. This can cause bubbles in the wall, blow holes through the wall, pock marks on the surface, rough inside surface, or overpressuring of the mold if adequate venting is not provided.

Table 10-4 shows the nominal physical properties of two rotational molding, crosslinkable high density polyethylene resins. When tested in a bent strip test (ASTM D1693), both Marlex® CL-100 and CL-50 have outstanding environmental stress cracking resistance (ESCR). The bent strip test is used for ESC characterization because it is presumed to be representative of the stresses and strains encountered in use. In this test, ten polymer samples (1½ in. x ½ in. x ⅛ in.) are bent 180° and immersed in a stress cracking agent. Each bar contains a longitudinal slit ¾ in. long x ^{2}⁄$_{100}$ in. deep down the center of the upper face. Resistance is defined as the length of time needed for five of the ten bars to show visible signs of cracking perpendicular to the slit. On the ASTM D1693, CL-100 has an F50 value greater than 1,000 hours. With a properly crosslinked CL-100 sample, not a single specimen has failed on this test. Marlex® CL-50 has a nominal ESCR F50 value of 200 hours. When properly molded, most CL-50 parts will have ESCR in excess of this. A more severe test under ASTM D1693 uses 10% solution of the stress cracking agent; the crosslinkable resins have similar values with this test as they do with less severe 100% solution.

Table 10-4: Nominal Physical Properties of Marlex® CL-100 and CL-50 High Density Polyethylene Resins

MARLEX® RESIN NUMBERS	**CL-50 Natural	*CL-100 Natural, white and standard colors
ASTM D1248 Classification	Type IV (1) Class A Category 1	Type IV (1) Class A Category 1
Meets Food and Drug Administration Regulation 177.1520 for Food Packaging	No	No
NOMINAL PHYSICAL PROPERTIES²		
Density, ASTM D1505, g/cm³	0.939-0.942³	0.937-0.940³
Melt Index, ASTM D1238, psi (MPa)	–	–
Flow Index, CIL, 375°F (190°C) 1500 psi (10.4 MPa), g/10 min.⁴	–	–
ROTATIONAL MOLDED PROPERTIES		
Environmental Stress Cracking Resistance, ASTM D1693, Condition A, F_{50}, h 100% Igepal / 10% Igepal	>200 / >175	>1000 / >1000
Tensile Strength @ Yield, ASTM D638 Type IV Specimen, 2 in. (50 mm) per min.	2600 (18)	2600 (18)
Elongation, ASTM D638 Type IV Specimen, 2 in. (50 mm) per min., %	400	450
Flexural Modulus, ASTM D790, psi (MPa)	110M (758)	100M (689)
Vicat Softening Temperature, ASTM D1525, °F (°C)	~255 (~124)	~260 (~127)
Mesh Size	35	35

* Available in either 35 mesh powder or pellets.
** Available in 35 mesh powder only.
(1) ASTM Classification for "Type" on base resin.
(2) Specimens molded in accordance with Procedure C of ASTM D1928.
(3) Density of natural crosslinked product.
(4) Data obtained using a gas operated extrusion plastometer based on a design by Canadian Industries, Ltd. with a die having an orifice diameter of 0.01925" (0.049 mm) and a land length of 0.176" (4.48 mm).

Data taken from Reference 17, reprinted with permission from the Phillips Chemical Co.

Long term hoop stress testing of the crosslinkable resins indicates that they are superior to other rotational molding resins and equal to high density polyethylene extrusion grade pipe resins. Both CL-100 and CL-50 rotational molded samples of 2 in. diameter pipe, 0.150 in. walls were used for long term hoop stress testing at both 80°F and 140°F. One set of CL-100 test samples at 1,750 psi hoop stress and lower has gone more than 50,000 hours without failure. Before testing, it was anticipated that the 1,700 psi sample would fail at approximately 100 hours. Because failure did not occur, design hoop stress could not be determined; however, it does indicate that a well crosslinked sample will have excellent long term stress at both 80°F and 140°F. The CL-50 data on 0.150 in. wall part indicated its hoop stress would be superior to conventional HDPE pipe resins.[18]

When properly cured, the crosslinked part has exceptional impact and overall toughness at both room and low temperatures. Good impact can be developed even at -20°F and -40°F. The impact trait has been demonstrated both by dart impact test and part drop tests. Tanks filled with liquid have been dropped 30 feet without failure at room and low temperatures (-20°F). A few noncrosslinked rotational molding resins might have similar dart impact but will not give the part drop performance of the crosslinkable resins. Another demonstration of the toughness of the crosslinked part is its ability to withstand repeated drop impacts of 30 feet without failure. Even parts which have been creased on previous drops can withstand repeated 30 foot drops.[18]

With plastics in general, it is difficult to correlate nominal physical properties with part performances. This is even more difficult when comparing nominal physical properties and part performance of crosslinkable to noncrosslinkable polyethylene. With the exception of environmental stress cracking resistance (ESCR) the nominal properties of the crosslinkable resins give little indication of the performance which crosslinked parts exhibit. In the early development stage of crosslinkable HDPE, special tests[18] were developed to illustrate and give a better understanding of what could be expected of crosslinked parts. These tests illustrated such properties as ESCR, long term hoop stress, gasoline resistance, impact resistance, and overall toughness. For one test, a 2 gallon jerry can (portable gasoline container) was 75% filled with gasoline, sealed and placed in a 130°F hot room. At 130°F the fuel has a vapor pressure of 7.25 psi. The crosslinked jerry can had a nominal wall thickness of $1/8$ in. and weighed 800 grams. This container was on test for three years and did not fail. Similar tests run on the same container molded from noncrosslinked resins available at that time had failures which varied from less than one hour to a maximum of four days. Some currently available resins would be expected to give better performance but still would not equal the crosslinkable resins.

Drop impact tests of molded containers were used to evaluate part impact strength and overall toughness. Containers varying in size from 2 gallons to 3,000 gallons have been drop tested. Containers, up to 80 gallons and filled with water or antifreeze solutions, have been dropped from heights of 30 feet at temperatures of 80°F. Larger containers containing only 50 gallons of water have been dropped 30 feet without failure. The capacity of the lift used for drop tests limited the volume of water used in the larger containers. A ½ in. wall, 3,000 gallon tank filled with water was dropped 10 feet without failure. This same tank filled with sand passed a similar drop impact test. For another drop, a 50 gallon tractor fuel tank was filled with water and dropped 30 feet. The part had a nominal wall thickness of 0.200 in. and weighed 30 pounds. The part and water had a total weight of 450 pounds. After the drop, no evidence of the deformation which occurred on impact could be detected.[18]

Low temperature impact testing is one of three quality control procedures for indicating the level of crosslinking in production parts. The others are percent gel and the bent strip test. Dart impact at -20°F is considered to be the most critical and comprehensive quality control test for parts molded from CL-100 and CL-50.[5c] As the compounds reach increasing degrees of crosslinking, the last property to reach maximum values is low temperature impact strength. Room temperature impact strength for crosslinkable resins can be misleading because values are high even for poorly crosslinked parts. Critical physical properties such as stress crack resistance, percent elongation, etc., develop their maximum values when low temperature impact strength of the molded part is equivalent to its room temperature impact strength. According to the manufacturer, particular attention should be given to failure patterns of impacted samples. Improperly crosslinked parts will crack or shatter when impacted. Such brittle type failures at -20°F have consistently been an indication of improper curing. When properly crosslinked specimens do fail, they exhibit ductile failures at temperatures of -20°F or higher. Ductile failures appear as a puncture through the specimen and show that the tensile strength of the material was exceeded. Should impact properties be poor a progressive increase in heating time and/or oven temperature is needed until impact properties improve.

The percent gel test is another method to indicate crosslinking levels in molded CL-100 and CL-50 parts.[5c] Refluxing ethylbenzene extracts the noncrosslinked portion of a specimen. The remaining insolubles are largely crosslinked polyethylene yielding a quantitative measure for degree of cure. Normally a high degree of crosslinking is indicated by gel levels of 85-90% but this can vary ±10% depending on wall thickness and molding conditions. The effect of wall thickness on percent gel is a phenomenon contrary to expected values. Wall thicknesses of less than ⅛ in. can produce 80-85% gel and be well crosslinked while a part in a ¼ in. wall may produce percent gels consistently above 90%. When wall thickness exceeds ¼ in., percent gel should be tested in the inner surface of the wall to assure good crosslinking throughout the wall. Other variations observed in this test are caused by the amount of surface area of the gel sample exposed to the refluxing ethylbenzene. Another consideration in interpreting percent gel data is its relationship to impact properties. A high percent gel is usually achieved before full development of maximum low temperature dart impact. Since low temperature dart impact is the last property to be improved in CL-100 and CL-50, it should be included with a gel test for the best quality control of production.

A bend test is a quick test for a rough estimate of the degree of cure.[5c] This method of determining whether the interior surface of rotational molded CL-100 parts is properly cured provides a means of checking for degree of crosslinking shortly after the part is molded. Due to the heat differential between the interior and exterior surfaces of a rotational molded part, the interior surface cures later than the exterior surface. The difference in cure means the interior portion of the part crosslinks last, so that stressing the interior portion will provide some indication of crosslinking. Conclusions from this test should be verified periodically by low temperature impact tests.

The producer[5c] concludes that the best production quality control test for crosslinking in Marlex® CL-100 and CL-50 is low temperature dart impact at -20°F or lower. Percent gel is also a good method but requires more time and expense than low temperature impact testing. The bend test may also be used but should be related to percent gel or dart impact testing. During initial molding trials for a new production part, low temperature impact levels and/or percent gel of proposed test areas should be correlated with acceptable performance of the overall part.

POST-IRRADIATION EFFECTS

One of the most interesting and novel features of crystalline olefin polymers is the irradiation-induced elastic memory phenomenon.[2b] When these polymers are crosslinked they behave as typical thermoplastics below their crystalline melting range and as elastomers above their melting range due to the crosslinked structure. Thus it is possible to deform these materials in their amorphous state and freeze the polymer in a deformed state. It will then remain in the deformed state until the material is heated above the crystalline melting range, whereupon it will return precisely to its crosslinked geometry.

When crosslinked polyethylene is heated above its crystalline melting point, the crystalline structure is destroyed and a rubbery, amorphous material results.[2b] In this state, its structure is quite similar to typical elastomers. Thus the material can be deformed by a force and will return to its original dimensions upon the removal of the force. If now the material is deformed while hot, the molecules will distort elastically. If the force is removed, the molecules will return to their original lower free energy state. If, however, the polymer is cooled in the elastically distorted state, the material will crystallize and remain in this deformed state. The molecules remain in this distorted state because the total crystalline forces have greater strength than the forces due to the relatively few crosslinks, and the molecules cannot relax to any extent until the crystals are remelted. The deformed condition is the form in which heat-shrinkable tubing is supplied to users.[19]

Although the initial work in the field of elastic memory was carried out with polyethylene, other polymers exhibit this phenomenom after irradiation. Table 10-5 provides a listing of various resins used for making shrinkable tubing and typical properties of the tubing after shrinking.[19] Specific applications for the radiation crosslinked thermoplastics include the insulation of a variety of electrical and electronic components including wires, lugs, terminals and connectors. In practice, the tubing is supplied in the form of an expanded diameter ranging from $3/64$ in. to 5 in. The user heats the tubing melting

Table 10-5: Heat-shrinkable Insulation and Encapsulation Tubings

	Product	Description
Flexible Polyolefins	RNF-100 Type 1	General purpose, flame retarded, flexible polyolefin
	RNF-100 Type 2	General purpose, flexible, transparent polyolefin
	RT-876	Highly flame retarded, very flexible polyolefin with low shrink temperature
	RT-102	Highly flexible, flame retarded, polyolefin with a very low shrink temperature
	RVW-1	Highly flame retarded, flexible polyolefin
Semirigid Polyolefins	CRN Type 1	General purpose, flame retarded, semirigid polyolefin
	RT-3	Semirigid flame retarded opaque polyolefin
Dual Wall Polyolefins	SCL	Meltable inner wall, selectively crosslinked, semirigid polyolefin
	TAT	Flexible dual wall adhesive tubing
	ATUM	Semi-flexible, high expansion, heavy dual wall, adhesive tubing
Fluoroplastics	Kynar	High temperature, flame resistant, clear, semirigid fluoroplastic
	Convolex	Convoluted, flexible irradiated polyvinylidene fluoride
	RT-218	Semirigid, white, high temperature, low-outgassing fluoroplastic tubing
Vinyl	PVC	Flexible, flame resistant, polyvinylchloride
Elastomers	NT (Neoprene)	Heavy duty, flexible, abrasion resistant, flame retarded elastomer
	SFR (Silicone)	Highly flexible, flame retarded, heat or cold shock resistant
	Viton (Fluoroelastomer)	Flexible, flame retarded, heat and chemical resistant fluoroelastomer
Caps	PD	Semirigid polyolefin, meltable inner wall

(continued)

Table 10-5: (continued)

	Product	Typical Applications
Flexible Polyolefins	RNF-100 Type 1	Insulation of wire bundles; cable and wire identification; terminal and component insulation, protection and identification.
	RNF-100 Type 2	Transparent coverings for components such as resistors, capacitors and cables where markings must be protected and remain legible.
	RT-876	Coverings for cables and components where excellent flexibility and outstanding flame retardance are needed.
	RT-102	Flexible material for general purpose protection and insulation. Especially effective for low temperature use.
	RVW-1	Lightweight harness insulation, terminal insulation, wire strain relief and general purpose component packaging and insulation where a UL recognized product with a VW-1 (FR-1) rating is needed.
Semirigid Polyolefins	CRN Type 1	Insulation and strain relief of soldered or crimped terminations; protection of delicate components; cable and component identification.
	RT-3	Particularly suited for automated application systems to insulate and strain relieve crimped or soldered terminals. Furnished in cut pieces.
Dual Wall Polyolefins	SCL	Encapsulation of components, splices, terminations, requiring moisture resistance, mechanical protection and shrink ratios as high as 6 to 1.
	TAT	Insulates and seals electrical splices, bi-metallic joints and components from moisture and corrosion.
	ATUM	Environmental protection for a wide variety of electrical components, including wire splices and harness breakouts.
Fluoro-plastics	Kynar	Transparent insulation, mechanical protection of wires, solder joints, terminals, connections and component covering.
	Convolex	Mechanical protection of cable harnesses. Excellent flexibility and chemical resistance. Good high temperature performance.
	RT-218	Insulation of splices and terminations in aircraft and mass transit markets; cable and wire identification.
Vinyl	PVC	Insulation and covering of cables, components, terminals, handles.
Elastomers	NT (Neoprene)	Insulation and abrasion protection of wire bundles and cable harnesses.
	SFR (Silicone)	Cable and harness protection requiring maximum flexibility and resistance to extreme temperatures; ablative protection for cables in rocket blast.
	Viton (Fluoro-elastomer)	Insulation and protection of cables exposed to high temperature and/or solvents such as jet fuel.
Caps	PD	Encapsulation of stub splices, especially fractional horsepower motor windings.

(continued)

Table 10-5: (continued)

	Product	Typical Values‡ Operating Temp. Range Continuous C (F)	Size Range, Expanded Diameter inches (mm)	Shrink Ratio (approx.)	Standard Colors
Flexible Polyolefins	RNF-100 Type 1	−55 to 135 (−67 to 275)	3/64 to 5 (1.2 to 127.0)	2:1	Black, white, red, yellow, blue
	RNF-100 Type 2	−55 to 135 (−67 to 275)	3/64 to 5 (1.2 to 127.0)	2:1	Clear
	RT-876	−55 to 135 (−67 to 275)	3/64 to 4 (1.2 to 101.6)	2:1	Black, white, red, yellow, blue
	RT-102	−75 to 125 (−103 to 257)	3/64 to 4 (1.2 to 101.6)	2:1	Black
	RVW-1	−30 to 135 (−22 to 275)	3/64 to 5 (1.2 to 127.0)	2:1	Black, white red, yellow, blue
Semirigid Polyolefins	CRN Type 1	−55 to 135 (−67 to 275)	3/64 to 1 (1.2 to 25.4)	2:1	Black (Available in clear as type 2)
	RT-3	−55 to 135 (−67 to 275)	.220 to .460 (0.6 to 1.2)	2.3:1	Black
Dual Wall Polyolefins	SCL	−55 to 110 (−67 to 230)	1/8 to 1 (3.2 to 25.4)	3:1	Black
	TAT	−55 to 110 (−67 to 230)	1/8 to 4 (4.7 to 101.6)	2:1	Black
	ATUM	−55 to 110 (−67 to 230)	1/8 to 1½ (3.2 to 38.1)	3:1	Black
Fluoro-plastics	Kynar	−55 to 175 (−67 to 347)	3/64 to 1 (1.2 to 25.4)	2:1	Clear
	Convolex	−55 to 150 (−67 to 302)	9/32 to 15/8 (7.1 to 41.3)	n/a	Black
	RT-218	−55 to 200 (−67 to 392)	3/64 to 4 (1.2 to 101.6)	2:1	White
Vinyl	PVC	−35 to 105 (−31 to 221)	3/64 to 2 (1.2 to 50.8)	2:1	Black
Elastomers	NT (Neoprene)	−70 to 121 (−94 to 250)	1/4 to 4 (6.4 to 101.6)	1.75:1	Black
	SFR (Silicone)	−75 to 180 (−103 to 356)	1/4 to 2 (6.4 to 50.8)	1.75:1	Black
	Viton (Fluoro-elastomer)	−40 to 200 (−40 to 392)	3/16 to 2 (4.7 to 50.8)	2:1	Black
Caps	PD	−55 to 110 (−67 to 230)	1/8 to ½ (3.2 to 12.7)	3:1	Black

‡ Specific minimum requirements and test methods are given in applicable Thermofit specifications.

(continued)

Crosslinked Thermoplastics

Table 10-5: (continued)

	Product	Minimum Shrink Temperature C (F)	Tensile Strength psi (kPa)	Elongation %	Dielectric Strength Volts/mil (Volts/mm)	Flame Resistance
Flexible Polyolefins	RNF-100 Type 1	121 (250)	1800 (12,410)	400	800 (32,000)	Flame-retarded
	RNF-100 Type 2	121 (250)	1800 (12,410)	400	1000 (40,000)	Not flame-retarded
	RT-876	100 (212)	2000 (13,790)	450	800 (32,000)	Highly flame-retarded
	RT-102	90 (194)	2500 (17,240)	400	800 (32,000)	Flame-retarded
	RVW-1	121 (250)	1500 (10,340)	300	700 (28,000)	Highly flame-retarded
Semirigid Polyolefins	CRN Type 1	135 (275)	2500 (17,240)	300	800 (32,000)	Flame-retarded
	RT-3	135 (275)	2500 (17,240)	300	1000 (40,000)	Flame-retarded
Dual Wall Polyolefins	SCL	135 (275)	2500 (17,240)	400	1000 (40,000)	Not flame-retarded
	TAT	121 (250)	1800 (12,410)	350	800 (32,000)	Flame-retarded jacket
	ATUM	125 (257)	2500 (17,240)	400	350 (14,000)	Flame-retarded jacket
Fluoro-plastics	Kynar	175 (347)	6000 (41,370)	300	900 (36,000)	Highly flame-retarded
	Convolex	n/a	*	*	*	Highly flame-retarded
	RT-218	175 (347)	6000 (41,370)	300	900 (36,000)	Highly flame-retarded
Vinyl	PVC	175 (347)	2500 (17,240)	300	700 (28,000)	Highly flame-retarded
Elastomers	NT (Neoprene)	175 (347)	1800 (12,410)	350	350 (14,000)	Flame-retarded
	SFR (Silicone)	175 (347)	600 (4,140)	300	500 (20,000)	Flame-retarded
	Viton (Fluoro-elastomer)	175 (347)	1500 (10,340)	300	400 (16,000)	Flame-retarded
Caps	PD	121 (250)	2500 (17,240)	350	900 (36,000)	Not flame-retarded

*Geometry of product precludes these test values.

(continued)

Table 10-5: (continued)

	Product	Fluid Resistance† (room temp.) Hydrocarbon Solvents	Aqueous Solutions Acids, Bases
Flexible Polyolefins	RNF-100 Type 1	Good	Excellent
	RNF-100 Type 2	Good	Excellent
	RT-876	Good	Excellent
	RT-102	Good	Excellent
	RVW-1	Good	Excellent
Semirigid Polyolefins	CRN Type 1	Good	Excellent
	RT-3	Good	Excellent
Dual Wall Polyolefins	SCL	Good	Excellent
	TAT	Good	Excellent
	ATUM	Good	Excellent
Fluoroplastics	Kynar	Excellent	Excellent
	Convolex	Excellent	Excellent
	RT-218	Excellent	Excellent
Vinyl	PVC	Excellent	Excellent
Elastomers	NT (Neoprene)	Good	Good
	SFR (Silicone)	Good	Good
	Viton (Fluoroelastomer)	Excellent	Excellent
Caps	PD	Good	Excellent

†Fluid resistance is presented only in general terms and is based on room temperature immersion of commonly encountered solvents.

Data taken from Ref. 19, permission to reprint from Raychem Corp.

the crystals. This allows the crosslinked material to return to its original shape and exhibit its elastic memory. Table 10-5 shows that the shrink ratio varies approximately from 1.75/1 to 3/1. After cooling, the crystals reform and the tubing retains its recovered form. Upon subsequent reheating, no further change in shape will take place unless a mechanical force is applied.

The heat needed to melt the crystals of a deformable crosslinked thermoplastic is, in certain instances, also sufficient to melt solder. This has led to the development of solder-sleeve devices consisting of fluxed solder and thermoplastic inserts at each end.[20] When placed over a cable shield and heated, the solder melts and flows, connecting the ground lead and shield (Figure 10-4). The outer sleeve shrinks and the thermoplastic inserts melt, encapsulating the termination to provide a sealed termination. Other devices are available for a connector that takes removable contacts. The contacts may terminate coaxial cables, shielded wires and twisted wire pairs. In addition, a rectangular, multicavity, heat-shrinkable device is available for permanent splicing of flat conductor cable to flat conductor cable or round wires.

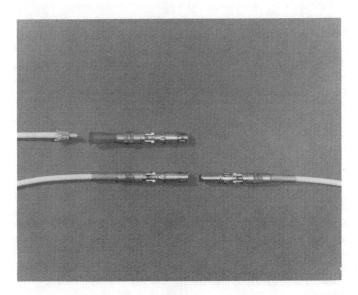

Figure 10-4: Soldertact® contact prior to and after heating. (Data taken from Reference 20, reprinted with permission from Raychem Corp.)

ACRYLATES

Crosslinking of thermoplastics by irradiation or with the aid of free radicals extends to acrylates also. Crosslinked acrylics are used as coatings or, in the form of sheets, in the instrument, aircraft and optical industries. Cell cast acrylic sheet is an optically clear, transparent material which is available in a variety of sheet sizes and thicknesses starting at about 0.06 inch; final thickness may depend on the supplier and the sheet size. The material is used in both outdoor and indoor applications and can be obtained in clear and transparent colors; general purpose sheets are available too in selected opaque colors. Table 10-6 shows some of the property differences between a thermoplastic acrylic and its crosslinked counterpart; the crosslinked acrylate exhibits superior solvent and craze resistance, higher impact strength as well as resistance to deformation at 122°F. While solvent cementing can be used to join a thermoplastic sheet to itself, special purpose cements are needed for the crosslinked material.

Coating systems are supplied usually as a combination of resins, crosslinking agents (oligomers), catalysts and various types of additives dissolved or dispersed in organic or aqueous solvents. Commercial processing of these formulations includes provisions for the safe and complete removal of the volatile solvents, usually with the application of heat. Spurred by government regulations to reduce substantially the amount of pollutants in the air, coating formulations containing no solvent are becoming available. In lieu of the solvent, the formulation consists of an oligomer containing unsaturation. The oligomer is used either by itself, or, more often, in combination with a monomer that contains unsaturation either along the side chain or in the backbone. The system is cured by the application of high-energy radiation with the monomer and all other organic components copolymerizing with the oligomer through a free radical mechanism. The so-called high solids systems make it easier to meet the non-polluting requirements since, now, there are no volatile, small organic molecule constituents.

An oligomer that contains groups which will undergo a free radical polymerization reaction on exposure to electron beam or ultraviolet radiation is an important component of the formulation. One system involves the use of oligomers capped at both ends with highly active acrylate groups, usually in combination with acrylate monomers. Table 10-7 shows examples of two formulations using, as oligomers, high molecular polystyryl polymers which are capped with methacrylate groups, Chemlink® 4545B and Chemlink® 4500B, with molecular weights of 4,500 and 13,000, respectively.[21] 1,6-Hexanediol diacrylate reportedly yields a clear, compatible solution with Chemlink® 4545B and with Chemlink® 4500B, up to concentrations of 60% and 50%, respectively; it is classified as a high-solvency monomer. Trimethylolpropane triacrylate dissolves Chemlink® 4500B providing a clear compatible solution up to a concentration level of 40%; it is categorized as a medium-solvency monomer. The radiation cured systems offer clear, hard films.

Table 10-6. Nominal Properties of Cell Cast and Crosslinked Acrylic Materials

Property	ASTM Method	Acrylic Sheet	Crosslinked Acrylic
Specific Gravity	D-792	1.19	1.19
Refraction Index	D-542	1.49	1.49
Transmission & Haze Measurements Parallel (%) Total (%) Haze (%)	D-1003	91 92 1	92 93 1
Tensile Strength—Rupture—p.s.i.	D-638	10,500	8,190
Flexural Strength—Rupture—p.s.i.	D-790	16,000	13,900
Compressive Strength—Yield-p.s.i.	D-695	18,000	18,200
Impact Strength-Izod Notched Ft. Lbs/in. of notch	D-256	3.5	4.2
Abrasion Resistance Falling Emery Test (x-methacrylate)	D-673	3	3
Deformation Under Load 2000 p.s.i. 122°F 24 Hrs. 4000 p.s.i. 122°F 24 Hrs.	D-621-485	0.2 0.5	0.4 2.2
Rockwell Hardness	D-785	M-104	M-94
Resistance to Stress Critical Crazing Stress Isopropyl Alcohol p.s.i. Toluene p.s.i.	Mil-P-6997 ARTC MOD.	1700 1300	4000 4000
Solvent Resistance—7 Days Immersion at 77°F Acetone Ethyl Acetate Ethylene Dichloride	D-543	Dissolved Dissolved Dissolved	Soft-Swollen Soft-Swollen Soft-Swollen
Coefficient of Linear Thermal Expansion x 10^{-5} per °F −40°F to −20°F −20°F to 0°F 0°F to 20°F 20°F to 40°F 40°F to 60°F 60°F to 80°F 80°F to 100°F Average −40°F to 100°F	D-696-44	2.9 3.1 3.3 3.6 3.9 4.2 4.6 3.6	2.8 3.0 3.3 3.8 4.2 5.3 6.0 4.0
Coefficient of Thermal Conductivity $\frac{BTU}{(Hr)(Sq. Ft.)(°F/In)}$ in	(Cenco-Fitch)	1.3	1.8
Dielectric Constant 60 Cycles 1 Kilocycle 1 Megacycle	D-150	3.7 3.3 2.5	3.5 3.2 2.7
Power Factor 60 Cycles 1 Kilocycle 1 Megacycle	D-150	0.050 0.040 0.030	0.002 0.001 0.004
Loss Factor 60 Cycles 1 Kilocycle 1 Megacycle	D-150	0.19 0.13 0.08	0.006 0.004 0.001
Arc Resistance	D-495	No Tracking	No Tracking
Surface Resistivity (ohm/square inch)	D-257	2×10^{16}	1.02×10^{16}
Flammability (Burning Rate) Inches per minute	D-635	Range .60 to 1.1	0.89
Ignition Temperature Flash Temperature Self Ignition Temperature	D-1929-62T	620°F 878°F	620°F 878°F
Water Absorption 24 hrs—25°C Water Absorbed—% Soluble Matter Lost—%	D-570	0.20 0.00	0.35 0.00
Odor	—	None	None
Taste	—	None	None
Ultra-Violet Transmittance 320 Nanometers	—	None	None

Table 10-7: Typical Formulations Containing Macromer™ Monomers as Oligomers

Component	Weight Percent
Chemlink® 4545B	50
Tetrahydrofurfuryl acrylate	37
Trimethylolpropane triacrylate	5
Ganex* V-516 Wetting agent	1
Irgacure** 184 Photoinitiator	2
Chemlink® 4500B	40
1,6-Hexanediol diacrylate	47
Tetrahydrofurfuryl acrylate	5
Trimethylolpropane triacrylate	5
Ganex* V-516 Wetting agent	1
Irgacure** 184 Photoinitiator	2

*GAF Corp.
**Ciba-Geigy Corp.

Data taken from Reference 21, reprinted with permission from Sartomer Company.

Acrylamide-functional oligomers are under evaluation.[22] They are derived from alkenyl azlactones and amine-terminated oligomers in what is described as a one-step, facile reaction at room temperature.

$$\underset{\text{Azlactones}}{\left[\begin{array}{c}CH_2=\overset{R}{\underset{}{\diagup}}\\ \diagdown N\diagdown Me\\ O\diagdown Me\\ \parallel\\ O\end{array}\right]_n} + (H_2N)_n \text{ oligomer} \longrightarrow \underset{\text{Acrylamide}}{(CH_2=\overset{}{\underset{\parallel}{C}}\overset{}{\underset{O}{C}}NH-\overset{MeO}{\underset{Me}{\overset{\mid}{C}}}-CNH)_n \text{ oligomer}}$$

In comparison with the isocyanate and epoxide groups, the polymeric amines react with azlactones at controlled and predictable rates that are intermediate between the very reactive isocyanate and the slow reacting epoxide moieties.

Electron beam-cured coatings based on oligomers capped with acrylate groups have been used for coating metal, wood and paper where film extensibility is not required generally. In textile applications, however, elongation at break of greater than 200% is needed for practically all end uses.[23] Elastomeric EB-cured coatings were obtained in films cast from polyester-urethane oligomers based on toluene diisocyanate that were capped by acrylate groups at both ends of the chain and monomer diluent. Increasing the chain length between the two acrylate groups resulted in an increase in ultimate elongation from 20% to 210% and a decrease in glass transition temperature from 50° to -25°C; breaking strength and Young's modulus decreased. The polyacrylo-urethane films have a one-phase morphology in which the hard urethane segments and the soft polyester segments are homogeneously mixed.[24] The films obtained from 1,000-molecular weight oligomer are hard and somewhat brittle due to their one-phase morphology in which hard glassy segments play a dominant role. The films obtained from 4,600- and 6,000-molecular weight oligomers are soft and tough, once again due to their one-phase morphology in which soft rubbery segments are more effective. The original crystalline structure of pure 6,000-molecular weight oligomer is retained in the precrystallized gamma-irradiated film, but the electron beam irradiated films show partial melting of the crystallites that was attributed to the heat of polymerization. The solid-state polymerized, gamma-irradiated films have a lower elongation, a higher modulus, and a higher breaking strength due to their much higher crystallinity.

TRADE NAMES

Trade Name	Product	Company
Chemlink® Macromer™ Series	High molecular weight monomers	Sartomer Corp.
Electroglas®	Acrylic sheet	Glasflex Corp.
Marlex®	Crosslinkable high density PE	Philips Chemical Corp.
Minicel®	Crosslinked polyolefin foam	Voltek Corp.
Photoglaze®	UV and EB curable coatings	Lord Corp.
Thermofit®	Crosslinked, heat shrinkable tubing	Raychem Corp.
Solder Tacts®	Solder preform encased in a crosslinked, heat shrinkable insulation sleeve	Raychem Corp.
Volara®	Closed-cell, radiation crosslinked polyethylene foam	Voltek Corp.

REFERENCES

1. *Polyethylene—The Technology and Uses of Ethylene Polymers*, Renfrew, A. and Morgan, P., Editors, Iliffe & Sons Ltd., London, 2nd edition (1960).

2. *Crystalline Olefin Polymers*, Raff, R.A.V. and Doak, K.W., Editors, Interscience Publishers, John Wiley & Sons, New York (1965). (a) Dole, M., Mechanism and chemical effects in irradiated polymers. (b) Lanza, V.L., Irradiation—properties changes.
3. Dole, M., Effect of radiation environments on plastics, in *The Effects of Hostile Environment on Coatings and Plastics*, Garner, D.P. and Stahl, G.A., Editors, American Chemical Society, Washington, D.C. (1983).
4. *Radiation Chemistry of Organic Compounds*, Vol. 2, Swallow, A.J., Pergamon Press, New York (1960).
5. Phillips Chemical Co., Plastics Technical Center (a) Technical Service Memorandum #243, *Engineering Properties of Marlex® Polyolefins*. (b) Technical Information Bulletin #17, *Rotational Molding*. (c) Technical Service Memorandum #291, *Quality Control Tests for Crosslinking Marlex® CL-100 and CL-50*.
6. Voltek Inc., (a) Technical Bulletin, *Closed Cell Foam Properties*; (b) Bulletin TB2 8/83, *Laminating Guide*; (c) Bulletin TB3 1/84, *Compression Molding Guide*.
7. Horng, P. and Klemchuk, P., *Plastics Eng.*, 40(4), 35 (1984).
8. Carlsson, D.J., Torborg Jense, J.P. and Wiles, D.M., *Polymer Preprints*, 25(1), 85 (1984).
9. Williams, J.L., *Polymer Preprints*, 25(1), 87 (1984).
10. Bowmer, T.N., Vroom, W.I. and Hellman, M.Y., *J. Appl. Polym. Sci.*, 28, 2553 (1983). (See references for previous papers by these authors and their associates).
11. Narkis, M., *Modern Plastics*, 57, 68 (November 1980).
12. Kolczynski, J.R., in *Modern Plastics Encyclopedia*, 46(10A), 316 (1969).
13. Yang, Y.T. and Phillips, P.J., *J. Appl. Polym. Sci.*, 28, 1137 (1983).
14. Kunert, K.A., *J. Macromol. Sci. Chem.*, A17(9), 1469 (1982).
15. Lem, K.W. and Han, C.D., *J. Appl. Polym. Sci.*, 27, 1367 (1982).
16. Borsig, E., Fiedlerova, A. and Lazar, M., *J. Macromol. Sci. Chem.*, A16(2), 513 (1981).
17. Phillips Chemical Co., Bulletin 842-84 TF, *Best Resins for Rotational Molding*.
18. Carrow, G.E., Crosslinkable polyethylene, the proven plastic for handling corrosive chemicals. *Corrosion/82*.
19. Raychem Corp. Technical Bulletin, *Thermofit® Heat-shrinkable Tubing Selection Guide*.
20. Raychem Corp. Technical Bulletin, *Solder Tacts Contacts*.
21. Sartomer Company Technical Bulletin, *Chemlink® 4545B/4500B High Energy Curable Macromolecular Monomers*.
22. Heilmann, S.M., Rasmussen, J.K., Krepski, L.R. and Smith, H.K. II, *Polymer Preprints*, 2, 35 (1984).
23. Oraby, W. and Walsh, J.K., *J. Appl. Polym. Sci.*, 23, 3227 (1979).
24. Wadhwa, L.H. and Walsh, J.K., *J. Appl. Polym. Sci.*, 27, 591 (1982).

11
Research Polymers and Future Directions

Kreisler S.Y. Lau

Materials Science Department
Technology Support Division
Hughes Aircraft Company
El Segundo, California

Commercialization of polymer resins has its origin in development research which initially establishes the synthetic feasibility of the process. The research resin is then evaluated for its potential in engineering applications. Optimization of reaction parameters ensues so that an efficient method can be realized for large scale preparations. With more materials that become available, more testing and more extensive evaluation can be pursued. Commercialization of a product thus represents the culmination of many years of persistent effort in developmental and application research.

Thermosetting polymers have long been in demand in producing composites for the fabrication of high performance aerospace engineering structural components. The success of epoxy resins as the most developed and most widely used polymer matrix materials rests in their excellent mechanical properties and processing characteristics. In fact, all new classes of matrix resins are measured against the epoxies in evaluating their processibility. Epoxy resins are not usable at temperatures beyond 347°F (175°C). On the other hand, the aerospace industry continuously demands new classes of high performance polymers as matrix resins and adhesives that have to survive for longer and longer times at higher and higher temperatures. Furthermore, the microelectronic industry has a need for an adhesive which must withstand processing temperatures of over 752°F (400°C) for a few hours in an inert atmosphere. Although the quantity requirement in the microelectronic industry is, in general, not critical at present, more material demands are foreseeable in the application of dimensionally stable high temperature polymer materials as structural matrix resins.

The search for thermally stable polymers in the past two decades has led to the conclusion that thermal stability and processibility are antithetical in nature. Although many classes of high temperature resistant polymers are in theory capable of reaching end use temperatures of 472° to 752°F (300° to 400°C), they are too rigid in nature and therefore too intractable to be useful. The research in high temperature polymers has always aimed at an optimal balance between processibility and high temperature stability. There are several classes of addition polymers developed which undergo chain extension and cure reactions at the reactive termini of shorter and therefore processible oligomers to yield crosslinked thermoset matrix resins applicable to advanced fiber composites. These addition polymers also remove the major processing problem associated with the evolution of volatiles.

Several classes of research thermoset polymers, particularly polyimides, have attained the stage of commercialization and have been discussed in Chapter 10. Other classes of research thermoset polymers will be discussed in this chapter. It is also the purpose of this chapter to present some new possibilities as future directions for research in thermosetting polymers.

NADIMIDE-TERMINATED THERMOSETTING POLYMERS

Investigations at the NASA-Lewis and NASA-Langley Research Centers have been instrumental in the development of nadimide-terminated addition-type polyimides. The requirement for high temperature polymers with enhanced processibility has been met with novel resins such as PMR, LARC-160 and LARC-13.

The PMR (for *in situ* polymerization of monomer reactants) technology was the subject of a recent review by its originator.[1] The development of the first generation PMR matrix resins has culminated in various forms of commercial applications. Selected properties and applications have been presented in Chapter 10. The key features of success in the PMR composite technology are the availability of monomers, the good solubility of monomer reactants in low boiling alcohols, the processibility of the resins and ultimately the excellent retention of properties at elevated temperatures.

The versatility of the PMR approach can be demonstrated with the tailoring of processing characteristics and properties by simple variations on the chemical nature of either the diester acid or the aromatic diamine, or both, and the stoichiometric proportion of the monomer reactants (see Chapter 10). Substituting m-phenylenediamine (MDA) with an aromatic polyamine (Jeffamine AP22), for example, yielded the LARC-160 family of thermoset polyimides.[2]

Second generation PMR resins were produced by partial or complete replacement of benzophenonetetracarboxylic ester acid (BTDE) with the dimethyl ester of 2,2-bis(3,4-dicarboxyphenyl)hexafluoropropane (6FDE).[3,4] The thermooxidative stability of the original PMR resin was improved. Although progress in developing this 6FDE-based PMR was slow due to an earlier lack of a commercial source for the corresponding dianhydride, 6FDA. Future outlook for this development is brightened with the upcoming 6FDA commercialization by American Hoechst and Hitachi Chemicals (Japan).

6FDE

6FDA

As the requirement for high temperature adhesives coincides with the requirement for structural matrix resins, the candidate high temperature polymeric materials are also screened for their suitability as high-temperature adhesives. Based on a structure-property relationship study of addition polyimides conducted at NASA-Langley,[5] LARC-13 was shown to exhibit the highest lap shear among two groups of nadimide-end-capped polyimide resins (Table 11-1) and was further developed as a structural adhesive for specialized bonding applications.[6]

Besides the demonstrated high lap shear strength, LARC-13 also has a high degree of flow during cure and is easily processible by conventional autoclave techniques. Its high crosslinking density provides dimensional stability during use at elevated temperatures in excess of its glass transition temperature of 518°F (270°C). Figure 11-1 shows the chemistry of LARC-13 adhesive.

To further enhance the potential application of LARC-13, researchers at Langley began work on elastomer-toughening of the inherently brittle LARC-13.[6] Some salient features are noteworthy from these studies: (a) a compromise exists between toughness improvement and elevated temperature adhesive strength; (b) novel high temperature resistant elastomers are needed; (c) a 50-50 bimodal distribution of long and short chained elastomers in LARC-13 can contribute to a significant improvement of adhesive properties.

Table 11-1: Adhesive Properties of Nadimide End-Capped Polyimide Resins

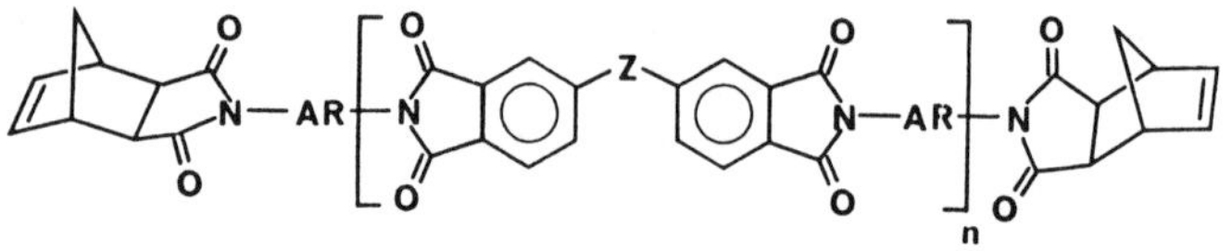

Amine Structure (AR)	Z	Amine Isomer	Lap Shear Str., psi (MPa)*
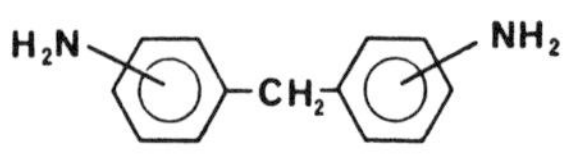 H_2N—⌬—CH_2—⌬—NH_2	—C(=O)—	3,3'**	2800 (19)
	—C(=O)—	4,4'	600 (4)
	—O—	3,3'	2500 (17)
	—O—	4,4'	1300 (9)
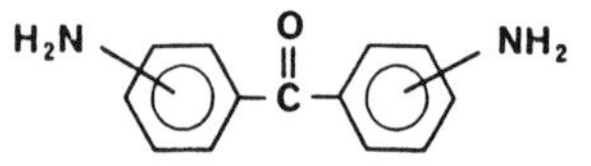 H_2N—⌬—C(=O)—⌬—NH_2	—C(=O)—	3,3'	2100 (14)
	—C(=O)—	4,4'	1300 (9)
	—O—	3,3'	3000 (21)
	—O—	4,4'	1300 (9)

*Titanium Adherends
**LARC-13

Reprinted by permission of Reference 6.

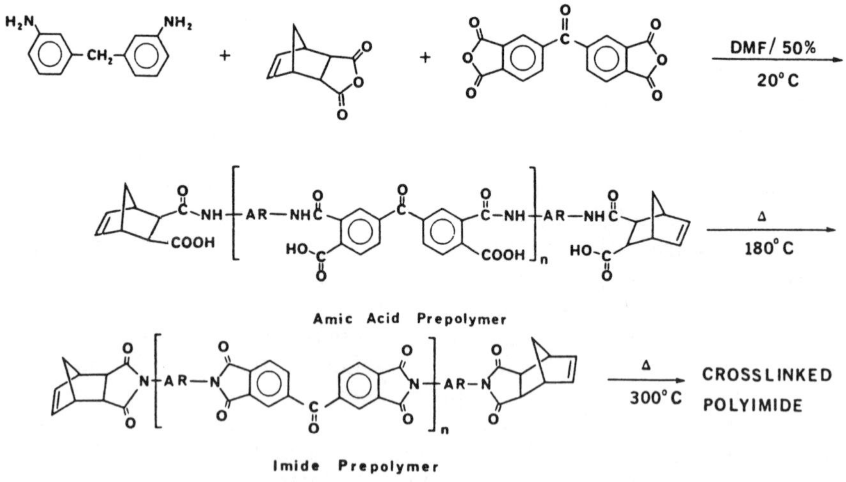

Figure 11-1: Chemistry of LARC-13 adhesive.

A further investigation was conducted to find a crosslinking end group which would improve the thermooxidative stability of LARC-13 adhesive. Table 11-2 shows that the acetylene-terminated LARC-13-based material has 40% higher adhesive lap shear strength than LARC-13 itself after aging 1,000 hours at elevated temperatures.

Table 11-2: Lap Shear Strengths of Titanium/Addition Polyimide Bonds

Oligomer End Group	LSS, psi (MPa) Unaged Sample		LSS, psi (MPa) after 1,000 Hours @ 450°F(232°C)	
	RT	450°F(232°C)	RT	450°F(232°C)
Nadic	3200 (22)	2600 (18)	2600 (18)	2000 (14)
Acetylene	2900 (20)	2500 (17)	2500 (17)	2800 (19)
N-Propargyl	3100 (21)	2800 (19)	800 (6)	1000 (7)

Reprinted by permission of Reference 6.

LARC-13 and its modified versions thus represent state-of-the-art polymeric materials suitable for applications as both structural matrix resins and high temperature adhesives. Owing to the high crosslink density, they are able to perform for short terms at temperatures up to 1112°F (600°C) where linear systems fail thermoplastically. On the other hand, further improvements are needed in terms of enhancing their thermooxidative stability and long term performance at elevated temperatures. As is well known in the field of high temperature resistant polymers, successful development of polymer systems possessing both ultrahigh thermal and thermooxidative stabilities while maintaining good processibility and a high level of toughness will represent a major technological breakthrough. A likely answer lies in extending the systematic resin toughening studies carried out at NASA-Langley to beyond the realm of organic polymers.

MALEIMIDE-TERMINATED THERMOSETTING POLYMERS

Maleimide-terminated polyimide oligomers have been commercially offered as the Kerimid® bismaleimide series.[7] Crosslinked poly(bismaleimides) were formed by free-radical catalyzed thermal polymerization of these bismaleimide oligomers. A melt processable bismaleimide copolymer (Kerimid 353) was developed using a ternary mixture of aliphatic and aromatic bismaleimides.[8-10] Further developments of bismaleimide technology using a combination of free radical cure and diamine addition had yielded new types of processible bismaleimide resins, the Kinels and Kerimid 601 series, which are suitable as molding and laminating resins, respectively.[11] Most of the bismaleimide resins do not show strong potential as high temperature structural matrix resins or high temperature adhesives. Some of the neat resin and composite properties of Kerimids® have been presented in Chapter 10.

Recent copolymerization studies have demonstrated that bismaleimide monomers or prepolymers can best be used as crosslinking agents to yield products with high glass transition temperatures.[12]

POLY(AMIDE-IMIDE) + [bismaleimide structure with CH$_2$ bridge]

3 parts   1 part

2-ethylimidazole (cat.)
⎯⎯⎯⎯⎯⎯⎯⎯⎯→   CROSSLINKED
2 hr, 190°C   RESIN
4 hr, 240°C   Tg >300°C

ACETYLENE-TERMINATED THERMOSETTING POLYMERS

As a result of the development of novel synthetic methods for attaching ethynyl and phenylethynyl groups on aromatic systems, more and more monomers, oligomers and prepolymers containing terminal ethynyl groups are available as basic units for a new class of addition curable resins. Similar to the development of nadimide-terminated polyimides and bismaleimides, the development of ethynyl-terminated, also called acetylene-terminated (AT) polyimides and other heteroaromatic polymers has the important feature of an addition cure mechanism which provides chain extension and crosslinking *without the evolution of volatile by-products*. Many of the cured resins have good solvent and moisture resistance, and exhibit outstanding physical and mechanical properties.

An upcoming comprehensive review will summarize the research activities that have been conducted on the development of acetylene-terminated monomers, oligomers and polymers.[13] The AT polyimides have been commercialized recently and have been discussed in Chapter 10. It is sufficient here to highlight some key features in the development of this novel class of addition thermoset resins.

The current view on the attachment of ethynyl groups to aromatic and heteroaromatic nuclei favors the application of organometallic chemistry and, in particular, organopalladium chemistry. The availability of a variety of organopalladium catalysts, the catalytic nature of the expensive palladium complexes for synthetic applications, the simple experimental procedure and the high yield of reaction product are the favorable factors.

An aromatic halide and a terminal acetylene compound can undergo coupling under the catalysis of palladium.[14] In principle, for the placement of an ethynyl group on an aromatic structure, a terminal acetylenic compound would be required having the protective R group easily removable after the palladium-catalyzed coupling reaction has been effected.

$$\text{Ar-X} + HC\equiv C-R \xrightarrow{Pd} \text{Ar-}C\equiv C-R$$

X = Br, I

In the synthesis of ethynylated aromatic compounds two key terminal acetylenes have been widely used. Both ethynyltrimethylsilane (variation A) and 2-methyl-3-butyn-2-ol (variation B) undergo facile palladium-catalyzed coupling with haloaromatics to give internal acetylene compounds. Removal of the trimethylsilyl group or the acetonyl group generates the corresponding terminal acetylenes.

$$HC\equiv C-Si(CH_3)_3 \qquad HC\equiv C-C(CH_3)_2-OH$$

A B

The highly basic reaction medium required in variation B for the removal of the acetonyl group precludes compounds containing base-sensitive functional groups. The removal of the trimethylsilyl group (variation A) requires only mild conditions.[15] Variation A, therefore, has specific application in ethynylation reactions of base-sensitive compounds.

As stated earlier, the most important characteristic of resins capable of addition cure is that no volatile by-products are released. Void-free components which can be produced with these thermoset resins are critical to large dimensional structural use. A variety of aromatic and heteroaromatic oligomers having terminal ethynyl groups have been prepared and evaluated. Some of them have the potential of meeting all the processing criteria of epoxies but are more thermally stable and moisture resistant.

Significant contributions to the synthesis and evaluation of acetylene-terminated reactive oligomers have been made by the Air Force Materials Laboratory. A recent review[16] illustrates that acetylene-terminated (AT) resins can be used in a wide range of processing temperatures and conditions. While resins with flexible backbones such as ATB[17,18] exhibit tack and drape characteristics, others with high molecular weight and rigid backbones possess only a short melt state, such as BATQ.[19]

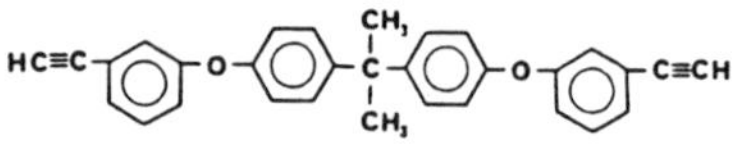

ATB

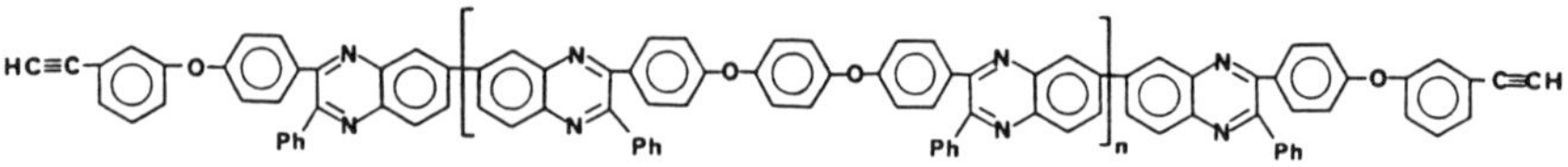

BATQ

Good mechanical properties are obtainable in AT cured systems. For example, the fully cured AT resin, ATQ, yielded a glass transition temperature of 610°F (321°C).[30] Table 11-3 summarizes tensile strength data obtained for this resin at both room temperature and 450°F (232°C). Mechanical properties at both temperatures are not affected by aging at 200°F (93.3°C), 95% humidity environment.

Table 11-3: Mechanical Properties of Net Resin ATQ

Temperature	Tensile Strength, MPa(KSI)	
	Dry	Wet
RT	98(14)	98(14)
450°F(232°C)	27(3.8)	32(4.5)

Reprinted by permission of References 17 and 18.

The ATQ resin has been evaluated as a matrix in graphite composites. The composite again showed good flexural strength and modulus.

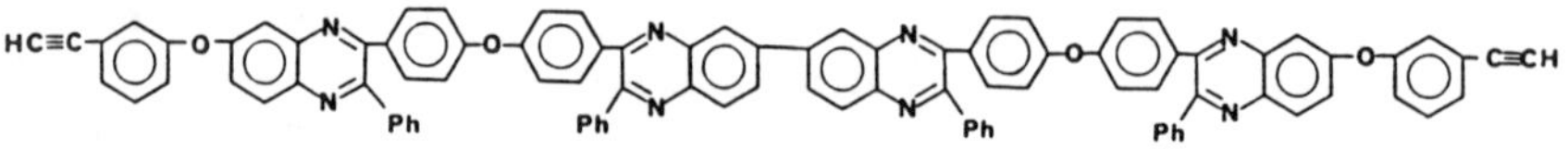

ATQ

It is important to point out that even AT systems with more flexible backbones, such as ATB, are capable of maintaining high temperature mechanical properties (Table 11-4).

Table 11-4: ATB Neat Resin Mechanical Properties

Property	RT dry	RT wet	212°F(100°C) dry	212°F(100°C) wet	350°F(177°C) dry	350°F(177°C) wet
Elongation at Break	3.4	3.1	4.4	3.4	5.9	6.9
Tensile Strength	67.9	58.4	57.2	49.4	39.7	37.9
Tensile Modulus, GPa	2.40	2.34	1.93	2.18	1.54	1.53
Shear Modulus, GPa	0.85	-	0.73	-	0.61	-

Reprinted by permission of References 17 and 18.

The acetylene-terminated sulfone (ATS) has been studied quite extensively. Depending on the method of synthesis, different forms of ATS were obtained. Some low molecular weight vinyl ether oligomers were present to the extent of 15-21%. The oligomers are responsible for the ATS' appearance as a tacky resin which can be melt impregnable. Neat resin mechanical properties have been determined for the ATS resins (Table 11-5).

$$HC{\equiv}C-\phi-O-\phi-SO_2-\phi-O-\phi-C{\equiv}CH$$

ATS

Table 11-5: ATS Neat Resin Mechanical Properties

Tensile Strength, MPa	48.9
Tensile Modulus, GPa	3.62
Tg	300°C
Elongation at Break, %	1.4

Reprinted by permission of Reference 13.

ATS has been evaluated as a matrix resin in graphite fiber reinforced composites. Although judged as brittle, the resin yielded composites that have good retention of mechanical properties at 350°F (177°C) before and after high humidity exposure.[21,22]

Successful experiments have been conducted with polymer blends. The objective is to achieve blends of resin materials that can retain all good properties originally ascribed to each neat material and reduce the undesirable properties at the same time. ATS has been evaluated as a reactive diluent with thermoplastic polysulfones. The high crosslinking density indigenous to ATS can help improve the T_g of the thermoplastic polysulfone while the brittleness associated with the high crosslinking density of ATS can be compensated by the presence of the thermoplastics. To this end, the polysulfone Radel® (Union Carbide), after being modified with various amounts of ATS, showed a significant improvement in its processibility. After cure, the resin blend showed a higher T_g (240°C versus 220°C for neat Radel® resin) and enhances solvent resistance.[23] Similar results were obtained for blends of ATS with UDEL® (Union Carbide).

Ethynyl-terminated sulfone oligomers other than ATS have also been prepared and used to modify properties of linear polysulfones such as UDEL® P-1700. While the toughness and thermoformability indigenous to the thermoplastic are preserved, the presence of crosslinked ethynyl-terminated sulfone oligomers improves the solvent resistance, expecially when under load.[13]

Acetylene-terminated diluents such as BADAB-BA, AA-BA and diethynyl-bisphenoxybenzene (ATPB) were developed for the improvement of processibility of BATQs so that they would become more amenable to melt processing. These diluents all plasticize the BATQs to give a significant lowering of the bulk viscosity. They then co-react with the BATQ oligomers to become part of the thermoset network.[24] The BATQ most extensively studied was the BATQ-H.

BATQ-H

BA-DAB-BA

AA-BA   **ATPB**

Acetylene-terminated imide oligomers were developed with similar objectives as for nadimide-terminated or maleimide-terminated polyimide resins (see also Chapter 10). The terminal ethynyl groups are capable of undergoing thermally induced addition reactions yielding a complex crosslinked network.[25] A significant development in acetylene-terminated imide oligomers was the commercialization of the Hughes HR600 resins as Thermid 600® (Gulf Chemicals). Beginning in 1983, various forms of Thermid 600® are being supplied by the National Starch and Chemical Corporation.

Although neat resin properties, composite properties and adhesive properties[26,27] of the HR600 resins (Tables 11-6 and 11-7) are indicative of their suitability as high temperature resistant matrix resins and adhesives, their poor flow characteristics and insolubility in common solvents make them difficult to process.

Table 11-6: Typical Properties of HR600

Tensile Strength, MPa	96.5
Tensile Modulus, GPa	3.79
Elongation, %	2.6
Flexural Strength, MPa	124
Flexural Modulus, GPa	4.48
Tensile Shear Strengths, MPa	22.1 (RT) 13.1 (232°C, 450°F) 8.3 (260°C, 500°F)

Reprinted by permission of References 26 and 27.

Table 11-7: Properties of Unidirectional HT-S Graphite Fiber Laminates made with Thermid 600® Resin

Test Condition	Flexural Strength, GPa (ASTM D-760)	Interlaminar Shear Strength, MPa (ASTM D-2344)
RT	1.28	83.4
200°C (after 500 hr. at 200°C in air)	1.17	60.0
288°C (after 500 hr. at 288°C in air)	0.99	51.0
316°C (after 500 hr. at 316°C in air)	1.04	41.4

Reprinted by permission of Reference 13.

Recent results showed that the HR resins can be processed through an isomeric form, the isoimide.

HR600P Isoimide Oligomer

These materials have substantially increased processability by virtue of drastically reduced melting points and prolonged gel time. Soluble and tractable isoimide oligomers have been prepared with degrees of polymerization (DPs) up to 25. By contrast, the Thermid 600® imide oligomers become insoluble and intractable at DP>3. At elevated temperatures, the ethynyl groups undergo the usual chain extension and crosslinking reactions while the isoimide functions rearrange to the imides. Preliminary results indicated that adhesive and composite properties obtained with cured acetylene-terminated isoimide oligomers are compatible to those of cured HR600 imide oligomers.[28,29] The high DP materials are excellent film formers and thus are useful for coating applications. The application of isoimide technology to acetylene-terminated polyimide precursors has brought to light the broad potential of this technology to yield easy-to-process high temperature polyimides. Especially important is the observation that the technology can be applied *without compromising* the thermomechanical properties of the final products. Previously, the concept of producing thermally stable addition curable polymers was severely limited by the intractability of the oligomers. With the introduction of isoimide functional groups into polyimide precursors, the processing window is significantly broadened, thereby making commercially available diamines and dianhydride monomers, which previously formed intractable polyimide prepolymers, viable candidates for future resin systems.

In addition to the melting point and gel time improvements, the isoimides impart to the precursors increased solubility in common low-boiling, noninteracting solvents, such as tetrahydrofuran. This is to be contrasted with state-of-the-art polyimide precursors, which generally are soluble only in strong aprotic solvents, i.e., N-methylpyrrolidone (NMP). The latter class of solvents is very difficult to remove during processing and its use often results in void formation in cured composites.

New families of processible ultrahigh temperature resistant polymers, suitable for long-term utility at >700°F (371°C), will likely be obtainable through the application of isoimide technology to the development of hybrid polymers containing polyheterocyclic structural units.

Studies with polyphenylquinoxalines showed that this class of high temperature thermoplastics were difficult to process, requiring temperatures in excess of 600°F (316°C). Subsequently, the AT oligomers were developed and they exhibited excellent potential as moisture-resistant adhesives and matrix materials.[30-32] It is noteworthy that the ease of processibility of these oligomers is inversely proportional to the oligomer chain length. On the other hand, thermooxidative stability and adhesive strength of the cured ATQ resin increase with longer linear segments in the oligomer. The cured ATQ resins are thermooxidatively less stable than the corresponding parent linear polymer. These observations have proven to be universal among AT polymers.[33] It has been suggested that the thermooxidative stability of the AT resins is related to the large percentage of polyene and enyne structures formed in the resin during cure. Solid state carbon-13 NMR[25] has demonstrated that only a small proportion of the acetylene terminations of AT resins undergo trimerization to yield thermal and thermooxidatively stable benzene rings.

Several acetylene-terminated quinoxalines containing only phenyl and quinoxaline units were synthesized specifically to measure their thermooxidative stability under isothermal aging conditions. Three systems, A,B,C, each being an isomeric mixture, were synthesized and cured prior to isothermal aging.[34]

A $\begin{cases} R_1 = \text{ethynyl}, \quad R_2 = H \\ R_1 = H, \quad R_2 = \text{ethynyl} \end{cases}$

B $\begin{cases} R_1 = R_3 = \text{ethynyl}, \quad R_2 = R_4 = H \\ R_1 = R_3 = H, \quad R_2 = R_4 = \text{ethynyl} \\ R_1 = R_4 = \text{ethynyl}, \quad R_2 = R_3 = H \end{cases}$

C $\begin{cases} R_1 = R_4 = \text{ethynyl} \quad R_2 = R_3 = H \\ R_1 = R_4 = H \quad R_2 = R_3 = \text{ethynyl} \\ R_1 = R_3 = \text{ethynyl} \quad R_2 = R_4 = H \end{cases}$

All three systems showed poor thermooxidative stability at 700°F (371°C). The cured acetylene linkages apparently were the predominant site of thermooxidative degradation. The rate of weight loss at 700°F (371°C) was directly proportional to the number of cured acetylene linkages (Table 11-8).

Table 11-8: Correlation of Curved Acetylene Linkages with
Thermooxidative Stability

System	Time Required to Give 90% Wt.Loss	% Ethynyl Component in System
A	30 hr.	15.2
B	50 hr.	9.4
C	85 hr.	8.2

Reprinted by permission of Reference 34.

These results adequately demonstrated that thermally cured AT resins do not have 700°F (371°C) utility. It is evident that a new cure mechanism must be sought for high temperature resins in order to achieve 700°F (371°C) applicability.

FUTURE DEMANDS IN ULTRAHIGH TEMPERATURE RESISTANT POLYMERS

Projected requirements for future high-performance jet engine, missile, and fighter aircraft structures will necessitate extensive use of advanced composite materials and high temperature structural adhesives. These requirements include high specific strength and stiffness at very high temperatures. For example, many advanced jet engine and tactical aircraft components will need to perform for hundreds of hours at 700°F (371°C) and above. For advanced air-to-air tactical missiles and air-launched stand-off missiles, composite airframe structures capable of maintaining strength for short periods (minutes) at 1000°F (538°C) and above will be needed. Other needs for composites and adhesives capable of performing at these high temperatures include structures for extended range cruise missiles, specialty materials for Stealth applications, and, potentially, space-plane structures.

The critical need for an easily processible resin capable of meeting these performance parameters cannot be satisfied through the implementation of state-of-the-art high temperature materials. Although polymers capable of such performance exist, none of them can be practically processed. Examples include polybenzimidazoles (PBIs), polyphenylquinoxalines (PBQs), polybenzothiazoles (PBTs), recently developed fluorine-containing TRW resins, polybisbenzimidazobenzophenanthrolines (BBBs), etc. The persistent problem is that the solubility, flow, gel time, and melting temperature of such linear and branched heteroaromatic polymer molecules are inadequate to achieve suitable processibility. Development of polymer systems possessing both ultrahigh thermal stability and good processibility while maintaining a high level of toughness will represent a major technological breakthrough.

Most of the state-of-the-art high temperature composite resin and adhesive systems are imide-based. The multifaceted engineering applications of various forms of polyimides are well known. High temperature resistance, in general, results from a high degree of aromaticity and crosslinking in the polymer structure.[35] The thermal stability of the imide ring is one of the major factors which would allow the use of polyimides as 700°F (371°C) resins, provided there were no structural units with lower temperature capability in the polymer backbone.

An ideal connecting linkage for the aromatics in high performance polymers must possess high thermooxidative resistance and good chemical resistance, comparable to the aromatics. The linkage must also facilitate processibility by state-of-the-art techniques and have the ability to promote high polymer formation either directly, via efficient reactions actually taking place at the group, or indirectly, by activiation of another functional group. The hexafluoroisopropylidene group (6F) is particularly suitable for connecting aromatic rings in macromolecules. It surpasses all other available groups in terms of high temperature and chemical resistance and also enhances processing properties. Incorporation of the 6F groups in the polymer backbone is one of the key features in the development of 700°F (371°C) polymers that possess good processibility.

6F-Containing Polymer Unit

As stated earlier, new crosslinking and chain extension techniques are required so that the ultimate structure, after thermal crosslinking, will be able to withstand long-term exposure at 700°F (371°C) in air. At the present time, this crucial requirement in achieving >700°F (371°C) stability has not been met. Some of the organic chemical techniques under study by various research groups are thermal and catalyzed biphenylene opening, and catalyzed acetylene trimerization. The potential 700°F (371°C) polymers thus have two characteristics in common: high thermal stability of the monomers and also stability of the groups or bonds linking the monomers in the polymers. Recent advancement in ceramic technology suggests the strong possibility of developing refractory (heat resistant) materials at low temperatures. It represents a viable entry to a new processing technique whereby polymer-ceramic composite (blend) materials can be made to achieve lightweight, dimensional stability and environmental resistance. It is interesting to note that a new structural material, recently commercialized as Quazite®, was molded from inorganic and organic polymer mixtures *(vide infra)*.

CHEMICAL STRUCTURES SUITABLE FOR ULTRAHIGH TEMPERATURE USE

Improved thermal stability in polymers can be achieved through the incorporation of rigid heterocyclic units, in particular ladder structures,[36] along the polymer chain. Chain scission in ladder structures is less likely to occur by a single bond-breaking reaction. Many of the potential 700°F (371°C) polymers have heterocyclic and/or ladder structures. Some well known examples are 1,4,5,8-naphthalenetetracarboxylic dianhydride-based poly[bis(benzimidazobenzophenanthrolines)], e.g., BTP,[37] BBB,[38-40] BBL,[41] BBL-DBS,[42] TAP-BB,[43] and bis(naphthalenedicarboxylic anhydride)-based poly[bis(benzimidazobenzoisoquinolinones)] (BBQ).[44] An important aspect of the BBB and BBL (ladder) polymers is that they show weight retention of 58% and 87%, respectively, after a 500-hour isothermal aging at 700°F (371°C).[45]

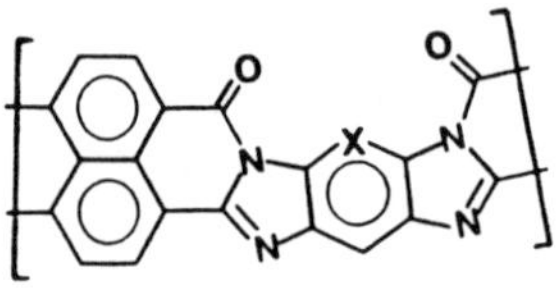

BBL (X=CH)
TAP-BB (X=N)

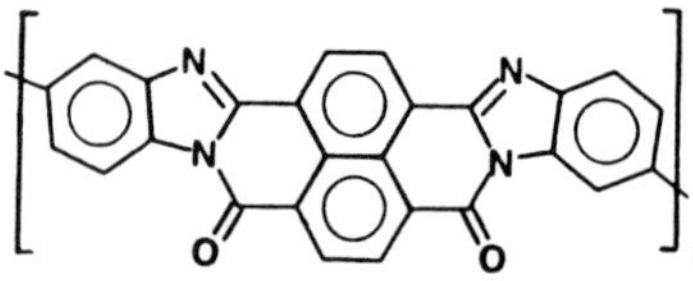

BBB

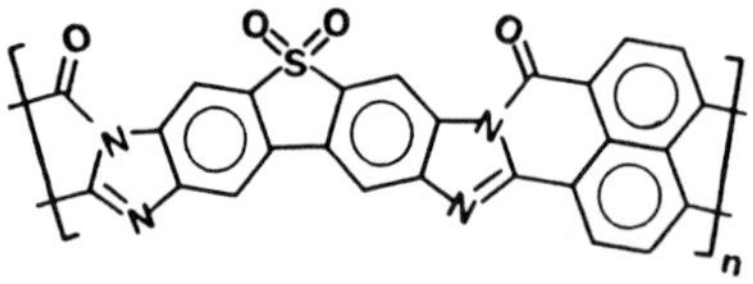

BBL-DBS

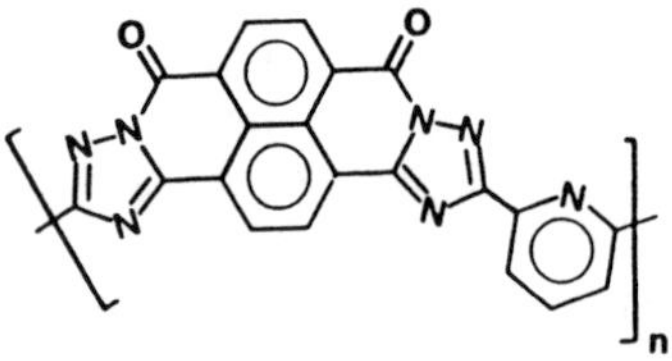

BTP

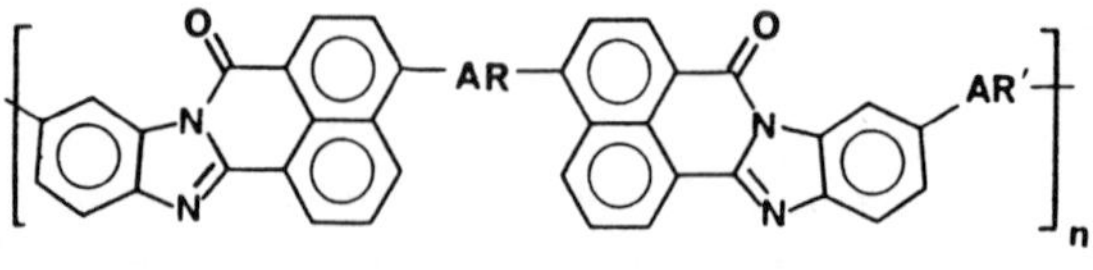

BBQ   AR, AR' = Aryl

Structures containing trigonal nitrogens have been known to impart outstanding thermal stability to polymer systems. It is noteworthy that *poly(benzimidazoquinazolines)* (PIQs) showed excellent retention of thermooxidative and thermomechanical properties after 200 hours exposure to air at 700°F (371°C).[46-48]

PIQ POLYMERS

High molecular weight *polybenzothiazoles* have been prepared and shown to have excellent thermal stability. Isothermal thermogravimetric analyses of these polymers at 600°F (316°C) for 200 hours have shown negligible weight loss and 50% weight loss at 700°F (371°C) after 200 hours.[49,50]

Poly(p-phenylenebenzobisthiazole) (PBT).

The *phthalocyanine* structure is recognized as one of the most thermally and thermooxidatively stable organic structures known and has been incorporated in the synthesis of thermally stable hybrid oligomers[51] and polymers, e.g., benzimidazoles.[52,53] The most relevant are phthalocyanine tetraamines and benzophenonetetracarboxylic dianhydride (BTDA)[54] and those derived from metal phthalocyanine tetraamines and pyromellitic dianhydride (PMDA).[55] Because of the apparent ease of synthesis of the phthalocyanine derivatives and the exceptional thermooxidative stability of the derived polymers (based on preliminary isothermal aging studies), these phthalocyanine polyimides can be considered as potentially useful 700°F (371°C) resins.

A parallel and apparently quite promising development in phthalocyanine-based polymers is the silicon-phthalocyanine polymers. They have been readily synthesized in large quantities, dissolved in acidic media and wet spun into fibers (either in pure form or as blends with Kevlar®).[56] This class of potential ultrahigh temperature resistant polymers is especially attractive because of its structural similarities relative to high temperature silane type coupling agents (needed for adhesive bonding) and silicate glassy networks produced by low temperature high-tech ceramic processing, which will find ultrahigh temperature utility.

M = Si

SILICON — PHTHALOCYANINE POLYMER

Incorporation of *adamantane* units in polyimides[57-61] is also expected to enhance processability without compromising their thermal and thermooxidative stabilities. Comparable studies, albeit less extensive, have also been carried out with *biadamantane*-based polymers.[62,63] Since an adamantane-based polybenzoxazole[64] has been developed as a useful 450°C material, a similar incorporation of adamantane units in polyimides is expected to yield processible resins for potential 700°F (371°C) applications. Recent literature also suggests that extremely high thermal stability can be attained through the incorporation of *dicarbadodecaborane* (carborane) units, as indicated by thermal studies.[65]

As thermooxidative stability (reflected by isothermal aging) is one important aspect in the design of a high performance polymer, consideration must be given to the presence of excessive aromatic hydrogens. Aromatic hydrogen atoms have been shown to be surprisingly labile at high temperatures and participate extensively in degradative and crosslinking reactions, thus decreasing the long term thermooxidative stability of the polymer. On this basis, the anhydride precursor *pyrazinetetracarboxylic dianhydride*, (PTDA), which does not contain any hydrogen, has been shown to impart extraordinary thermal and thermooxidative stability (25 hours at 400°C in air) to polyimides containing this skeletal moiety.[66]

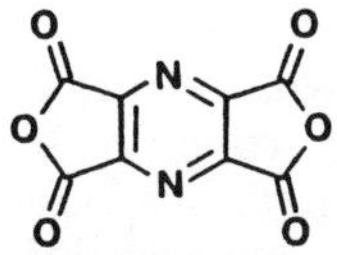

PTDA 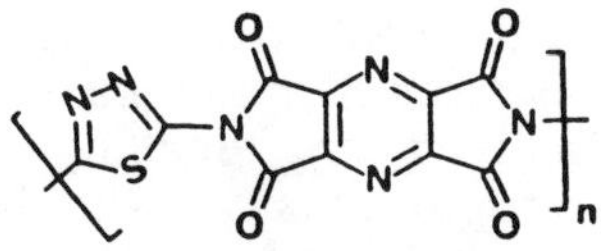   PTDA-DATD

In summary, imide-based polymers incorporating the various classes of ultrahigh temperature resistant structures are potential 700°F (371°C) resin materials. Good processibility, however, must also be achieved.

The introduction of fluorine-containing monomers, particularly those containing the hexafluoroisopropylidene (6F) group, has been shown to yield polymers with increased processibility while maintaining excellent oxidative stability and improved moisture resistance.[67] Many 6F-containing monomers are or can be readily available.[68,69]

The fundamental properties of the monomers and also of the polymers in which they are incorporated have indicated improved thermooxidative stability (c.f. DuPont's NR-150®, which incorporates the 6F-containing dianhydride 6FDA). The application of 2,2-bis[4-(4-aminophenoxy)phenyl]hexafluoropropane (4-BDAF) in polyimide coatings for 700°F (371°C) exposure is also illustrative.[70] The hydrophobic nature of the hexafluoroisopropylidene group is expected to enhance the moisture resistance of the resin systems of which it is a part. Furthermore, the flexible nature of the 6F-linkage is conducive to improvement of fracture toughness.[21]

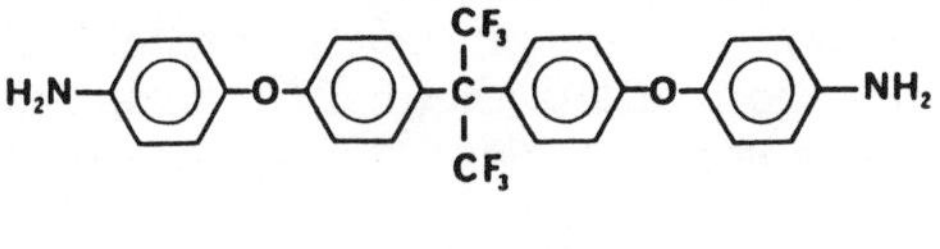

4-BDAF

CURE AND CROSSLINKING MECHANISMS AT ULTRAHIGH TEMPERATURES

The harsh environmental conditions at 700°F (371°C) in air pose severe limitations on the choice of a suitable crosslinking and/or chain extension mechanism. The ultimate structure after thermal crosslinking must be able to withstand long-term exposure up to 700°F (371°C) in air. It has been demonstrated that ethynyl and phenylethynyl end groups on polyphenylquinoxalines undergo thermally induced chain extension and crosslinking, but the final cured resins have less thermooxidative stability than the corresponding polymers which do not contain the ethynyl groups.[71,72] The thermooxidative instability is ascribed to the presence of non-aromatic end products as a result of the ethynyl groups undergoing thermally induced crosslinking.

An attractive crosslinking mechanism presents itself in the thermally induced ring opening of biphenylene, to give dimeric and polymeric products via diradical intermediates.[73-77] Such a crosslinking mechanism for high temperature polymers containing these biphenylene units along the polymer chain provides thermally stable crosslinks and yields no volatile by-products.

Bisbiphenylene compounds having the corresponding skeletal structures to the respective imide, quinoline and phenylene oxide polymers were synthesized. The presence of the *bisbiphenylene crosslinking agent* in the polymer initially plasticizes the polymer and, on curing, this plasticizer forms a crosslinked network with the polymers.[77] The addition of an organometallic reagent to catalyze the crosslinking reaction did not seem to affect the ultimate thermal and thermooxidative stability of the resins. Although the mechanism looks promising, certain technical problems persist. For example, the control over completeness in ring opening reaction is lacking. Furthermore, it is not certain that the recombination of the radical species generated necessarily produces thermally stable products. More extensive studies are necessary.

A recent extension of the biphenylene recombination cure mechanism is the application of the thermal (or nickel-catalyzed) cycloaddition of biphenylene to acetylene[78,79] as the cure mechanism.[80]

Organometallic complex-catalyzed trimerization,[81] intramolecular cyclization,[82,83] and polymerization[84] of many acetylenic compounds often lead to well-defined highly condensed aromatic structures. Of particular interest are complexes based on cobalt, nickel, palladium, rhodium, and most recently, tungsten, tin, niobium, and tantalum. The catalyzed reactions are often facile and proceed at relatively low temperatures.

In the light of the generally accepted catalytic effect of organometallic compounds in the reactions of aromatic acetylene compounds the addition of organometallic reagents is expected to *catalyze trimerization specifically* and at the same time lower the temperature required for the reaction. Candidate reagents for achieving this include rhodium[I] carbonyl, nickel[0] carbonyl, and palladium[II] benzonitrile complexes. The ultimate product after the acetylene cure would have thermally stable crosslinks.

The inherent high temperature instability in thermally cured acetylene-terminated resins is due to the low percentage of stable aromatic and/or heterocyclic ring structures *(vide supra)*. Besides the use of metal catalysts to promote the ring formation reaction of the terminal ethynyl groups (so that the final resin may attain applicability at 700°F), two interesting diazine compounds[85,86] capable of undergoing 1,3-dipolar addition reactions with various dipolarophiles, also can be considered in the context of chain extension and thermal cure for acetylene-terminated polymers.

$$R_1R_2C=N-N=CR_1R_2$$
$$+$$
$$R_3C\equiv CH$$

$\xrightarrow{\Delta}$

a. $R_1 = R_2 = CF_3$
b. $R_1 = Ph, R_2 = H$

Recently, the possibility of preparing refractory materials at very low temperatures has been demonstrated, and these materials have been shown to be stable well beyond 700°F (371°C). For example, aluminum phosphate-based glass material is refractory up to 1600°C, at which point aluminum phosphate begins to decompose. The processing is typically carried out at low temperature and heat treatment requires only temperatures as low as 100°C.[87] The related silicon alkoxide-based glass materials, made also at low temperatures, have been the subject of intense studies.[88-90]

Physical blending of an inorganic (ceramic) precursor and a high temperature polymer has been studied in the development of a high temperature adhesive-sealant composition.[91] The composition comprised a mixture of an aluminous cement and a poly(amic acid) precursor. A rectangular glass fiber sheet was formed into a cylinder and the edges were joined with the adhesive system. After drying for 3 hours at 400°F (204°C) and heating for 3 hours at 800°F (426°C), the joint was shown to have good cohesive and adhesive strength.

The recent commercialization of Quazite® as an improved structural material is another related testimony to the utility of inorganic polymer composites. Quazite® is molded from 95% inorganic material and 5% high performance polymers.[92] The process, which involves controlled mixing, molding and curing, is proprietary to Quazite Corp. The ultimate structure is an intertwined crosslinked network, monolithic in nature and having tailorable characteristics. Quazite® is advertised to have the formability of fiberglass, twice the bending strength of and more abrasion resistance than granite, excellent impermeability to liquids and chemical resistance equal to titanium-clad steel. It can also be fiber-reinforced, polished or gel coated. Commercial applications of Quazite® can be illustrated by the ability of a 20-foot x 1-1/2 inch thick Quazite® panel to withstand a 50,000-amp, 10,000-volt arc. Temperatures above 2000°F (1093.5°C) were contained. Chemically, it resisted temperatures up to 450°F (232°C) and pressures to 500 psi in a highly concentrated, corrosive salt solution.

The utilization of ceramic materials in matrix resins and structural adhesives affords still another beneficial feature. Chemical compounds which undergo *expansive phase transformations* such as the one occurring in partially stabilized zirconium oxide (PSZ), can contribute to the *toughening* of the structural adhesive matrix. The PSZ is a matrix of cubic zirconium oxide containing 20% to 50% of the metastable tetragonal form. As the adhesive matrix sustains a crack, the crack tip induces the metastable tetragonal particles to transform to the monoclinic form with a net volume increase. Such a volume increase exerts a compressive stress to the crack tip thus halting further propagation. While PSZ is known to impart toughening to alumina-based ceramics, other silicate-based expansive toughening agents, such as calcium silicate, can perform well in silicon-derived ceramic matrices.[93] The interdisciplinary research in polymer-ceramics has the potential of offering a novel solution to the quest for ultra-high temperature (>700°F or 371°C) processible polymer resins, particularly thermosetting resins.

REFERENCES

1. Serafini, T.T., in *Polyimides: Synthesis, Characterization, and Applications*, ed. K.L. Mittal, Vol. 2, Plenum Press, New York and London, pp 957 (1984).
2. St. Clair, T.L. and Jewell, R.A., *NASA TM-74994* (1978).
3. Vannucci, R.D. and Alston, W.B., *NASA TMX-71816* (1975).
4. Serafini, T.T., Vannucci, R.D. and Alston, W.B., *NASA TMX-71894* (1976).
5. St. Clair, A.K. and St. Clair, T.L., *Polym. Eng. Sci.*, 16, 314 (1976).
6. St. Clair, A.K. and St. Clair, T.L., in *Polyimides: Synthesis, Characterization, and Applications*, ed. K.L. Mittal, Vol. 2, Plenum Press, New York and London, pp 977 and references herein (1984).
7. Mallet, M.A.J. and Darmory, F.P., *Amer. Chem. Soc. Org. Coatings Plast.*, 34, 173 (1974).

8. Stenzenberger, H.D., *Appl. Polym. Symp.*, 22, 77 (1973).
9. Stenzenberger, H.D., *Appl. Polym. Symp.*, 31, 91 (1977).
10. Hummel, D.O., *J. Appl. Polym. Sci.*, 18, 2015 (1974).
11. Critchley, J.P., *Angew. Makromol. Chem.*, 109/110, 41 (1982).
12. Japan Kokai Tokkyo Koho JP 57/44626, 13 Mar. 1982, issued to Mitsubishi Gas Chemical Company, Inc.; *Chem. Abst.*, 97, 39809y (1982).
13. Hergenrother, P.M., *Encyclopedia of Polymer Science and Engineering*, to be published.
14. Dieck, H.R. and Heck, F.R., *J. Organomet. Chem.*, 93, 259 (1975).
15. Austin, W.B., Bilow, N., Kelleghan, W.J. and Lau, K.S.Y., *J. Org. Chem.*, 46, 2280 (1981).
16. Lee, C.Y.C., Goldfarb, I.J., Arnold, F.E. and Helminiak, T.E., *Org. Coat. Appl. Polym. Sci. Proc.*, 48, 904 (1983).
17. Denny, L.R., Lee, C.Y.C. and Goldfarb, I.J., *Polym. Prepr.*, 25, 183 (1984).
18. Lee, C.Y.C., Denny, L.R. and Goldfarb, I.J., *Polym. Prepr.*, 24, 139 (1983).
19. Hedberg, F.L. and Arnold, F.E., *AFWAL Technical Report*, TR-78-142 (1978).
20. Browning, C.E., Wereta, A. Jr., Hartness, J.T. and Kovar, R.F., *SAMPE Series*, 21, 83 (1970).
21. Abrams, F.I. and Browning, C.E., *Org. Coat. Appl. Polym. Sci., Preprints*, 48, 909 (1983).
22. Maximovich, M.G., Lockerby, S.C., Arnold, F.E. and Loughran, G.A., *Soc. Adv. Mat'l. Proc. Eng. Ser.*, 23, 490 (1978).
23. Sikka, S. and Goldfarb, I.J., *Org. Coat. Plast. Chem., Preprints*, 43, 1 (1980).
24. Reinhardt, B.A., Jones, W.B., Helminiak, T.E. and Arnold, F.E., *Polymer Preprints*, 22(2), 100 (1981).
25. Sefcik, M.D., Stejskal, E.O., McKay, R.A. and Shaefer, J., *Macromolecules*, 12, 423 (1979).
26. Bilow, N., Landis, A.L. and Aponyi, T.J., *Sci. Adv. Mat'l. Proc. Eng. Ser.*, 20, 618 (1974).
27. Bilow, N., Landis, A.L., Boschan, R.H. and Fasold, J.G., *SAMPE Journal*, 18, 8 (1982).
28. Landis, A.L. and Naselow, A.B., *SAMPE Series*, 14, 236 (1982).
29. Landis, A.L. and Naselow, A.B., in *Polyimides: Synthesis, Characterization and Applications*, ed. K.L. Mittal, Vol. 1, Plenum Press, New York and London, pp 39 (1984).
30. Kovar, R.F., Ehlers, G.F.L. and Arnold, F.E., *Polymer Preprints*, 16(2), 246 (1975); *J. Polym. Sci., Polym. Chem. Ed.*, 15, 1081 (1977).
31. Maximovich, M.G., Lockerby, S.C., Kovar, R.F. and Arnold, F.E., *Adhesives Age*, 11, 40 (1977).
32. Kovar, R.F. and Arnold, F.E., *Sci. Adv. Mat. Process. Eng. Ser.*, 8, 106 (1976).
33. Hergenrother, P.M., *Polymer Preprints*, 25, 97 (1984).
34. Hedberg, F.L. and Arnold, F.E., *Polymer Preprints*, 21, 176 (1980).
35. Arnold, C. Jr., *J. Polym. Sci., Macromol. Rev.*, 14, 379 (1979).
36. Bailey, W.J. and Feinberg, B.D., *Polymer Preprints*, 8, 229 (1967).
37. Evers, R., *Polymer Preprints*, 12(1), 240 (1971).
38. Coleman, J. and Van Deusen, R.L., *Air Force Technical Report*, AFML-TR-69-289 (March 1970).
39. Van Deusen, R.L., Goins, O.K. and Sicree, A.J., *J. Polym. Sci.*, A-1, 1776 (1968).
40. Waddella, W.H. and Younes, U.E., *Polymer Preprints*, 21(1), 195 (1980).
41. Arnold, F.E., *Polymer Preprints*, 12(2), 179 (1971).

42. Sicree, A.J. and Arnold, F.E., *Air Force Technical Report*, AFML-TR-73-256 (February 1974).
43. Gerber, A.H., *J. Polym. Sci., Polym. Chem. Ed.*, 11, 1703 (1973).
44. Jedlinski, Z.J., Kewalski, B. and Gaik, U., *Macromolecules*, 16, 522 (1983).
45. Arnold, F.E. and Van Deusen, R.L., *Macromolecules*, 2, 497 (1969).
46. Milligan, R.J., Delano, C.B. and Aponyi, T.J., *J. Macromol. Sci. Chem.*, A10, 1467 (1976).
47. Korshak, V.V. and Rusanov, A.L., *Izv. Akad. Nauk SSR, Ser. Khim.*, 1917 (1970).
48. Korshak, V.V. and Rusanov, A.L., *J. Macromol. Sci.-Rev. Macromol. Chem.*, C21, 275 (1982).
49. Wolfe, J.F. and Arnold, F.E., *Macromolecules*, 14, 909 (1981).
50. Wolfe, J.F., Loo, B.H. and Arnold, F.E., *Macromolecules*, 14, 915 (1981).
51. Achar, B.N., Fohlen, G.M. and Parker, J.A., *J. Polym. Sci., Polym. Chem. Ed.*, 20, 269 (1982).
52. Achar, B.N., Fohlen, G.M. and Parker, J.A., *J. Polym. Sci., Polym. Chem. Ed.*, 20, 1785 (1982).
53. Achar, B.N., Fohlen, G.M. and Parker, J.A., *J. Polym. Sci., Polym. Chem. Ed.*, 20, 2073 (1982).
54. Achar, B.N., Fohlen, G.M. and Parker, J.A., *J. Polym. Sci., Polym. Chem. Ed.*, 20, 2773 (1982).
55. Achar, B.N., Fohlen, G.M. and Parker, J.A., *J. Polym. Sci., Polym. Chem. Ed.*, 20, 2781 (1982).
56. Dirk, C.W., Inabe, T., Schoch, K.F. Jr. and Marks, T.J., *J. Amer. Chem. Soc.*, 105, 1539 (1983).
57. Novikov, S.S., Khardin, A.R., Radchenko, S.S., Novakov, I.A. and Pershin, V., *Vysokomol. Soedin, Ser. B*, 18, 35 (1976); *Chem. Abst.*, 84, 136091x (1976).
58. Novikov, S.S., Khardin, A.P., Novakov, I.A. and Radchenko, S.S., *Vysokomol. Soedin, Ser. B*, 16, 155 (1974); *Chem. Abst.*, 81, 50081v (1974).
59. Novikov, S.S., Khardin, A.P., Novakov, I.A. and Radchenko, S.S., *Vysokomol. Soedin, Ser. A*, 18, 1146 (1976); *Chem. Abst.*, 85, 47135c (1976).
60. Novikov, S.S., Khardin, A.P., Novakov, I.A. and Radchenko, S.S., *Vysokomol. Soedin, Ser. B*, 18, 462 (1976); *Chem. Abst.*, 85, 78598a (1976).
61. Korshak, V.V., Novikov, S.S., Vinogradava, S.V., Khardin, A.P., Vygodoskii, Y.S., Novakov, I.A. Orlinson, B.S. and Radchenko, S.S., *Vysokomol. Soedin, Ser. B*, 21, 248 (1979); *Chem. Abst.*, 91, 57587e (1979).
62. Burreson, B.J. and Levine, H.H., *J. Polym. Sci., Polym. Chem. Ed.*, 11, 215 (1973).
63. Reinhardt, H.F., *J. Org. Chem.*, 27, 3258 (1962).
64. Bellman, G.G., Groult, A.M. and Arendt, J.H., *Ger. Offen.* 2,330,452; *Chem. Abst.*, 83, 179909s (1975).
65. Raubach, H., Schumann, A., Oehlert, J., Korshak, V.V., Krongauz, E.S. and Bekasova, N.I., *Ger. (East DD)* 155,075 (issued 12 May 1982); *Chem. Abst.*, 97, 216953p (1982).
66. Hirsch, S.S., *J. Polym. Sci.*, A-1, 15 (1969).
67. Hedberg, F.L. and Arnold, F.E., *Air Force Technical Report*, AFML-TR-76-198 (March 1977).
68. Lau, K.S.Y., Landis, A.L., Kelleghan, W.J. and Beard, C.D., *J. Polym. Sci., Polym. Chem. Ed.*, 20, 2381 (1982).
69. Fusaro, R.L., *NASA Technical Memorandum 82832* (April 1982).
70. Jones, R.J., Chang, G.E., Powell, S.H. and Green, H.E., *Abstracts, First Technical Conf. Polyimides, SAMPE*, Ellenville, New York, p 38 (November 10-12, 1982).

71. Hergenrother, P.M., *Macromolecules*, 14, 891 (1981).
72. Hergenrother, P.M., *Macromolecules*, 14, 898 (1981).
73. Garapon, J. and Stille, J.K., *Macromolecules*, 10, 627 (1977).
74. Recca, A., Garapon, J. and Stille, J.K., *Macromolecules*, 10, 1344 (1977).
75. Swedo, R.J. and Marvel, C.S., *J. Polym. Sci., Polym. Lett. Ed.*, 15, 683 (1977).
76. Sutter, A., Schmutz, P. and Marvel, C.S., *J. Polym. Sci., Polym. Chem. Ed.*, 20, 609 (1982).
77. Vancraeynest, W. and Stille, J.K., *Macromolecules*, 13, 1361 (1980).
78. Friedman, L. and Rabideau, P.W., *J. Org. Chem.*, 33, 451 (1968).
79. Eisch, J.J., Pietrowski, A.M., Han, K.I., Krueger, C. and Tsai, Y.H., *Organometallics*, 4, 224 (1985).
80. Stille, J.K., submitted to *Macromolecules* (1985).
81. Udovich, C.A. and Fields, E.K., U.S. Patent 4,339,595 (issued 13 July 1982); *Chem. Abst.*, 97, 183025g (1982).
82. Müller, E., Heiss, J., Sauerbier, M., Streichfuss, D. and Thomas, R., *Tetrahedron Lett.*, 1195 (1968).
83. Dougherty, T.K., Lau, K.S.Y. and Hedberg, F.L., *J. Org. Chem.*, 48, 5273 (1983).
84. Masuda, T., Kawai, H., Ohtori, T. and Higashimura, T., *Polym. J.*, 11, 813 (1979).
85. Wagner-Jauregg, T., *Synthesis*, 349 (1976).
86. Armstrong, S.E. and Tipping, A.E., *Perkin Trans. I*, 1411 (1975).
87. Birchall, J.D. and Kelly, A., *Scientific American*, 248(5), 104 (1983).
88. Yoldas, B.E., *J. Mat. Sci.*, 12, 1203 (1977).
89. Yoldas, B.E., *J. Mat. Sci.*, 14, 1843 (1979).
90. Yoldas, B.E., *J. Non-Cryst. Solids*, 51, 105 (1982).
91. Holloway, J.G., Barch, H.W. and Fahey, D.M., U.S. Patent 3,990,409 (issued 7 December 1976); *Chem. Abst.*, 86, 91245y (1976).
92. Available from Quazite® Corporation, Houston, Texas.
93. Sanders, H.J., High-tech ceramics, *Chem. Eng. News*, p 26 (July 9, 1984).

Index

Abrasion resistance - 215, 264, 341, 388
   crosslinked acrylics - 363
   Taber, polyetherimide - 290
   urethane elastomer - 234
Abrasives - 41
Accelerators - 144, 151, 157
Accoustical insulating properties - 42
Acetamides - 84
Acetic acid - 62
Acetic anhydride - 254, 255
Acetoacetates - 84
Acetophenone - 347
Acetoxy cure - 324
   of silicones - 324, 325
Acetylene terminated quinoxalines (see ATQ)
Acetylenes, terminal - 373
Acid anhydride - 144
Acid number - 223
   unsaturated polyester alkyds - 70
Acids, organic - 150, 151
Acrylate (see Acrylics)
Acrylate monomers (see Acrylic monomer)
Acrylic
   crosslinked, properties of - 363
   thermoplastic - 362
   vacuum formed - 100
Acrylic acid - 76, 77
Acrylic backup - 99
Acrylic fabric - 129

Acrylic fibers - 115
Acrylic monomer - 81, 362
Acrylic sheet, cell cast - 362
Acrylics - 160, 326, 364
   crosslinked - 361, 362
   waterborne - 52
Acrylonitrile - 239
Acrylonitrile-butadiene-styrene (ABS) - 326
Activator - 141
Active hydrogen equivalent - 142
Acyl peroxide - 84
Adamantane - 384
Addition cure - 327
   RTV rubber, properties of - 329
   RTV silicone rubber - 328
   systems - 328
Addition of glycol to maleate and fumarate double bonds - 62
Additives - 114, 157, 164
Adduct - 191, 222
Adduct resins - 139
Adducting - 16
Adduction - 192
Adhesion - 56, 135, 161
Adhesion of polyimide and silicone-polyimide to semiconductor wafers - 287
Adhesion promoters - 157, 164, 206, 217
   silane - 218
Adhesive strength - 387

Adhesives
    high strength - 369
    structural - 43, 369
Adipic acid (AA) - 66, 69, 72, 74, 210
A/E ratio - 152
Age hardening - 146
Aging, urethane elastomers, effects of - 197
Alcohol - 212
Aliphatic acid anhdrides (see Anhydrides, acid)
Aliphatic acids, long chain - 72
Aliphatic amines (see Amines, aliphatic)
Aliphatic amines, primary and secondary (see Amines, primary and Amines, secondary)
Alkoxy cure - 324
    of silicones - 324, 325
Alkyd ester linkages - 76
Alkyd-styrene incompatibility - 74
Alkyd ureas - 52
Alkyds - 3, 52, 62
    oil modified - 56
Alkylated phenolic novolacs - 43
Alkylated phenols - 21
Allophanates - 186, 187, 192, 211, 224, 259, 260
Allyl diglycol carbonate - 112
Allyl esters - 344-345
Allyl formulation - 115
Allyl glycidyl ether - 158, 161
Allyl
    properties - 116
    trade names - 130
Allyls - 112
Alpha cellulose - 37, 53, 54, 56
Alpha methyl styrene (AMS) - 71
Alpha relaxation point - 347
Alumina - 162, 163
Alumina, hydrated [see Alumina trihydrate (ATH)]
Alumina trihydrate (ATH) - 88, 89, 98, 163, 165, 167
Aluminum - 164, 165
    chloride - 150
    oxide - 41, 165
    trihydrate [(see Aluminum trihydrate (ATH)]
Amic acid - 267
Amic ester intermediate - 267

Amidopolyamines - 149
Amine adducts - 145
Amine equivalent - 253, 254
Amine hydrogen equivalent - 142
Amine terminated products - 239
Amines - 187, 212
    aliphatic - 144, 145, 146, 152
    aromatic - 134, 146, 147, 152, 156, 212, 237
    aromatic adducts - 147, 149
    bisimide - 149
    cyclic - 146
    cycloaliphatic - 146
    heterocyclic - 156
    polyoxypropylene - 146
    primary - 144, 156, 218, 224
    secondary - 144, 218
    tertiary - 77, 144, 151, 152, 156, 212, 236
Amino molding compounds, properties of - 54-55
Amino wood products - 49
N-Aminoethylpiperazine (AEP) - 146
3-Aminophenylacetylene (APA) - 313
Aminoplasts - 151
Aminopolyamides - 143, 149, 150
Aminos - 3
Aminosilane finish - 306
Ammonium chloride catalysts - 51
Anhydride - 156
Anhydride/epoxy ratio (see A/E ratio)
Anhydride reactivity, relative - 155
Anhydrides
    acid - 150, 151, 268
    cyclic - 151
    eutectic - 155
Antifoams - 190, 227, 233, 235
Antimony oxide - 167
Appearance - 34
Applicability at 700°F - 380
Arc resistance - 115, 153, 163
    crosslinked acrylics - 363
    DAP molding compounds - 119-120, 121-125, 126-128
    polyetherimide - 292, 294
Aromatic diols - 75
Aryl sulfonic acids - 64
Asbestos - 31, 34, 37, 40, 55, 115
    defibrillated - 168
A-stage - 14
ASTM test methods - 185
    for polyurethanes - 264
ATB - 374

ATQ - 374, 379
ATS - 375, 376
Auto body putties - 89
Axial fatigue, *Vespel* polyimide - 276
Azelaic acid - 66
Azlactones, alkenyl - 364

Baekeland, L. - 3, 19, 31
Balance sheet - 37
Barium - 83
Barium sulfate (see Barytes)
Barytes - 116, 162
BATQ - 374, 376
BATQ-H - 376
Batt - 34
Bending strength - 388
Bent strip test - 351, 353
Bentonite (see Clays, bentonite)
Benzidine - 269
Benzimidazoles [see Polybenzimidazoles (PBI)]
Benzophenones, substituted - 86
3,3',4,4'-Benzophenonetetracarboxylic acid, dimethyl ester of (see BTDE)
Benzophenonetetracarboxylic dianhydride (see BTDA)
Benzophenonetetracarboxylic ester acid (see BTDE)
Benzotriazole - 224
    substituted - 86
Benzoyl alcohol - 149
Benzoyl chloride - 225
Benzoyl peroxide - 85, 100, 113, 331
    catalyzed room temperature cures - 84
Benzyldimethylamine (BDMA) - 144, 152
Benzyltrimethylammonium chloride - 77, 84
Beryl - 163
Beryllium oxide - 165
Berzelius - 59
Bis-Amine A (Japanese MOCA) - 213
1,3-Bis(aminoethyl)cyclohexane - 262
1,3-Bis(3-aminophenoxy)benzene (APB) - 313
2,2-Bis[4-(4-aminophenoxy)phenyl]-hexafluoropropane (4-BDAF) - 385
1,2-Bis(2'-aminophenylthio)ethane - 213, 262

Bis-beta-hydroxyethyl ether of hydroquinone - 215
Bisbiphenylene
    compounds - 386
    crosslinking agent - 386
Bis(2,4-dichlorobenzoyl) peroxide - 331
Bisimide monomer - 289
Bismaleimide
    copolymer, melt processable - 371
    modified, processing of - 305
    modified, properties of - 305-308
    N-substituted - 304
    two component - 312
4,4'-Bismaleimide-diphenylmethane - 312
Bismaleimides - 268, 372
Bisphenol A - 76, 78, 79, 80, 132
    derived diols - 76
    diglycidyl ether of (see DGEBA)
    hydrogenated - 75
    propoxylated - 75
Bisphenol F - 139
Bisphenols - 162
Biuret - 186, 192, 211, 214, 259, 260
Blending, physical - 387
Blistering - 217
Blowing agents, fluorocarbon - 107
BMC - 82, 88, 89, 90, 100, 104, 105
Bonding - 80
Boron trifluoride
    adducts - 157
    complexes - 152
    etherate - 150
    monoethylamine - 150
Bounce - 219
Bouyancy - 342
Bowen, R.L. - 77
Branching - 260
Brittleness - 174
    low temperature - 264
Bromine - 75
Bromophenol blue indicator - 253, 255, 256
B-stage - 14, 16, 36, 40, 148
    laminates (see Laminates, B-staged)
    paper - 36, 37
BTDA - 154, 282, 285, 313
BTDE - 309, 369
Bubble release agents - 82
Bulk factor, DAP molding compounds - 121-125
Bulk molding compounds (see BMC)

Bulk resistivity, silicone polyimide - 287
Butadiene dioxide - 159
Butadiene-acrylonitrile
   resins - 160
   rubber, carboxy terminated (see CTBN)
1,4-Butanediol - 219, 237, 240
Butanol - 254, 255, 282
Butyl acrylate - 71
Butyl glycidyl ether (BGE) - 158, 160, 161
t-Butyl hydrazinium chloride - 107
Butyl methacrylate - 71
t-Butyl perbenzoate - 85, 100, 105, 113
t-Butyl peroctoate - 85, 100, 105
t-Butyl phosphonium acetate - 80
Butyl stannoic acid - 64

Cage effect - 338
Calcium - 83
   carbonate - 88, 89, 98, 105, 116, 162, 163
   cyanamide - 48
   oxide - 216, 217
   silicate - 116, 388
   stearate - 39
   sulfate - 163, 216
Calendaring - 40
Canvas - 38
Capping - 191
ε-Caprolactam - 224
Caprolactones - 264
Carbamic acid - 186
Carbodiimides, polymeric - 249
Carbon black - 164, 331
Carbon powders - 162
Carbonyl dipoles - 346
Carborane - 384
Carboxy group functionality - 78
Carboxy terminated polybutadiene-acrylonitrile (see CTBN)
Carboxylic acid anhydrides (see Anhydrides, acid)
Carboxylic acids - 212, 224
Carcinogen - 187, 214
Carothers, W. - 59
Casein glue - 27
Cashew nut oil - 139
Cast resin, properties of - 72
Castable liquids - 190

Castan, P. - 132
Casting, unsaturated polyester - 95
Castor oil
   derivatives - 164
   hydrogenated - 86
Catalysts - 8, 104, 115, 141, 157, 235
   for room temperature cobalt promoted resins - 84
   handling of - 85
   indicators - 82
   injection spray equipment - 93
   latent - 150
   organopalladium - 373
   proprietary, polyurethane - 242
   urethane - 212
Catalytic curatives - 142
Catalyzed acetylene trimerization - 381
Catalyzed biphenylene opening - 381
Cellulose (see *Alpha*-cellulose)
Cellulose fibers - 41
Cellulosics - 52, 115
Ceramic processing - 384
Ceramic spheres - 163
Chain
   extension - 372, 381, 386
   scission - 338, 339, 344, 382
   stoppers - 223, 227
   transfer agent - 343
Chalk, precipitated - 89
Chalk resistance - 161
Change in molecular weight of irradiated polypropylene and polystyrene - 344
Chemical resistance - 34, 56, 57, 73, 74, 75, 76, 77, 78, 82, 89, 99, 112, 116, 133, 136, 144, 145, 147, 150, 175
   heat shrinkables - 360
China wood oil - 38
Chlorendic acid (CA) - 66, 75
Chlorendic anhydride - 154
Chlorine - 75
Chopped fabric - 34
Chopped twisted cord - 34
Chrome green - 164
Cinnamic acid - 77
Cis isomer - 63
Cis maleate isomer - 63
Clay - 88, 89, 116, 163
   bentonite - 163, 164
   bentonite, modified - 86
   treated - 168

Coal tar - 160, 179
Coatings
    elastomeric, electron beam
        cured - 364
    high solids - 175
    moisture cured - 205
    powder - 175
    radiation curable - 344
    solventless - 175
    ultraviolet curable, epoxy - 175
    water based - 175
Cobalt
    naphthenate - 83
    neodecanoate - 83
    octoate - 83
    promoter - 107
    soaps - 82
Co-cob (see Oat hulls)
Coefficient of expansion (see Co-
    efficient of thermal expansion)
Coefficient of friction
    Kapton polyimide - 271
    Vespel polyimide - 277
Coefficient of linear thermal expan-
    sion (see Coefficient of
    thermal expansion)
Coefficient of thermal expansion -
    163, 275
    acetylene terminated polyimide -
        315
    amino molding compounds - 54-55
    crosslinked acrylics - 363
    DAP molding compounds - 121-125
    epoxy - 170-173
    PMR polyimide - 310, 311
    polyamide-imide - 300
    polyetherimide - 291, 293
    Pyralin polyimide - 281
    urethane elastomer - 234
    Vespel polyimide - 278
Cohesive strength - 387
Color - 53, 134, 137, 147, 153
    irradiated polyethylene - 340
    polyurethane elastomers - 194
    RTV silicone rubber - 326
Colorants - 32, 90, 157, 163
    for epoxy resins - 166-167
Common unsaturated polyester
    resin synthesis raw materials -
    65
Composition board - 29, 30, 49
Compounding
    epoxy - 185

fundamentals - 185
materials - 60
polyurethane elastomers - 223, 228
unsaturated polyesters and vinyl
    ester resins - 81
Compressibility - 321
    dimethylpolysiloxane fluids - 321
Compression fatigue - 264
Compression molding - 33, 40, 114
Compression ratio, amino molding com-
    pounds - 54-55
Compression set - 214, 215, 218, 233,
    264, 331
    crosslinked polyethylene foam - 341
    urethane elastomer - 234
Compressive modulus
    amino molding compounds - 54-55
    epoxy - 170-173
    polyetherimide - 290
    Vespel polyimide - 276
Compressive strength - 153, 163
    acetylene terminated polyimide - 315
    anhydride cured epoxy - 155
    aromatic amine cured epoxy - 148
    crosslinked polyethylene foam - 341
    cycloaliphatic amine cured epoxy -
        147
    DAP molding compounds - 115, 119-
        120, 121-125
    DETA/TETA cured epoxy - 145
    epoxy - 170-173
    modified bismaleimide - 308
    PMR polyimide - 310, 311
    polyamide-imide - 298
    polyetherimide - 290, 293
    unsaturated polyesters - 108
    Vespel polyimide - 276
Compressive yield strength
    crosslinked acrylics - 363
    PMR polyimide - 310
Compton scattering - 337
Concrete - 202
Condensation cure - 327, 328
    RTV silicone rubber - 327, 328
    RTV silicone rubber, properties of -
        329
Condensation polymers - 61
Conductivity, electrical (see Electrical
    conductivity)
Conductivity, thermal (see Thermal
    conductivity)
Containers - 189

Continuous heat resistance,
    DAP molding compounds -
    121-125
Continuous operating temperature,
    heat shrinkables - 358
Continuous service temperature,
    polyetherimide - 291, 293
Continuous use temperature, DAP
    molding compounds -
    119-120
Copolymerization - 66
Copolymerize - 60
Copolymers, graft - 345
Copper - 164, 165
Core binder - 38
Corona treatment - 342
Corrosion potential - 325
Cost - 163, 165
    reduction - 162
    volume - 165
Cotton canvas - 37
Cotton floc - 34
Crack growth resistance - 176
Crack resistance - 335
Cracking - 163
Craze resistance - 362
Creep - 335
Creep resistance - 32, 350, 351
Cresols - 21, 38
Cresyl glycidyl ether (CGE) -
    149, 158, 160
Croning process - 38
Crosslink density - 133, 154, 224,
    369, 376
Crosslinked
    polymer - 66
    thermoplastics - 335
    thermoplastics, trade names -
        365
Crosslinking - 4, 5, 192
    agent - 114, 327
    chemical - 335, 336, 346
    degree of - 354
    irradiation - 335, 336
    rate - 345
    reactions, epoxy - 141
Crosslinks, thermally stable - 386
Crotonic acid - 77
Crystallinity, degree of - 336
C-stage - 14
CTBN - 80, 139
Cumene - 22
Cumene hydroperoxide - 83

Curative
    catalytic - 142
    concentration, effect on urethane
        elastomers - 215
    level - 232
Curatives - 8, 169, 184
    aromatic amine - 233
    epoxy - 141
    latent - 177
Cure - 169, 226
    and crosslinking mechanisms at ultra-
        high temperature - 385
    degree of - 354
    flow - 369
    rate - 137
    schedules - 11, 169
    shrinkage - 82
    time - 325, 327
    time, RTV silicone rubber - 326
    time, urethane elastomer - 234
Curing - 4, 8, 11, 38
    agents - 8, 31, 134, 141, 142, 143,
        145, 157, 169, 215, 327
        acid anhydrides - 144
        acid, epoxy - 150
        alkaline, epoxy - 144
        aromatic amine - 144
        paste - 327
        peroxide, for silicones - 331
    behavior - 87
    catalysts - 61
    compounds, and prepolymers - 232
    free radical - 76
    heat - 85
    room temperature - 76, 85
    unsaturated polyesters - 82
Curtain coating - 29
Cyanamide - 48
Cycloaddition of biphenylene to acetyl-
    ene - 386
Cycloaliphatic diols - 75
Cyclodehydration - 268
Cyclohexanedimethanol (CHDM) - 66,
    74
Cyclohexene vinyl monoxide - 158

DABCO - 212, 225, 252, 263
DAIP - 71, 82, 113, 116
DAP - 71, 82, 112, 113, 116, 118, 129,
    130
    molding compounds, properties of -
        115, 119-128
    polyesters - 130

398 Handbook of Thermoset Plastics

Dart impact
  low temperature - 355
  test - 353
Dearating agents - 157
Decabromobiphenyl - 167
  oxide - 167
Decarboxylation - 73
Dechlorane - 164, 167
Decomposition temperature
  MDI - 250
  Pyralin polyimide - 281
Deflection temperature (see Heat deflection temperature)
Degree of condensation - 69, 72
Dehydrochlorination - 344, 345
Delaminating strength, modified bismaleimide - 308
Demold time, urethane elastomer - 234
Density - 163, 250, 335
  control - 162
  crosslinked high density polyethylene - 352
  crosslinked polyethylene foam - 341
  irradiated polyethylene - 340
  Kapton polyimide - 271, 274
  modified bismaleimide - 305
  polyamide-imide - 300
  polyethylene - 336
  Pyralin polyimide - 281
Department of Transportation red label - 71
Dermatitic potential - 150
Dermatitis - 159
DETA - 53, 143, 145, 149
DGEBA - 76, 77, 79, 80, 133, 134, 137, 139, 142, 145, 149, 151, 159, 161
  brominated - 140
  chlorinated - 140
o,o'-Diallyl bisphenol A - 312
Diallyl chlorendate - 113
Diallyl fumarate - 113
Diallyl isophthalate (see DAIP)
Diallyl maleate (DAM) - 71, 113
Diallyl orthophthalate (see DAP)
Diallyl phthalate (see DAP)
Diallyl tetrabromophthalate - 71
Diaminodiphenylsulfone (DADS) - 148, 149
Di-(4-aminophenyl)ether (ODA) - 269

Diatomaceous
  earth - 330
  silica - 34
Diazides - 225, 260
Diazine compounds - 387
Dibromoneopentyl glycol (DBNPG) - 66, 75
Dibutyl phthalate (DBP) - 149, 159, 162
Dibutyl tin
  dilaurate (DBT) - 212, 263, 327
  oxide - 64
Di-n-butylamine - 252, 254, 255, 256
Dicarbododecaborane (see Carborane)
Dicumyl peroxide - 113, 268
Dicyandiamide - 150
Dielectric breakdown, DAP molding compounds - 121-125
Dielectric constant - 116, 118, 148
  acetylene terminated polyimide - 315
  addition cured RTV silicone - 329
  anhydride cured epoxy - 155
  comparison of two polyimide films - 304
  condensation cured RTV silicone - 329
  crosslinked acrylics - 363
  DAP molding compounds - 115, 119-120, 121-125, 126-128
  Kapton polyimide - 272, 273, 274
  methylphenylpolysiloxane fluids - 323
  methylpolysiloxane fluids - 320
  modified bismaleimides - 307
  phenolic molding compounds - 52
  polyamide-imide - 299
  polyetherimide - 292, 294
  polyimide film - 302
  Pyralin polyimide - 281
  RTV silicone rubber - 326
  silicone polyimide - 287
  Skybond polyimide - 284
  urethane elastomers - 196, 197
  Vespel polyimide - 277
Dielectric loss - 346
Dielectric properties - 238
Dielectric strength - 116
  addition cured RTV silicone - 329
  amino molding compounds - 54-55
  condensation cured RTV silicone - 329
  DAP molding compounds - 119-120, 121-125, 126-128
  epoxy - 170-173
  heat shrinkables - 359

Dielectric strength (Cont'd)
   Kapton polyimide - 272, 273, 274
   methylphenylpolysiloxane fluids - 323
   methylpolysiloxane fluids - 320
   modified bismaleimide - 307
   polyamide-imide - 299
   polyetherimide - 292, 294
   polyimide film - 302
   Pyralin polyimide - 281
   RTV silicone rubber - 326
   silicone polyimide - 287
   Skybond polyimide - 284
   urethane elastomers - 196, 197
   Vespel polyimide - 277
Diepoxide resins - 76
Diethylaminoethanol (S-2) - 144
Diethylaminopropylamine (DEAPA) - 145
Diethylammonium triflate - 151
Diethylaniline - 84
Diethylene glycol (DEG) - 63, 66, 69, 72
Diethylene triamine (see DETA)
Diethyltoluenediamine - 239
Diethynyl-bisphenoxybenzene (ATPB) - 376
Difunctional
   acids - 61
   alcohols - 61
   reactants - 62
Diglycidyl ether of Bisphenol A (see DGEBA)
Dihydric alcohols - 63
m-Dihydroxybenzene (see also Resorcinol) - 21
Dihydroxystearic acid - 238
Diisocyanates - 185, 191, 192
   aliphatic - 211, 239
   polymeric aromatic - 250
   saturated - 211, 238
Diluent efficiencies - 160
Diluents - 159, 164, 217
   acetylene terminated - 376
   nonreactive - 149, 161
   reactive - 159, 376
Dimensional stability - 32, 34, 105, 112, 116, 275, 289, 369
   DAP molding compounds - 121-125
Dimer acids - 66
Dimerized fatty acids - 139

Dimethacrylates - 345
Di-(2-methoxyethyl phthalate) - 218
Dimethyl ester of 2,2-bis(3,4-dicarboxyphenyl)hexafluoropropane (6FDE) - 369
Dimethyl phenols (see also Xylenols) - 21
N,N-Dimethylacetoacetamide - 84
Dimethylaminoethanol (S-1) - 84, 144, 218
Dimethylaniline - 84
2,5-Dimethyl-2,5-di(t-butyl peroxyl)-hexane - 331
Dimethylethanolamine [see Dimethylaminoethanol (S-1)]
Dimethylol urea - 47
Dimethylpolysiloxanes - 319
Dimethylsilicone fluids - 321, 322
Dimethyl-p-toluidine - 84
Dioctyl phthalate - 162
Diol, polypropylene polyether - 229
Diols - 222, 229, 235
Dipentene monoxide - 158
Dipropylene glycol (DPG) - 63, 66, 72, 225
   dibenzoate - 218, 263
Dissipation factor - 116, 118, 175
   addition cured RTV silicone - 329
   anhydride cured epoxy - 155
   comparison of two polyimide films - 304
   condensation cured RTV silicone - 329
   DAP molding compounds - 115, 119-120, 121-125, 126-128
   Kapton polyimide - 272, 273, 274
   modified bismaleimide - 307
   phenolic molding compounds - 32
   polyamide-imide - 299
   polyetherimide - 292, 294
   polyimide film - 302
   Pyralin polyimide - 281
   RTV silicone rubber - 326
   Skybond polyimide - 284
   urethane elastomers - 196
   Vespel polyimide - 277
Diundecyl phthalate (DUP) - 345
Divinyl benzene - 345
DMP-30 - 80, 144, 152, 212
   tri-2-ethyl hexoate - 144
Dodecenylsuccinic anhydride (DDSA) - 153
Dolomite - 88

Dose rate - 339
Double promoted polyester reactivity - 84
Double promoted system - 84
Drape - 374
Driers - 216
Drop impact test - 353
Ductile failure - 354
Durability - 115, 175, 205
Durene - 269
Dyes - 157, 163

EEW (see Epoxy equivalent weight)
Eicosanoic acid - 238
Elastic memory - 355, 361
Elasticity - 230
Elastomers - 193, 222, 226
  high temperature - 369
  room temperature curing - 235
  toughening - 369
Electrical conductivity - 162, 163, 164
Electrical insulation - 275, 342
Electrical loss - 112
Electrical properties - 34, 53, 60, 74, 89, 115, 116, 130, 134, 150, 154, 155
  properties of urethane elastomers - 196
  properties of *Vespel* polyimide - 277
Electrical resistance - 145, 150, 153, 162, 179
Electrons, high energy - 337
Ellis, C. - 60
Elongation - 72, 77, 78, 80, 87, 135, 187, 215, 218, 219, 224, 226, 233, 235, 236, 264, 275, 313, 348, 354, 364
  acetylene terminated polyimide - 314, 315, 377
  amino molding compounds - 54-55
  anhydride cured epoxy - 155
  aromatic amine cured epoxy - 148
  ATB - 375
  crosslinked high density polyethylene - 352
  crosslinked polyethylene foam - 341
  cycloaliphatic amine cured epoxy - 147

DETA/TETA cured epoxy - 145
  epoxy - 170-173
  heat shrinkables - 359
  irradiated polyethylene - 340
  Kapton polyimide - 271, 274
  polyamide-imide - 298
  polyetherimide - 290, 293
  polyimide film - 303
  polyurethane elastomers - 194, 197, 234
  Pyralin polyimide - 281
  RIM urethane - 240
  RTV silicone rubber - 326
  Skybond polyimide - 283
  unsaturated polyesters - 108
  Vespel polyimide - 276
  vinyl ester resin, cast - 81
Elongation (strain)
  addition cured RTV silicone - 329
  condensation cured RTV silicone - 329
Elongation, tensile (see Elongation)
Emulsifiers - 190
Emulsifying agents - 94
Endomethylenetetrahydrophthalic anhydride - 66
Endotherm - 222
Endothermic reaction - 224
Ene-yne structure - 315
Engineering thermoplastics - 112
Engineering thermosets - 4
Environmental stress crack resistance (see Stress crack resistance)
Environmental stress cracking - 340
Epichlorohydrin - 132, 133
  methacrylated - 79
Epoxide - 132
Epoxide containing reactive diluents - 157
Epoxidized
  linseed oil - 137
  polybutadiene - 137
  polyglycols - 137
  soya oil - 137
Epoxy - 16, 40, 43, 46, 52, 71, 76, 77, 79, 112, 132, 157, 367, 373
  acrylates - 76
  anhydride cured, properties of - 155
  applications - 174
  aromatic amine cured, properties of - 148
  coal tar - 175
  containing reactive diluents - 158

Epoxy (Cont'd)
  crosslinking reactions - 141
  curatives - 141
  cycloaliphatic amine cured,
    properties of - 147
  cyclohexyl-spiro-epoxy cyclo-
    hexane dioxide - 137
  equivalent weight - 157
  ethynyl terminated - 141
  flexible - 139
  formulation principles - 157
  halogenated - 164
  lignin resins - 141
  methacrylates - 76
  molecular weight - 133
  novolacs - 76, 78, 133
  peracid - 133, 137
  properties - 169, 170-173
  resin formulation - 162
  resin types - 133
  silicone hybrid resins - 141
  stoichiometry - 141
  trade names - 180
  waterborne - 52
3,4-Epoxycyclohexylmethyl-3,4-
  epoxycyclohexane carboxylate - 137
Equivalent number - 223
Equivalent weight - 142, 153, 185, 223
Ester linkage shielding - 80
Ester linkages - 78
Esterification catalysts - 64
Ethanol - 62, 282
Ether glycols - 72
Etherified melamine - 52
Etherified urea - 52
Ethyl acetoacetate - 84
Ethyl fumarate - 63
Ethyl maleate - 63
N-Ethyl morpholine - 212, 263
Ethyldiethanolamine - 212
Ethylene diamine - 149
Ethylene glycol - 63, 66, 74, 282
Ethylene oxide - 132, 145, 219, 324
  DETA adduct - 146
Ethynyl groups to aromatic and heteroaromatic nuclei, attachment of - 372
Ethynylation - 373
Ethynyltrimethylsilane - 373

Eutectic blends - 148
Eutectic solders - 164
Exotherm - 8, 68, 70, 83, 84, 87, 135, 145, 146, 147, 152, 153, 161, 162, 163, 193, 222, 229, 230, 232, 259, 260, 328
Exothermic heat - 29, 258
Exothermic reaction - 191
Expansive phase transformations - 388
Exposure, long term at 700°F, 381
Extenders - 28, 35, 160, 165, 217, 235, 236, 330
Extrusion rate, RTV silicone rubber - 326

Fatty acids
  dimerized - 149, 151
  trimerized - 151
6FDE - 309
Ferric chloride - 107, 150
Fiber glass (see Fibers, glass)
Fiberboard - 29
  low density - 30
  medium density - 30, 46, 49, 51
Fibers
  glass - 34, 53, 91, 115, 162, 165
  graphite - 162
  polyamide - 91
  polyaramid - 91, 162, 164
  polyester - 91
  reinforcing - 162, 164
Filler dispersion and mixing equipment - 88
Filler loading - 87
Fillers - 28, 31, 40, 41, 61, 82, 87, 104, 114, 115, 157, 159, 162, 163, 165, 217, 236
  extending - 330
  fluorocarbon - 275
  low density - 87
  nonreinforcing - 162
  particulate - 116
  reinforcing - 330
Film
  electron beam irradiated - 365
  ethylene-vinyl acetate - 343
  formers - 378
  gamma irradiated - 365
  polyacrylo-urethane - 364
  polyethylene - 343
Fire resistance - 34, 42
Fire retardance - 70

Flaking - 65
Flame resistance (see also Flame retardance)
    DAP molding compounds - 121-125
    heat shrinkables - 359
Flame retardance - 75, 76, 82, 89, 90, 105, 140, 154, 169, 331
Flame retardant
    additives - 157
    additives for epoxy resins - 167
    synergists - 167
Flame retardants - 164
Flame treating - 342
Flammability
    crosslinked acrylics - 363
    DAP molding compounds - 119-120
    polyamide-imide - 300
    Pyralin polyimide - 281
    rating - 289
    Skybond polyimide - 283
Flammable liquids - 71
Flash - 37
Flash point - 71, 81, 248, 319
    methylphenylpolysiloxane fluids - 323
    methylpolysiloxanes fluids - 330
    Pyralin polyimide - 280
    styrenated GP polyester liquid resin - 70
    vinyl ester resins - 81
Flex resistance - 215
Flexibility - 56, 72, 137, 139, 146, 150, 153, 157
    low temperature - 233, 236, 238
    Pyralin polyimide - 281
Flexibilization - 72, 73
Flexibilized resins - 73
Flexural fatigue, *Vespel* polyimide - 276
Flexural modulus - 177, 313, 374
    acetylene terminated polyimide - 315, 377
    amino molding compounds - 54-55
    anhydride cured epoxy - 155
    aromatic amine cured epoxy - 148
    crosslinked high density polyethylene - 352
    cycloaliphatic amine cured epoxy - 147
    DAP molding compounds - 115, 119-120, 126-128
    DETA/TETA cured epoxy - 145
    epoxy - 170-173
    modified bismaleimides - 305, 306, 308
    phenolic molding compounds - 32
    PMR polyimide - 311
    polyamide-imide - 298
    polyetherimide - 290, 293
    Skybond polyimide - 283
    unsaturated polyesters - 108
    Vespel polyimide - 276
    vinyl ester resin, cast - 81
Flexural strength - 130, 153, 154
    acetylene terminated polyimide - 315, 377
    amino molding compounds - 54-55
    anhydride cured epoxy - 155
    aromatic amine cured epoxy - 148
    crosslinked acrylics - 363
    cycloaliphatic amine cured epoxy - 147
    DAP molding compounds - 115, 119-120, 121-125, 126-128
    DETA/TETA cured epoxy - 145
    epoxy - 170-173
    modified bismaleimides - 305, 306, 308
    phenolic molding compounds - 32
    PMR polyimide - 311
    polyamide-imide - 298
    polyetherimide - 290, 293
    Skybond polyimide - 283
    unsaturated polyesters - 108
    Vespel polyimide - 276
    vinyl ester resin, cast - 81
Flooring, monolithic - 95, 99
Flow - 34, 89, 380
    control - 82, 116
    control agents - 85
    index, crosslinked high density polyethylene - 352
    length - 289
Fluorinated ethylene propylene (FEP) - 269
Fluoroelastomer - 356
Foam
    closed cell, polyethylene, radiation crosslinked - 341

Foam (Cont'd)
  copolymer, polyethylene, crosslinked - 341
  crosslinked polyethylene - 342
  crosslinked polyolefin - 342
  crosslinked polypropylene - 342
  low density urethane - 343
  microcellular - 193
  phenolic - 42
  polyester - 107
  polyolefin - 343
  polypropylene - 341
  radiation crosslinked polyolefin - 343
  syntactic - 162
  urethane - 342
  urethane, flexible - 195
  urethane, rigid - 195
Folding endurance, *Kapton* polyimide - 271
Formaldehyde - 18, 22, 23, 26, 42, 45, 47, 49
Formica - 36, 56
Formulation - 169
Foundry resins, phenolic - 38
Fracture toughness - 385
Franck-Rabiniwitch effect - 338
Free isocyanate (see Isocyanate, free)
Free radical
  curing - 76
  formation - 343
Free radicals - 339, 346, 350
Freezing point - 248, 319, 322
  methylphenylpolysiloxane fluids - 323
  methylpolysiloxane fluids - 320
Friction materials, phenolic - 39
Frictional properties - 34
Fumarate ester - 63
Fumarate unsaturation - 63
Fumaric acid (FA) - 66, 74
Functionality - 5, 62, 213, 222, 233
Functionality theory - 62
Furan - 45, 56
  asbestos filled - 57
  molding compounds, properties of - 55
  polymers - 56
Furfural - 18, 21, 22, 56, 57, 160
Furfural-phenolic resins - 22
Furfuryl alcohol - 56, 57

phenolic - 39
urea - 39

Galvanic corrosion - 116
Gamma radiation - 117
Gamma rays - 337
Gas permeability, *Kapton* polyimide - 275
Gasoline resistance - 353
Gel coating - 85
Gel coats - 87, 89, 92
Gel filtration chromatography - 26
Gel permeation chromatography - 26, 73, 74
Gel phase chromatography - 23
Gel point - 7, 68
Gel time - 7, 25, 83, 84, 146, 305, 314, 378, 380
  drift - 83
  modified bismaleimide - 305
  urethane elastomer - 234
Gelation, premature - 189
General chemistry of unsaturated polyesters - 62
General purpose resins, unsaturated polyesters - 72
GFC (see Gel filtration chromatography)
Glass - 179, 202
  aluminum phosphate based - 387
  chopped - 163
  cloth - 38, 305, 332
  fibers (see Fibers, glass)
  frit - 98
  hollow spheres - 165
  reinforced polyetherimide - 289
  silicon alkoxide based - 387
  solid spheres - 165
  spheres, silver coated - 162
Glass transition temperature (see T_g)
Gloss resistance - 56
Glue line - 27
Glutaric acid - 66
Glycerin - 186, 219
  glycidyl ethers of, water-soluble - 141
Glycerol - 139
Glycidyl
  acrylate, esterified - 77
  ester of tert-carboxylic acid - 158
  ethers of glycerin, water-soluble - 141
  groups - 133, 136

404 Handbook of Thermoset Plastics

Glycidyl (Cont'd)
  methacrylate - 158, 161
    esterified - 77
Glycol - 63, 186, 191, 192, 222
  loss - 64
  maleates - 59
  polypropylene polyether - 264
Gold - 164
  powdered - 162
GPC (see Gel permeation chromatography)
Graft copolymers - 345
Granular molding powders - 129
Graphite - 34, 57, 163, 164, 179, 275
  cloth - 305
  composites - 373, 375
  fibers (see Fibers, graphite)
  PMR composites - 310
  powders - 162
Greases, silicone - 217
Green strength - 78, 101, 202, 239, 331
Greenlee, S.O. - 132
Group II metal hydroxides - 76, 91, 104
Group II metal oxides - 76, 91, 104
GR-S rubber - 60

Halogenated intermediates - 69, 75
Hand lay-up - 85, 91
Hardboard - 29
  wet process - 30
Hardener - 8, 136, 141
Hardness - 56, 143, 187, 193, 213, 218, 224, 226, 230, 233, 237, 264, 340
  Barcol - 306
    acetylene terminated polyimide - 315
    DAP molding compounds - 126-128
    epoxy - 170-173
    Skybond polyimide - 283
    vinyl ester resin, cast - 81
  cured - 72, 87
  Norton - 41
  RIM urethane - 240
  Rockwell
    amino molding compounds - 54-55
    anhydride cured epoxy - 155

    aromatic amine cured epoxy - 148
    crosslinked acrylics - 363
    cycloaliphatic amine cured epoxy - 147
    DAP molding compounds - 115
    DETA/TETA cured epoxy - 145
    epoxy - 170-173
    polyamide-imide - 300
    polyetherimide - 290
  Shore - 215
    acetylene terminated polyimide - 315
    addition cured RTV silicone - 329
    condensation cured RTV silicone - 329
    epoxy - 170-173
    irradiated polyethylene - 340
    polyurethane elastomers - 194, 197, 234
    RTV silicone rubber - 326
  Vespel polyimide - 279
Hazard
  factor - 187
  of IPDI - 252
  of MDI - 250
  of TDI - 249
  of saturated MDI - 251
Haze - 64
HDT (see Heat deflection temperature)
Heat conduction - 194
Heat conductivity - 34
Heat deflection point - 72, 73, 78
Heat deflection temperature - 134, 135, 137, 143, 145, 148, 150, 153, 155, 161, 213, 214, 233, 264, 289
  amino molding compounds - 54-55
  anhydride cured epoxy - 155
  aromatic amine cured epoxy - 148
  crosslinked acrylics - 363
  cycloaliphatic amine cured epoxy - 147
  DAP molding compounds - 115, 119-120, 121-125, 126-128
  DETA/TETA cured epoxy - 145
  epoxy - 170-173
  polyamide-imide - 300
  polyetherimide - 291, 293
  Vespel polyimide - 278
  vinyl ester resin, cast - 81
Heat distortion temperature (see Heat deflection temperature)
Heat resistance - 34, 53, 56, 76, 116, 136, 187

Heat resistance (Cont'd)
  continuous, DAP molding
    compounds - 115
Heat shrinkable insulation and
  encapsulation tubings - 356
Heat tempering - 226
Heterocyclic units, rigid - 382
Hexa - 24, 25, 31, 32, 38, 39,
  41, 42
Hexafluoroisopropylidene group -
  381, 385
Hexahydrophthalic anhydride
  (HHPA) - 153
Hexamethylene tetramine (see
  Hexa)
1,6-Hexanediol - 219
  diacrylate - 362, 364
1,2,6-Hexanetriol - 219
High maleic content resins - 74
High temperature properties -
  130, 266
High temperature stability - 331
Homopolymerization - 60, 304
Hoop stress - 352, 353
Hot melt - 8
Hot strength - 101, 104
Humidity - 116, 117, 174
  and temperature, effects of,
    urethane elastomers - 197
  exposure - 115, 375
  resistance - 144
Hycar CTBN - 80
Hydantoin resins - 138
Hydrogen bonds, shear labile - 86
Hydrogen peroxide - 85
Hydrogenated Bisphenol A - 66, 75
Hydrolysis prone ester linkages - 80
Hydrolysis resistance - 80, 238
Hydrolytic stability - 73, 74, 264
Hydroperoxides - 339, 344
Hydrophilic groups - 94
Hydroquinone - 65, 79, 80, 314
  di(beta-hydroxyethyl)ether -
    215, 262
Hydroxyl equivalents - 222
Hydroxyl number - 184, 220, 221,
  223, 226, 231, 242, 255
  determination of - 254
N-(2-Hydroxypropyl)-ethylenedi-
  amine (see Quadrol)
Hydroxypropylmelamine - 239
Hygroscopic agents - 95

Hysteresis - 193
  factor - 194

Ignition temperature, crosslinked
  acrylics - 363
Imidazole
  2-ethyl - 150
  2-ethyl-4-methyl - 150
  3-methyl - 150
Imide oligomers, norbornenyl termi-
  nated - 308
Imide structure - 266
Immersion - 264
Impact resistance - 157, 163, 353
Impact strength - 34, 87, 115, 135,
  146, 150, 153, 161, 193, 350,
  351, 362
  DAP molding compounds - 126-128
  Gardner, polyetherimide - 290
  Izod
    amino molding compounds -
      54-55
    anhydride cured epoxy - 155
    aromatic amine cured epoxy -
      148
    crosslinked acrylics - 363
    cycloaliphatic amine cured
      epoxy - 147
    DAP molding compounds - 115,
      119-120, 121-125
    DETA/TETA cured epoxy - 145
    epoxy - 170-173
    modified bismaleimide - 308
    phenolic molding compounds - 32
    PMR polyimide - 311
    polyamide-imide - 298
    polyetherimide - 290, 293
    Vespel polyimide - 276
    vinyl ester resin, cast - 81
    Kapton polyimide - 271, 274
    low temperature - 354, 355
Impermeability - 388
Induction - 193
Induction period - 191
Initiator - 347
Injection molding - 33
Insulation resistance - 116, 117
  Skybond polyimide - 284
Interlaminar bond strength - 107
Interlaminar shear strength - 310
  acetylene terminated polyimide -
    377

Interlaminar shear strength (Cont'd)
   PMR polyimide - 311
   short beam, *PMR* polyimide - 311
Internal mold release (IMR)
   formulations - 239
Intramolecular cyclization - 386
IPDI - 205, 210, 252, 262
   hazards of - 252
Irradiation - 338, 339, 343, 344
   beta - 339, 340
   effects on properties of polyethylene - 340
   gamma - 339, 343, 344
   pile - 337
Isocyanate - 79, 184, 186, 188, 190, 335, 364
   equivalent - 184, 220, 222, 256
   equivalent, determination of - 255
   equivalent weight - 252
   free - 184, 190, 192, 228, 230, 231, 235, 236, 258, 261
   index - 184, 211, 221, 222, 230, 232, 233, 235, 236, 246, 247, 249, 255
   percent - 221, 248, 250, 251, 252
   percent free - 221, 228, 229, 237, 238, 254
   percent free, determination of - 252
   polymeric - 210
3-Isocyanatomethyl-3,5,5-trimethyl-cyclohexyl isocyanate (see IPDI)
Isoimide - 377
   functional groups - 378
   thermally unstable - 268
Isomerization - 63
   and glycol structure - 64
   and temperature - 64
   of maleate to fumarate - 62, 63
   variations - 63
   vs. glycol type - 63
   vs. reaction temperature - 64
Isophorone diisocyanate (see IPDI)
Isophthalic
   acid (IPA) - 66, 69, 72
   polyester, SMC formulation - 105
   polyesters - 74, 100, 101, 104, 105
   resins - 73
Isopropyl alcohol - 254
Isothermal aging - 379

Isothermal thermogravimetric analysis - 383

Jackson, R.J. - 78

Kapton - 269
   physical properties of - 271
Karl Fisher method for testing water content - 216
Ketenes - 268
Ketone peroxides - 82
K-factor, crosslinked polyethylene foam - 341
Kienle, R.H. - 62
Kinetics of maleate-fumarate isomerization - 64
Kirksite dies - 100
Kraft paper - 22, 37

Ladder structures - 382
Laminate prepregs - 115
Laminates - 129, 332
   B-staged - 42
   decorative, phenolic - 36
   industrial - 36
   industrial, phenolic - 37
   paper - 42
Lap shear - 369
Lap shear strength - 177, 312, 371
   nadimide polyimides - 370
   polyimide - 285
   RTV silicone rubber - 326
LARC - 368, 369, 371
   chemistry of - 370
Latent catalysts - 150
Latent thermosets - 8
Lead naphthenate - 212, 263
Lead octoate - 212, 252, 263
Leveling aids - 190
Lewis acids - 150, 156, 238, 268
Lewis bases - 144, 151
Light stabilizers - 61
Lignin - 35
   humidified - 29
Lime - 32
Limestone, ground - 89, 98
Limiting oxygen index - 169
   polyamide-imide - 300
   polyetherimide - 291, 294
   Vespel polyimide - 279
Linear polymer - 4
Linear primary glycol structure - 74

Linear shrinkage
  amino molding compounds - 54-55
  epoxy - 170-173
Linoleic acid - 238
Linolenic acid - 238
Linseed oil, epoxidized - 137
Lithium aluminum silicate - 163
Logarithmic decrement - 347
LOI (see Limiting oxygen index)
Loss factor - 60, 148
  crosslinked acrylics - 363
Loss of glycol - 62
Loss regions
  alpha - 346
  beta - 346
  gamma - 346
Loss tangent, acetylene terminated polyimide - 315
Lost wax process - 38
Low angle laser light scattering (LALLS) detector - 73
Low profile
  additive - 105
  agent - 89
  resin - 104
Lubricants - 31, 32
Lubricity - 162

Machinability - 163
Magnesium oxide - 32, 78, 216, 217
  thickener - 105
Maleic acid - 61, 62, 78, 79, 151
Maleic anhydride (MA) - 62, 63, 65, 66, 72, 153, 154
Manganese - 83
Marble
  cultured - 89, 95, 99
  cultured, matrix formulation - 98
  synthetic - 88
Masking tapes - 43
Masonite - 29
Mass, influence of - 7
Mass transfer area - 64
Master batch - 89
Mat molding - 89, 100
Matched die molding - 85, 100, 114
Mathematical principles of polyurethanes - 242

Mathematics, industrial for polyurethanes - 220
Maturing - 104
May, C.A. - 79
MDI - 187, 192, 194, 205, 210, 211, 221, 224, 226, 227, 231, 236, 237, 238, 239, 247, 250, 261
  hazards of - 251
  prepolymers, preparation of - 260
  saturated - 251, 262
  saturated, hazards of - 251
Mechanical properties - 130, 138, 153, 155
  Vespel polyimide - 276
Medium density fiberboard - 46, 49, 51
MEK peroxide - 81, 83, 85, 92, 95, 98, 107
Melamine - 37, 48, 49
  etherified - 52
  molding compounds, properties of - 54-55
  molding powders - 56
  resin chemistry - 48
  resins - 45, 46, 53
Melamine-formaldehyde - 136
  molding compounds, properties of - 54
Melamine-phenolic molding compounds, properties of - 55
Melt index - 350, 351
  crosslinked high density polyethylene - 352
Melt strength - 351
Melting point - 250, 336, 378
  alkyd - 69
  crystalline - 339
  Pyralin polyimide - 281
  unsaturated polyester alkyds - 70
Melting temperature - 54, 55, 314, 380
Menthane diamine - 146, 147
Mer - 62
Mercaptans - 155, 157, 187, 212
Mercury octoate - 212
Mercury oleate - 212
Meta-dihydroxybenzene - 42
Meta-tetramethylxylene diisocyanate - 241
Metal adhesion - 144
Metallic soaps - 327
Metals, powdered - 162, 163
Metastable tetragonal form - 388

Methacrylate
 groups - 335
  methyl group - 80
  termination - 80
Methacrylic acid - 76, 77, 79, 80
Methanol - 255, 256
Methyl acetoacetate - 84
Methyl ethyl ketone - 305, 314
 peroxide (see MEK peroxide)
Methyl ethyl ketoxime - 224
Methyl methacrylate (MMA) - 62, 71, 82, 93
N-Methyl morpholine - 212, 255, 256, 263
N-Methyl pyrrolidone (NMP) - 214, 263, 282, 301, 313, 314, 378
α-Methylbenzyldimethylamine - 144
2-Methyl-3-butyn-2-ol - 373
Methylchlorophenylpolysiloxane - 322
Methylene bis-methyl anthranilate - 213, 262
4,4'-Methylene bis(o-chloroaniline) (see MOCA)
Methylene chloride - 301, 305
Methylene dianiline (MDA) - 148, 149, 213, 214, 262, 285, 309
Methylene dicyclohexane-4,4'-diisocyanate (see MDI, saturated)
Methylene diphenyl diisocyanate (see MDI)
Methylol phenol - 25, 26
Methylphenylpolysiloxanes - 319, 322
Methylpolysiloxanes - 319, 322
Mica - 31, 34, 162, 163, 164
Michael addition - 304
Microballoons - 87
 ceramic - 87
 epoxy 87
 glass - 87, 162
 phenolic - 87, 88, 162, 165
 polyvinylidene chloride - 87
Microspheres, glass - 88
Millable gums - 190, 192
Mine bolt resins - 81
Mineral fillers, fibrous - 115
Mix ratio - 14, 16, 142, 143, 150, 185, 327

urethane elastomer - 234
Mixed viscosity - 7, 28
Mixers
 dough - 88, 98
 high speed - 189
 propeller - 89
 sigma blade - 88
MOCA - 187, 212, 214, 215, 232, 233, 237, 262
Moduli - 87
Modulus - 214, 218
 of elasticity, *Skybond* polyimide - 283
 storage - 347, 348
 Young's - 348, 349, 364
Moisture - 191
 content - 188
 cured systems - 190, 192, 238
 degradation - 220
 protection - 342
 resistance - 149, 372, 385
 sensitivity - 46
Molar ratio of phthalic anhydride to maleic anhydride - 72
Mold release - 104
 agents - 82, 157, 217, 289
 internal - 331
Mold shrinkage
 amino molding compounds - 54-55
 DAP molding compounds - 115
 epoxy - 170-173
 polyetherimide - 292, 294
Molding compounds - 70, 81, 82
 urea - 53
Molding powders - 115
Molding pressure DAP molding compounds - 121-125
Molding pressure range, amino molding compounds - 54-55
Molding shrinkage, DAP molding compounds - 121-125
Molding temperature, DAP molding compounds - 121-125
Molecular sieves - 216, 227, 255, 263
Molecular weight - 7, 29, 69, 72, 73, 77, 78
 distribution - 343, 344, 349
 number average - 343, 344, 349
 ortho- and isophthalic unsaturated polyesters - 73
 unsaturated polyester alkyds - 70
 weight average - 343, 344, 349
Molybdic oxide - 167

Monoepoxides - 158, 159
Monofunctional reactants - 62
Monomer - 223
    used in unsaturated polyesters - 71
Monomethyl phenols (see also Cresols) - 21
Monomethyl urea - 47
Mono-N-methylacetamide - 84
Monomethylol urea - 47
Monotertiary butyl hydroquinone - 65
Morpholine - 263
Multifunctional coreactants - 142
Mutagen - 241

Nadic methyl anhydride (NMA) - 153
Najvar, D.J. - 79
Naphthylene-1,5-diisocyanate - 241, 262
Natural rubber - 3, 60
Neopentyl glycol (NPG) - 66, 74
    diglycidyl ether - 160
Neoprene - 43, 356
Nepheline syenite - 88
Network polymer - 1, 5
Nickel - 164
Nitrogen, dry - 189, 190
Nitrophthalic monomer - 289
No-bake binders - 57
No-bake systems - 39
Nonreactive diluents (see also Diluents, nonreactive) - 149, 157
Nonreactive flexibilizers - 146
Nonyl phenol - 21
5-Norbornene-2,3-dicarboxylic acid (NE) - 309
Novolac - 24, 25, 31, 32, 38, 39, 40, 41, 42, 57, 136, 137, 151
    epoxy - 133
    phenol-formaldehyde - 133
Nuclear magnetic resonance - 23
Number average molecular weight - 66
Nylon - 115
    fabric - 38

Oat hulls - 28, 29
Ocher - 164
Octabromobiphenyl - 167
Octyl acrylamide (OAA) - 71
Octylene oxide - 158, 161
Odor - 325
    crosslinked acrylics - 363
Oil absorption - 89
Oil resistance - 223
Oils, hydrocarbon - 162
Olefin
    oxides - 158
    polymers - 335, 337
    polymers, crystalline - 355
Oleic acid - 238
Oligomers
    acrylamide functional - 364
    imide, acetylene terminated - 376
    isoimide - 378
    polyester-urethane - 364
    reactive, acetylene terminated - 373
    sulfone, ethynyl terminated - 376
One can systems - 8, 150, 157, 177
One component systems - 324, 325
One-shot polyurethane prepolymers - 217
One-shot system - 190, 191
One-step phenolic resins - 35
One-step resins - 31
Onium salts - 77
Onyx
    cultured - 90, 95, 99
    cultured, production - 98
    synthetic - 88, 90
Opacity - 162
Oriented strandboard - 49
Ortho cresol - 38
Ortho resins - 72
Orthophthalic polyesters - 100
    unsaturated - 74
Ortho-toluidine diisocyanate - 241, 262
OSB (see Oriented strandboard)
Overlay sheet - 37, 53
Oxalic acid - 151
Oxidation per unit of radiation dose - 339
Oxidative destruction of double bonds - 62
Oxidative stability - 266, 267, 275, 385
Oxirane ring - 133, 143, 151
Oxygen index, DAP molding compounds - 126-128
Ozone resistance - 193, 195

Package life - 85

Palmitic acid - 238
Paper - 41
  laminates - 42
Para-butyl phenyl glycidyl ether - 158
Para-cresol - 38
Paraffin - 337
  crosslinked - 346
  waxes - 107
Para-methyl styrene (PMS) - 71, 82, 93
Para-phenyl phenol - 22
Para-phenylene diamine - 309
Para-phenylene diisocyanate - 241, 262
Para-t-butyl phenol - 21
Para-tetramethylxylene diisocyanate - 241
Para-t-octyl phenol - 21
Para-toluene sulfonic acid - 64
Parker, E.E. - 62
Particleboard - 29, 30, 46, 49, 50, 51, 52, 129
Parts per hundred of resin (see phr)
Pasa-Jell - 312
Peak exotherm (see Exotherm)
Peanut meal - 27, 161
Peel strength, RTV silicone rubber - 326
3-(Pentadecyl)phenyl glycidyl ether - 158, 160
Pentaerythritol - 219, 224
  triallyl ether - 350
Pentanedione peroxide - 85
Peracid epoxies - 133, 137
Peracid resins - 136
Percent gel - 355
  test - 354
Percent solids - 28
Perfluorinated isopropylidene analogue of BTDE (see 6FDE)
Perfluoroisopropylidene containing acid anhydride - 288
  group - 287
Peroxide - 339
  decomposer - 343
  dimer content - 85
  organic - 346
Peroxyl radicals, trapped - 344
Phenol - 18, 21, 22, 26, 38, 57, 149, 150, 151, 152, 186, 212, 224
  formaldehyde novolacs - 133

formaldehyde polymers - 18
furfural resins - 57
Phenolic
  decorative laminates - 36
  fabric - 42
  foam - 42
  foundry resins - 38
  friction materials - 39
  industrial laminates - 37
  molding compounds - 31
    properties of - 32
  nitrile adhesives - 43
  paper - 42
  polyvinyl butyral - 43
  raw materials - 20
  reaction mechanisms - 23
  resin - 3, 18, 28, 29, 34, 35, 37, 39, 40, 51, 57, 136, 151, 160
    one-step - 35
    water-soluble - 35
  thermal and sound insulation - 34
  trade names - 43
Phenolphthalein - 254, 255
Phenyl ether linkages - 78
Phenyl glycidyl ether (PGE) - 149, 158, 159, 161
Phenyl isocyanate - 255, 256
Phenyl mercuric oleate - 212
Phenyldiethanolamine - 84
m-Phenylene diamine (MPDA) - 148, 149, 269, 282, 368-369
2-Phenyl-2-propanol - 347
Phosphine - 77
  oxides - 249
Phospholines - 249
Phosphorous trichloride - 268
Photon absorption - 337
Phr - 142, 228
Phthalate ester plasticizers - 162
Phthalic acid - 61, 72, 151
Phthalic anhydride (PA) - 62, 65, 66, 72, 134, 152
Phthalocyanine - 383
  tetraamines and benzophenone-tetracarboxylic dianhydride - 383
  tetraamines and pyromellitic dianhydride - 383
Physical properties
  of *Kapton* polyimide - 271
  of unsaturated polyesters - 108
Pigments - 32, 61, 82, 90, 104, 115, 157, 164

Pigments (Cont'd)
  inorganic - 90
  organic - 90
  phthalocyanine - 90
  veining - 98
α-Pinene oxide - 158
Piperidine - 146
Plasticization - 56, 162
Plasticizer - 218, 235, 237, 344, 345, 349, 386
  bleeding - 204
  external - 73
  phthalate ester - 162
  secondary - 161
Platinum - 164
Plywood - 27, 28, 49, 129
  phenolic bonded - 27
  urea - 52
  urea resin bonded - 27, 51
PMR - 368, 369
  composite technology - 368
Poisson's ratio, *Vespel* polyimide - 276
Polyamic prepolymer - 269
Polyamide-epichlorohydrin polymers - 53
Polyamide-imides - 295
  properties of - 298-300
Polyamides - 134, 135, 149, 156, 159
Polyamines - 135, 159
  aliphatic - 134
Polyaramid - 179
  fibers (see Fibers, polyaramid)
Polybenzimidazoles (PBI) - 380, 383
Poly(benzimidazoquinazolines) (PIQ) - 383
Polybenzothiazoles (PBT) - 380, 383
Polybenzoxazole, adamantane based - 384
Poly[bis(benzimidazobenzoisoquinolines)] - 382
Poly[bis(benzimidazobenzophenanthrolines)] (BBB) - 380, 382
Polybutadiene - 338
  epoxidized - 137
  rubber, carboxyl terminated - 79
Poly(n-butyl acrylate) - 176
Polybutylene terephthalate (PBT) - 116

Polycaprolactone polyols - 238
Polycarbonates - 202, 326
Polycarboxylic acids - 61
Polydimethylsiloxanes - 324
Polyester - 67, 112, 160, 169
  alkyd - 81
  carboxy terminated - 151
  fibers - 115
  resins - 142
Polyesterification - 61
  reaction - 62
  reaction speed - 64
Polyether polyol prepolymers - 214
Polyetherimide - 214
  properties of - 290-294
Polyethylene - 335, 337, 338, 339, 346, 350, 351
  crosslinked - 347, 348, 354, 355
  glycol - 139
  high density - 336, 339, 340, 349, 351
  high density, crosslinked, properties of - 352
  linear low density - 347
  low density - 336, 338, 347, 349
  post-irradiation oxidation of - 339
  terephthalate - 61
Polyglycols, amine terminated - 146
Polyimide - 266, 384
  acetylene terminated - 268, 313, 372, 378
  addition - 304
  adhesives - 282
  ethynyl terminated (see Polyimide, acetylene terminated)
  from condensation reactions - 268
  LARC types - 310
  maleimide terminated - 376
  nadimide terminated - 376
  PMR types - 308, 309
  thermoplastic - 267, 286
  thermosetting - 267
  trade names - 288
  Vespel
    electrical properties of - 277
    mechanical properties of - 276
Polyisobutylene - 338
Polyisocyanates - 39, 187
Polyisoimide - 313
Polymer - 223
  blends - 376
  ceramic - 388
  composite - 381

Polymer (Cont'd)
   concrete - 9, 95, 99
   resins, high temperature processable - 388
Polymercaptans - 157
Polymerization inhibitors - 65
Polymers
   700°F - 382
   addition type - 368
   biadamantane based - 384
   blocked - 224
   high temperature with enhanced processability - 368
   hybrid containing polyheterocyclic structural units - 378
   thermally stable - 368
   thermosetting - 367
      acetylene terminated - 372
      maleimide terminated - 371
      nadimide terminated - 368
Polyolefin
   flexible - 356
   resins - 342
   semirigid - 356
Polyols - 39, 61, 63, 151, 184, 186, 191, 218, 222, 235, 237
   castor oil derived - 238
   epoxidized - 137
   graft - 239
   organophosphorous - 164
   polycaprolactone (see Polycaprolactone polyols)
   polyester based - 220, 238
   polyether polyethylene - 224
   polyether with polyethylene backbone - 220
   polypropylene polyether - 219, 224, 238, 258, 259
Polyphenylene sulfide (PPS) - 116
Poly(p-phenylenebenzobisthiazole) (PBT) - 383
Polyphenylenes - 267
Polyphenylquinoxalines (PBQ) - 378, 380, 386
Polypropylene - 338, 343, 346, 349, 350
   glycol - 139
      maleate - 64
   hydroperoxide - 344
   isotactic - 350
   polyether - 226
Polysiloxane - 319, 324
Polystyrene - 164, 326, 343

Polysulfides - 155
Polysulfone - 376
   thermoplastic - 376
Polytetramethylene oxide - 237
   diols - 264
Polyurethane (see also Urethane) - 16, 71, 112, 160, 174, 185, 207
   adhesives - 202
   chemistry - 186
   coatings - 204
      moisture cured - 204
      two component - 207
   components, handling of - 189
   elastomers - 194
      applications of - 198
   ingredients - 238
   products, types of - 193
   raw materials and moisture - 188
   reversion - 220
   systems, types of - 190
   trade names - 261-264
   unsaturated - 241
Polyvinyl butyral (PVB) - 202
Polyvinyl chloride (PVC) - 326, 338, 344, 346, 356
Polyvinylidene chloride (PVDC) microspheres - 162, 165
Polyvinylidene fluoride - 326
   irradiated - 356
Post cure - 226
Post curing - 12, 152
Post irradiation effects - 355
Post irradiation thermal treatment - 345
Postmold shrinkage, DAP molding compounds - 115
Pot life - 11, 88, 134, 135, 136, 145, 146, 147, 148, 152, 153, 161, 162, 210, 215, 232, 327
   curing agent/DGEBA - 156
Power factor, crosslinked acrylics - 363
Precipitated silicas - 86
Preform molding - 89, 100
Premix gunk molding compounds - 129
Premix molding - 100
Prepolymer - 16, 66, 114, 184, 185, 190, 191, 192, 213, 222, 228, 229, 231, 232, 233, 235, 238, 249, 260
   DAP - 129
   polyether polyol - 214
Prepolymerization - 16
Prepreg - 70, 81, 82
Prepromoted resins, commercial - 84

Press cycles - 100
Pressure sensitive tapes - 43
Primary alcoholic hydroxyl
    groups - 224
Primers, polyamide-epoxy - 218
Print sheet - 37, 53
Processibility - 368, 371
Processing
    aids - 115, 157
    characteristics - 367
    temperature range, amino
        molding compounds - 54-55
    window - 378
Promoters - 61
    elevated temperature cure - 81
    handling of - 85
    primary - 82
    room temperature cure - 81
    secondary - 83, 84
1,2-Propane diol - 72
Properties of
    addition cured RTV silicone - 329
    amino molding compounds -
        54-55
    anhydride cured epoxy - 155
    aromatic amine cured epoxy -
        148
    condensation cured RTV silicone - 329
    crosslinked acrylic - 363
    crosslinked high density polyethylene - 352
    cycloaliphatic amine cured
        epoxy - 147
    DAP molding compounds -
        115, 119-128
    epoxy - 170-173
    furan molding compounds - 55
    melamine molding compounds -
        54-55
        formaldehyde molding compounds - 54
        phenolic molding compounds - 55
    modified bismaleimide - 305-308
    phenolic molding compounds -
        32
    polyamide-imides - 298-300
    polyetherimide - 290-294
    RTV silicone rubber - 326
    styrenated GP polyester liquid
        resins - 70

unsaturated polyester alkyd - 69
unsaturated polyester alkyd GP - 70
urea molding compounds - 54
urethane elastomers - 197, 234
Property promoters - 157
Propoxylated Bisphenol A - 66, 75
Propylene glycol (PG) - 62, 63, 65, 66,
    69, 72, 74, 192
    maleate - 62
    polyesters - 62
Propylene oxide - 145, 219, 324
Polytetramethyldisiloxane (GAPD) - 285
Prototype molding - 100
Pseudoplastic behavior - 319
Pultrusion - 85, 101
Pyrazinetetracarboxylic dianhydride
    (PTDA) - 385
Pyridine - 254
Pyrogenic silicas - 86
Pyromellitic dianhydride (PMDA) - 154,
    268, 269

Quadrol - 145-146, 212
Quartz - 163
    ground - 330
Quaternary ammonium chlorides - 84
Quaternary salts - 84
Quenching - 227

Radar transparency - 60
Radiation
    electron beam - 362
    exposure - 116
    ultraviolet - 362
Radical scavenger - 343
Radomes - 60
Rate of isomerization - 64
Raw material cost - 87
Reaction injection molding (RIM) - 107,
    176, 199, 217, 218, 227, 239
Reaction temperature - 258, 261
Reactive diluent - 159
Reactive number - 223
Reactivity - 72, 145
Rebound (see also Resilience) - 264
    RIM urethane - 240
Recovery - 195
Reduced styrene emission resins - 107
Refractive index - 90
    cast resin - 72
    crosslinked acrylics - 363
    Kapton polyimide - 271

Refractive index (Cont'd)
   Pyralin polyimide - 281
Refractory materials - 381, 387
Reinforcement - 61
   fiber - 82, 91
Reinforcing fillers - 330
Resilience (see also Rebound) - 153,
      195, 215, 226, 234, 237, 341
   urethane elastomer - 234
Resin - 8
   content - 306
   cooking temperatures - 69
   efficiency - 35
   laminating - 87
   liquid - 169
   thixotropic - 100
   transfer molding (RTM) - 93,
      94, 176
Resinous modifiers - 157, 160
Resole - 31, 40, 42
   phenol-formaldehyde - 151
Resorcinol - 21, 42
   diglycidyl ether - 159
Reversion - 174
   polyurethanes - 220
Rheogoniometer, *Weissenberg* -
      347, 348, 349
Rheological additives - 157, 164
Rheology
   pseudoplastic - 86
   thixotropic - 86
Ricinoleic acid - 238
Rigidity - 32, 232
Rock wool - 34, 35
Roller coating - 29
Room temperature curing - 76
   systems - 82
Room temperature vulcanizate
   (see RTV)
Rope molding compounds - 129
Rosin - 38
Rotational molding - 350, 351
Roving choppers - 93
RTV - 324
   silicone rubber - 325, 328
   properties of - 326
Rubber - 40, 41
   natural - 195
   Neoprene - 195, 202
Rubber Reserve Corporation - 60

Safety regulations - 179
Sag

RTV silicone rubber - 326
   values - 168
Sagging - 82
Salicylic acid - 149
Sand - 163
   binder - 38
   molds - 38
Saturated polyester from tartaric acid
   and glycerin - 59
Sawdust - 163
Scratch resistance - 56
Scuff resistance - 187, 193, 205, 233,
      238
Sealants - 204
   polyurethane - 202
   vinyl plastisol - 202
Secondary
   alcoholic hydroxy groups - 224
   glycol - 63, 74
   hydroxyls - 79, 80
Service temperature - 177, 233, 236,
      237, 251
Settling - 163
Shear
   modulus
      ATB - 375
   strength - 40
      polyamide-imide - 298
      polyetherimide - 298
      Vespel polyimide - 276
Sheet molding compounds (see SMC)
Shelf life - 10, 130, 331
   urethane elastomer - 234
Shelf stability - 249
Shell molding - 38, 39
Shock absorption - 342
Shock resistance - 115
Shrink control - 89
   additive - 104
Shrink ratio - 361
   heat shrinkables - 358
Shrink temperature, minimum, heat
   shrinkables - 359
Shrinkage - 14, 114, 157, 163, 169,
      177, 192, 230, 335
   cure - 87, 104
   DAP molding compounds - 119-120
   linear, urethane elastomer - 234
   mold - 67, 112, 130
Sienna, natural - 164
Sieves, molecular (see Molecular sieves)
Silanes - 39
   amino functional - 217

Silanes (Cont'd)
  coupling agents, high
    temperature - 384
  organofunctional - 326
Silanol terminated polymers - 327
Silica - 162, 163
  aerogel - 163
  fumed, hydrophilic - 168
  fumed, hydrophobic - 164, 168
  precipitated - 86
  pyrogenic - 86, 330
  pyrolitic - 164
  sand - 88, 99
  wet process - 331
Silicon carbide - 41
Silicon-phthalocyanine polymers - 383
Silicone - 160, 174, 318, 356
  compounding - 330
  fluids - 319
  heat cured systems - 330
  laminates - 331
  polyimides - 282
  rubber - 95, 324, 330, 332
    compounded - 331
    RTV - 325
  trade names - 333
Silver - 164, 165
  powdered - 162
Simultaneous interpenetrating networks (SIN) - 176
Single package system - 8
Single promoted resin - 84
Sisal - 91
Size exclusion chromatography - 73
Skin over time, RTV silicone rubber - 326
Skin sensitizers - 145, 159, 174
Sloughing - 101
Slump, RTV silicone rubber - 326
SMC - 76, 82, 89, 90, 104, 107
Smoke density - 90
Smoke generation - 42
Sodium aluminum silicate, synthetic - 162
Softening temperature - 72, 310
Softness - 224
Solid state carbon-13 NMR - 379
Solubility - 309, 314, 322, 340, 380
  irradiated polyethylene - 340
  modified bismaleimide - 305

unsaturated polyester alkyds - 70
Solvent cage - 337
Solvent resistance - 159, 362, 372, 376
  crosslinked acrylics - 363
Solvents, aprotic - 378
Solvents for urethanes - 227
Soya oil, epoxidized - 137
Soybean meal - 27
Sparging - 64
Specialty thermosets - 4
Specific gravity - 28, 76, 87, 162, 164, 165, 234, 306
  addition cured RTV silicone rubber - 329
  amino molding compounds - 54-55
  condensation cured RTV silicone rubber - 329
  crosslinked acrylics - 363
  DAP molding compounds - 115, 119-120, 121-125, 126-128
  epoxy - 170-173
  methylphenylpolysiloxane fluids - 323
  methylpolysiloxane fluids - 320
  polyetherimide - 292, 294
  RTV silicone rubber - 326
  unsaturated polyester alkyds - 70
  urethane elastomer - 234
  Vespel polyimide - 279
Specific heat
  Pyralin polyimide - 281
  Vespel polyimide - 278
Specific stiffness - 380
Specific strength - 380
SPI gel time - 70
  vinyl ester resins - 81
Spray lay-up - 85, 93
Stabilizers - 343
Staging - 14
Stain resistance - 56, 95
Stannous octoate - 212, 263
Stannous tin octoate - 327, 328
Stearic acid - 238
Steel - 164, 165
  titanium clad - 388
Stereoregularity, alkyd chain - 74
Steric shielding - 76
Stiffness - 105
Stiffness to weight ratio - 32
Stoichiometric
  analysis - 158
  balance - 142, 149
  calculations - 16

Stoichiometric (Cont'd)
  considerations - 14
  excess of glycol - 64
  ratio - 15
Stoichiometry - 145, 152, 187,
  211, 222, 231, 232, 246,
  309
  epoxy - 141
Strandboard - 29, 30
Strength-to-weight ratio - 179
Stress crack resistance - 350, 351,
  354
  crosslinked acrylics - 363
  crosslinked high density polyethylene - 352
Stress cracking agent - 351
Structural adhesives - 43
Styrenated GP polyester liquid
  resin, properties of - 70
Styrenated unsaturated polyester - 71
Styrenation tank - 65
Styrene - 62, 65, 66, 67, 76, 78,
  79, 129, 162, 239
  butadiene rubber - 60
  compatibility - 74
  content, vinyl ester resins - 81
  in the air or workplace - 71
  monomeric - 60
  oxide - 149, 158, 161
Substituted hydroquinone - 65
Sucrose - 224
Sulfone, acetylene terminated
  (see ATS)
Sulfuric acid - 64
Surface activators - 164
Surface charge, silicone polyimide - 287
Surface resistance, DAP molding
  compounds - 121-125
Surface resistivity
  crosslinked acrylics - 363
  polyamide-imide - 299
  Pyralin polyimide - 281
  Skybond polyimide - 284
  urethane elastomers - 196, 197
  Vespel polyimide - 277
Surface tension - 324
  methylphenylpolysiloxane
    fluids - 323
  methylpolysiloxane fluids - 320
Surfactants - 107, 216
  silicone - 217

Szmercsanyi, V. - 63

Tack - 27, 202, 374
Tack free time - 325
  RTV silicone rubber - 326
Tackifiers - 21, 43
Talc - 88, 89, 162, 163, 217
Tapes, masking - 43
Tapes, pressure sensitive - 43
TDI - 187, 192, 195, 205, 210, 221,
  223, 225, 226, 229, 231, 237,
  238, 239, 246, 248, 249, 252,
  258, 259, 260, 261, 364
  free - 211
  hazards of - 249
  prepolymers, preparation of - 256
Tear resistance - 213, 214, 215, 220,
  237, 238
  crosslinked polyethylene foam - 341
Tear strength - 218, 220, 239, 264
  Kapton polyimide - 271, 274
  RIM urethane - 240
  RTV silicone rubber - 326
  urethane elastomer - 197, 234
Tego film - 27
Telechelic polymers - 139
Temperature, effects of, electrical
  properties of urethane elastomers - 197
Temperature, influence of - 7
Temperature resistance - 133, 138,
  147, 148, 149, 150, 153, 381
Tensile modulus - 177, 313
  100% & 300% - 264
  100%, urethane elastomer - 197, 234
  acetylene terminated polyimide -
    314, 315, 377
  amino molding compounds - 54-55
  ATB - 375
  comparison of two polyimide films -
    304
  epoxy - 170-173
  Kapton polyimide - 271, 274
  PMR polyimide - 310, 311
  polyamide-imide - 298
  polyetherimide - 290
  polyimide film - 303
  RIM urethane - 240
  unsaturated polyesters - 108
  vinyl ester resin, cast - 81
Tensile shear properties, polyimide -
  285

Tensile shear strength, acetylene
  terminated polyimide - 377
Tensile strength - 34, 78, 116, 117,
  153, 154, 214, 215, 218,
  224, 233, 264, 313, 340,
  348, 349, 354, 374
  acetylene terminated polyimide -
    314, 315, 377
  addition cured RTV silicone
    rubber - 329
  amino molding compounds -
    54-55
  anhydride cured epoxy - 155
  aromatic amine cured epoxy -
    148
  ATB - 375
  comparison of two polyimide
    films - 304
  condensation cured RTV sili-
    cone rubber - 329
  crosslinked acrylics - 363
  crosslinked high density poly-
    ethylene - 352
  crosslinked polyethylene
    foam - 341
  cycloaliphatic amine cured
    epoxy - 147
  DAP molding compounds -
    115, 119-120, 121-125
  DETA/TETA cured epoxy -
    145
  epoxy - 170-173
  heat shrinkables - 359
  irradiated polyethylene - 340
  Kapton polyimide - 271, 274
  modified bismaleimide - 308
  phenolic molding compounds -
    32
  PMR polyimide - 310, 311
  polyamide-imide - 298
  polyimide film - 303
  polyurethane elastomers - 194,
    197, 234
  Pyralin polyimide - 281
  RIM urethane - 240
  RTV silicone rubber - 326
  Skybond polyimide - 283
  unsaturated polyesters - 108
  Vespel polyimide - 276
  vinyl ester resin, cast - 81
Tensile yield strength
  amino molding compounds -
    54-55

epoxy - 170-173
Kapton polyimide - 271, 274
polyetherimide - 290, 293
TEPA - 145, 149
Terephthalic acid (TPA) - 66, 72
Terminal unsaturation - 76
Tertiary alcoholic hydroxy groups - 224
Tertiary amines - 77, 84, 144
TETA - 142, 143, 145, 149, 159, 161
Tetrabromobisphenol A - 78
Tetrabromophthalic anhydride (TBPA) -
  66, 75, 85
Tetrabutyl titanate - 64, 238
Tetrachlorophthalic anhydride (TCPA) -
  66, 75, 85
Tetraethylene pentamine (see TEPA)
Tetraglycidyl methylene dianiline - 138
Tetrahydrofuran - 57, 301, 378
Tetrahydrofurfuryl acrylate - 364
Tetrahydrophthalic anhydride (THPA) -
  153
N,N,N-Tetrakis(2-hydroxypropyl)-
  ethylene diamine (see *Quadrol*)
1,2,4,5-Tetramethylbenzene (see Durene)
Tetramethylbutanediamine (TMBDA) -
  218
Tetramethylene oxide - 224, 226, 238
Tetramethylol urea - 47
T_g - 54, 55, 267, 268, 288, 301, 313,
  314, 364, 369, 372, 374, 376,
  acetylene terminated polyimides -
    314
Thermal biphenylene opening - 381
Thermal conductivity - 162, 163, 164,
  165
  amino molding compounds - 54-55
  crosslinked acrylics - 363
  epoxy - 170-173
  polyamide-imide - 300
  polyetherimide - 291
  Pyralin polyimide - 281
  Vespel polyimide - 278
Thermal expansion, coefficient of (see
  Coefficient of thermal expansion)
Thermal insulating properties - 42
Thermal insulation - 342
Thermal resistance - 179
Thermal shock resistance - 163
Thermal stability - 266, 322, 368, 381,
  382
  crosslinked polyethylene foam - 341
  polyimide film - 302

Thermally induced ring opening of
 biphenylene - 386
Thermoformability - 376
Thermooxidative properties - 267
Thermooxidative stability - 287,
 288, 310, 369, 371, 379,
 383, 384, 386
Thermoplastic - 223
 pellets - 192
Thermoset polyurethanes - 183
Thermosetting - 223
 resins - 388
Thick molding compounds (see
 TMC)
Thickener - 104
Thickening - 104
 agents - 82, 90, 164
  SMC, nonmetal oxide/hydroxide - 91
 response - 78
Thinning tank - 65, 69
Thiols - 150
Thixotropes - 157, 164
Thixotropic
 agents - 82, 85, 86, 164
 index - 87
Time, influence of - 7
Tires, low speed - 194
Titanium dioxide - 98, 162
Titanium oxides - 90
TMC - 104
Toluene - 254, 255, 256
 diisocyanate (see TDI)
Toluhydroquinone - 65
Tooling - 199
Torsion pendulum - 347
Toughness - 72, 73, 77, 80, 157,
 161, 175, 187, 193, 213,
 220, 232, 238, 275, 351,
 353, 369, 371, 376, 381
Toxicity - 71, 145, 146, 159
Track resistance - 89
Trade names
 allyls (DAP) - 130
 aminos - 58
 crosslinked thermoplastics - 365
 epoxy - 180
 phenolics - 43
 polyimide - 288
 polyurethanes - 261-264
 silicones - 333
 unsaturated polyesters - 108-109
 vinyl ester resins - 108-109

Trans fumarate isomer - 63
Trans isomer - 63
Trans-1,4-cyclohexane diisocyanate -
 241, 262
Transfer molding - 33, 114
Transition metal soaps - 82
Transition temperature, second order -
 319
Translucence - 90
Transmission and haze, crosslinked
 acrylics - 363
Triacrylates - 345
Triallyl cyanurate (TAC) - 68, 71
 polyesters - 130
Triallyl isocyanurate (TAIC) - 71, 345
Triamide of cyanuric acid (see Melamine)
2,4,6-Triamino-1,3,5-triazine (see
 Melamine)
Triazides - 225, 260
Triethanolamine - 218
Triethyl amine - 144, 255
Triethylene diamine (see DABCO)
Triethylene glycol
 dicaprate - 218, 263
 dicaprylate - 218, 263
Triethylene tetramine (see TETA)
Triflic acid - 151
Trifluoromethane sulfonic acid (see Triflic acid)
Triglycidyl p-aminophenol - 140
Trigonal nitrogens - 382
Trimellitic anhydride (TMA) - 154, 295
Trimerization, organometallic complex
 catalyzed - 386
Trimethacrylates - 345
Trimethyl ammonium chloride - 79
Trimethylamine - 144
Trimethyleneglycol di-p-aminobenzoate - 213, 262
Trimethylol urea - 47
Trimethylolpropane - 219, 224
 polyoxypropylene derivatives - 219
 triacrylate (TMPT) - 71, 362, 364
 trimethacrylate (TMPTMA) - 345
Trimethylpentanediol (TMPD) - 66
2,2,4-Trimethyl-1,3-pentanediol
 (TMPD) - 76
Triol - 222, 229, 235
 polypropylene polyether - 229, 230
Triphenyl methane - 140
Triphenylphosphine - 167
Triphenylsulfonium chloride - 84
Tris-β-chloroethyl phosphate - 167

Tris(dibromopropyl)phosphate - 164
Tris-dimethylaminomethylphenol
 (see DMP-30)
Tubing, heat shrinkable - 355
Tung oil - 38
Turbidity, time to - 25
Two-component
 compounds, RTV silicone
  rubber - 327
 polyurethanes - 204
  coatings - 207
  systems - 190, 324
Two-pot method - 84
Typical general purpose (GP)
 unsaturated polyester
 resin - 65

UL flammability, DAP molding
 compounds - 121-125,
 126-128
UL rating, DAP molding compounds - 119-120
Ultraviolet
 absorbers - 85, 86
 energy - 174
 resistance - 74
 transmission, crosslinked
  acrylics - 363
Underwater cure - 150
Unsaturated polyesters - 59, 64,
 77, 114, 129
 alkyd - 65, 67
  properties of - 70
 chemistry - 61
 common raw materials for - 66
 compounding of - 81
 emulsions, water-in-oil - 94
 halogenated - 85
 history - 59
 isophthalic - 74
 monomers used in - 71
 ortho-phthalic - 74
 physical properties of - 108
 properties and chemical composition - 71
 resin alkyd, properties of - 69
 trade names - 108-109
Unsaturation
 terminal - 339
 trans-vinylene - 339
 vinylidene - 339
Urea - 35, 47, 186, 187, 188, 211,
 212, 224, 254, 259

bonded plywood - 51
etherified - 52
formaldehyde - 56, 136
 furfuryl alcohol resins - 56
 molding compounds - 53
 properties of - 54
 plywood - 52
 resin - 45, 46, 49, 50, 53, 57
  chemistry - 47
Urethane (see also Polyurethane) - 176,
 186, 211, 212, 365
elastomer,
 electrical properties of - 196
 properties of - 197, 234
hybrid resins - 107
hydrogen - 224
polyester - 195, 209, 210
polyether - 195
raw material sources - 261
Uretidione - 249
Use ratio - 232, 235
Useful life of silicone rubber at elevated
 temperatures - 333
Useful temperature range
 addition cured RTV silicone rubber -
  329
 condensation cured RTV silicone
  rubber - 329

Vacuum - 189, 190
Vacuum bag laminates - 114
Vanadium - 83
 oxide - 269
Vapor pressure
 methylphenylpolysiloxane fluids -
  323
 methylpolysiloxane fluids - 320
Vibration damping - 342
Vicat softening point, polyetherimide -
 291, 293
Vicat softening temperature, crosslinked
 high density polyethylene - 352
Vinsol - 35
Vinyl coatings - 52
Vinyl ester resins - 59, 76, 93, 94, 99,
 100
 acid modified - 78
  thickening of - 79
 cast, properties of - 81
 chemistry - 76
 compounding of - 81
 flame retardant - 78
 impact modified - 79

Vinyl ester resins (Cont'd)
  liquid, styrenated, properties of - 81
  one-step - 79
  rubber modified - 79
  SMC, maleic acid modified, ingredients of - 79
  specialty - 78
  structure and properties - 78
  thickening for SMC - 78
  trade names - 108-109
  urethane modified - 79
Vinyl toluene - 71, 81, 82
Vinyl unsaturation - 78
  terminal - 80
Vinylcyclohexane dioxide (VCHD) - 137, 159
Vinyls - 169
Viscosity - 28, 87, 134, 146, 147, 149, 152, 153, 159, 162, 169
  addition cured RTV silicone rubber - 329
  bulk - 376
  condensation cured RTV silicone rubber - 329
  depressants - 164
  false - 86
  methylphenylpolysiloxane fluids - 323
  methylpolysiloxane fluids - 320
  mixed - 161
  modified bismaleimide - 305
  reducers - 134
  suppressants - 157
  thinned resin - 74
  urethane elastomers - 234
  vs. styrene content for ortho- and isopolyesters - 92
Viscosity-temperature coefficient - 319, 322
  methylphenylpolysiloxane fluids - 323
  methylpolysiloxane fluids - 320
Volatility - 145, 150
Volume coefficient of expansion
  methylphenylpolysiloxane fluids - 323
  methylpolysiloxane fluids - 320
Volume resistance, DAP molding compounds - 121-125

Volume resistivity - 164
  Kapton polyimide - 272, 273, 274
  modified bismaleimide - 307
  polyamide-imide - 299
  polyetherimide - 292, 294
  polyimide film - 302
  Pyralin polyimide - 281
  RTV silicone rubber - 326
  Skybond polyimide - 284
  urethane elastomers - 196, 197
  Vespel polyimide - 277
Vorlander - 59
Vulcanization - 3, 8
Vulcanizing agents - 330
  peroxide - 331

Waferboard - 30, 49
Wash-and-wear clothing - 53
Water absorption - 32, 56, 161
  acetylene terminated polyimide - 315
  amino molding compounds - 54-55
  crosslinked acrylics - 363
  crosslinked polyethylene foam - 341
  DAP molding compounds - 119-120, 121-125, 126-128
  epoxy - 170-173
  polyamide-imide - 300
  polyetherimide - 292, 294
  Skybond polyimide - 283
  Vespel polyimide - 279
Water abstract conductance, DAP molding compounds - 121-125
Water dilutability - 25
Water extended polyester (WEP) - 94, 95
Water immersion, effects of, urethane elastomers - 197
Water resistance - 34, 56, 150
Water soluble phenolic resins - 35
Water white castings - 147
Waxes, silicone - 217
Wear rate, *Vespel* polyimide - 277
Wear resistance - 193, 205, 238
Weather degradation - 141
Weathering - 67, 112, 351
  outdoor - 81
Web - 36, 37
Weight loss - 161, 379, 380
  acetylene terminated polyimide - 315
  modified bismaleimide - 306
  Pyralin polyimide - 281
  Skybond polyimide - 283

Weight per gallon - 81
    Pyralin polyimide - 280
    urethane elastomers - 234
Wet dough process - 40
Wettability - 342
Wetting - 80
    agents - 82, 157
Wheat flour - 28, 29, 52
Wollastonite - 163
Wood - 27
    adhesives - 46
    flour - 19, 31, 32, 34, 55, 163
        molding - 30
Work life - 325
Work time - 328
Working life - 11
Working time, RTV silicone
        rubber - 326

X-rays - 337
Xylenols - 21
m-Xylylene diamine (MXDA) -
        146, 147, 262

Yellowing resistance - 74
Yield stress - 348

Zinc borate - 167
Zinc octoate - 252
Zinc stearate mold release - 105
Zinc sulfide white - 164
Zirconium oxide
    cubic - 388
        partially stabilized (PSZ) - 388